THE STUDY OF LANDFORMS

A TEXTBOOK OF GEOMORPHOLOGY

BY

R. J. SMALL

Senior Lecturer in Geography
University of Southampton

CAMBRIDGE
AT THE UNIVERSITY PRESS
1970

Published by the Syndics of the Cambridge University Press
Bentley House, 200 Euston Road, London N.W.1
American Branch: 32 East 57th Street, New York, N.Y. 10022

Library of Congress Catalogue Card Number: 74–92254
Standard Book Number: 521 07449 5

Printed in Great Britain
by Jarrold & Sons Ltd, Norwich

CONTENTS

PREFACE

This book is designed both for advanced work in schools and as a background to introductory courses in geomorphology at universities and colleges of education. Some sections are doubtless more appropriate to study in schools than others; for instance, the chapters on Weathering, transport and erosion (2), Geological structure and landforms (3), Rock-type and landforms (4), and Drainage development (7) cover what I regard as the branches of geomorphology which can be most profitably studied at this stage. The chapters on The cycle of erosion (5), Slope development (6), and The study of planation surfaces (8) are perhaps of greater interest to university students, though many older pupils in schools could be introduced to these topics providing their reading is strictly supervised by teachers.

In writing this book I have been guided by five main principles. Firstly, I have not attempted a fully comprehensive coverage of landform study; there are some branches and problems of geomorphology that are deliberately not touched upon or discussed in any detail. I have preferred to aim for greater depth in the material I have chosen to present here. However, I have also tried to achieve a fair balance, so that despite the lacunae the reader of this book will gain a reasonably competent all-round knowledge of the more important aspects of landform study. Secondly, I have attempted to bring out the problematical nature of most landforms. The subject has not been presented as a series of known facts (which it is very far from being), but attention has constantly been focused on the inherent difficulties of landform study, the equivocal nature of much geomorphological evidence, the existence of alternative hypotheses and so on. Moreover, apparently simple processes, such as river capture, are shown to be far more complex and less well understood than is commonly supposed. Thirdly, I have aimed at introducing or developing concepts, methods of study and evidence which is either new or is not in my opinion sufficiently emphasised in some other geomorphological texts. In particular, I have discussed the decline of the cycle concept, its possible replacement by the dynamic equilibrium theory, the current emphasis on the quantification of processes and landforms, and have devoted a whole chapter (10) to periglaciation, which is recognised by professional geomorphologists as a major factor in landform development but which is hardly touched upon in schools. Fourthly, I have wherever possible or appropriate geared this book to the detailed working out of examples. I have deliberately chosen those that I know well (the Chalk country of southern England, south Wales, Dartmoor, the Grands Causses of southern France, the Val d'Hérens of Switzerland and so on) and which I consider repay such detailed study. It is true that concentration on regional case-studies again militates against comprehensive coverage, but in my view a real knowledge of one area (which may or may not contain all the so-called 'typical' landforms) is infinitely preferable to a wider—and vaguer—knowledge of examples which are neither described nor discussed in sufficient depth for the student to gain any true understanding of them. Fifthly, I have tried to emphasise that landforms result from the complex interaction of many factors and processes. One cannot relate landforms merely to, say, the type of rock in an area; climate, structure, erosional history and so on must all be considered if a realistic assessment is to be made. The 'one cause—one effect' approach to landform study has in my opinion been a serious weakness of much geomorphological teaching in the past.

Finally, I should like to express my deep gratitude to Carson Clark (formerly Chief Cartographer in the Department of Geography, University of Southampton), Alan Burn (Experimental Officer in the same Department), Robert Smith and Barbara Manning, who have drawn with patience and skill the maps and diagrams that illustrate this text.

Southampton
November 1968

R. J. S.

ACKNOWLEDGEMENTS

Thanks are due to the following for supplying photographs. Aerofilms Ltd: Figs 140 and 200. M. J. Clark: Figs 10, 155, 156, 163, 174 and 190. French Government Tourist Office: Fig. 61. F. Gay: Fig. 64. Eric Kay: Figs 3, 8, 9, 16, 19, 26, 29, 31, 35, 45, 47, 51, 52, 115, 147, 148, 149, 153, 157, 171, 173, 176, 179, 181, 187, 191, 194 and 199. G. R. Siviour: Figs 72, 74 and 127. M. F. Thomas: Fig. 75. H. Roger Viollet: Fig. 63.

1

THE AIMS AND METHODS OF LANDFORM STUDY

In its strictest sense geomorphology, the science of landform study, is concerned with the study of the form of the earth. In practice, however, this simple definition is too all-embracing. For instance, the actual shape of the earth body is taken to lie within the field of geodesy and geophysics. The form and origin of mountain systems, among the most important of the earth's surface features, are studied more by geologists than geomorphologists. Problems concerning the shape and distribution of the continental land-masses are at present being investigated by physicists, who are constructing theories of continental drift based on the evidence of the magnetic properties of rocks. Geomorphologists have undertaken the study of very large-scale landforms, such as planation surfaces of continent-wide extent, but for the most part their attention has been focused on the smaller-scale phenomena of the earth's surface, such as drainage basins, areas of uniform rock-type, individual river valleys or the hill-side slopes of a small region. The reason for this is no doubt partly a practical one: geomorphology at the research level is essentially a field study, and only comparatively restricted parts of the landscape can be adequately investigated by one worker or a small group of workers.

It has been said that geomorphology is concerned with the study of the evolution of landforms, particularly those produced by the processes of erosion. Clearly this statement must not be taken too literally, for geomorphologists have contributed many valuable studies of depositional forms, especially in coastal and glaciated areas. However, there is some measure of truth in it; and the narrowing of the field of study that it implies seems to be the result of a deliberate preference on the part of a majority of geomorphologists. There are, too, other relevant factors. In the first place, there is no denying the great influence that the American geographer, W. M. Davis, has had on geomorphological thought and research. Davis, whose opinions and arguments will be referred to at many points in this book, was primarily concerned with the problem of formulating a genetic system of landform description, or in other words the classification of landscapes according to their mode of origin. Davis achieved his goal in the concept of the cycle of erosion, in which he attempted to show that the most fundamental erosional landforms (river valleys and the slopes bordering them) experience a gradual change of form with the passage of time (or, in short, evolve). Secondly, the fact that many of the most influential geomorphologists have been geographers, rather than geologists, may help to account for the lack of emphasis on deposits and depositional landforms. In 1949, the American geomorphologist, R. J. Russell, castigated what he termed 'the cult of pure morphology', or attempts to reconstruct stages in landform development by an analysis of form and relief alone, without reference to available geological evidence. Obviously the point is an important one, for erosion at one place necessarily produces detritus which is ultimately deposited elsewhere; and when two lines of evidence exist, it is foolhardy to ignore one. However, it must be added that Russell's criticism has to some extent lost its validity in the intervening years. Our knowledge of the vital role of periglaciation in shaping the landscapes of present-day temperate areas is based almost wholly on analysis by geomorphologists of the distribution and character of solifluxion gravels, 'head', 'coombe rock' and the like.

It is apparent, then, that landform study has grown up, largely in the present century, with certain well-marked emphases and some notable omissions. It is probably fair to add that geomorphology has also exhibited failings on the methodological side. It can hardly be disputed that the geomorphologist is confronted by three main tasks, and that these need to be approached in a logical order. Firstly, he must attempt to describe, as accurately as possible, the various forms of the physical landscape. Secondly, it is usually necessary for him to classify the landforms and processes that he encounters, if only with the object of rendering more assimilable a large body of information. Thirdly, he must suggest hypotheses as to the origin of landforms, and if possible test these hypotheses by further field study as a prelude to the formulation of theories. By and large, however, geomorphologists have in the past been reluctant to concentrate sufficiently on careful landscape description, and over-anxious to explain the origin of landforms about which insufficient basic information has been obtained. An oft-quoted example lies in the work of W. M. Davis, who in the context of his cycle concept was prepared to make suggestions as to the manner of slope evolution without detailed measurement of slope profiles in the field or actual study of the processes of weathering and transport at work on slopes. Davis's hypothesis is, in fact, an instance of the type of deductive reasoning which has been very popular in geomorphology; he made certain assumptions (as regards uplift, form and process) and argued logically from these. If the assumptions are wrong, however, the hypothesis is also incorrect. Other similar examples of the deductive approach are the work of Penck (1953) and Wood (1942) on slope development and Lawson's (1915) explanation of rock pediments in deserts.

THE DESCRIPTION OF LANDFORMS

Description of the physical landscape can be attempted with varying degrees of precision and with different objectives in mind. One approach may be illustrated by the following quotation from A. E. Trueman's *Geology and scenery in England and Wales.*

> The Chalk areas of England, which Huxley thought so suggestive of mutton and pleasantness, are perhaps the most easily recognisable, for the wide expanses of grassy downland and the smooth rounded curves of the hills are as characteristic of Salisbury Plain as of the Chilterns or the Downs. In many parts the short grass barely covers the white rock, for the soil is extremely thin, and the white gashes of chalk pits on the hillsides and the pale cream of the rough flinty tracks give to these areas a lightness, a delicacy of colouring, which is unlike that of any other type of hill country.

Such a description, which is largely personal or subjective, is very useful in introducing the layman to the general appearance of a 'typical' Chalk area, but is of little use to the geomorphologist who is concerned with form rather than vegetation or colouring. Davis, with a more specifically geomorphological purpose in mind, would doubtless have described such a landscape in quite different terms. To him the Chalk areas of England would have been in the stage of 'late-youth' or 'early-maturity', phrases which imply that the valley-side slopes show some evidence of 'grading' and the interfluves have been lowered by 'wasting'. Such a description is not only more technical, in the terms used, but also introduces the notion of landform origin. The Chalk is, in fact, seen as having evolved through the stage of youth, and as now evolving towards the stage of late-maturity and old age. Furthermore, 'grading' and 'wasting' relate to the way in which the slopes are weathered and the resultant detritus transported away.

The Davisian method of genetic description has, as stated above, been very widely applied in geomorphology, but it is unfortunately a decidedly clumsy tool, lacking in any real precision. The main weakness is that Davis recognised only three basic stages of landscape evolution (youth, maturity and old age), with some refinements added by use of the prefixes 'early' and 'late'. In reality, the landscape exhibits infinite gradations of form, and is not readily classifiable into such a simple set of categories. Furthermore, a particular landscape may be characterised by some features which Davis regarded as symptoms of youth, and by others that were supposed by him to indicate maturity. It is not therefore surprising that many modern geomorphologists have become dissatisfied with the Davisian approach to landform description. Indeed, the American geomorphologist Strahler (1950) has stated: 'a generalised overall scheme of landscape evolution stated in terms of youth, maturity and old age [can] contribute next to nothing to the understanding of factors determining the mechanism and intensity of erosion on slopes'.

Strahler himself has in part been responsible for initiating the most significant development that has taken place in geomorphology during the last fifteen years: the replacement of subjective or unscientific descriptions of the landscape by more objective and scientific methods. Precise measurements of dimensions and forms are derived from accurate topographical maps or from survey in the field, and the data so obtained are carefully analysed using accepted statistical methods. The value of such 'quantification' in the study of slope angles and forms must be immediately obvious, and some of the results obtained—and their sometimes surprising implications—are discussed in chapter 6. Its application to the analysis of drainage patterns and basins is briefly described in chapter 7. Quantification is applied not only to landscape forms, giving rise to the branch of modern geomorphology known as 'morphometry', but also to processes such as river flow, movement of sediment, types and rates of weathering, soil creep, solifluxion and so on.

One important result of the new objective approach to landforms has been the development of interest in morphological mapping. Geomorphologists have always used maps as a means of illustrating the features they are studying and the conclusions they have drawn, but these have often been highly selective and with a genetic emphasis. For instance, many of the older geomorphological maps depict remaining fragments of old planation surfaces, even though these may constitute a small part of the existing landscape (Fig. 1). In short, features presumed to have a special significance, as recording the main stages in the

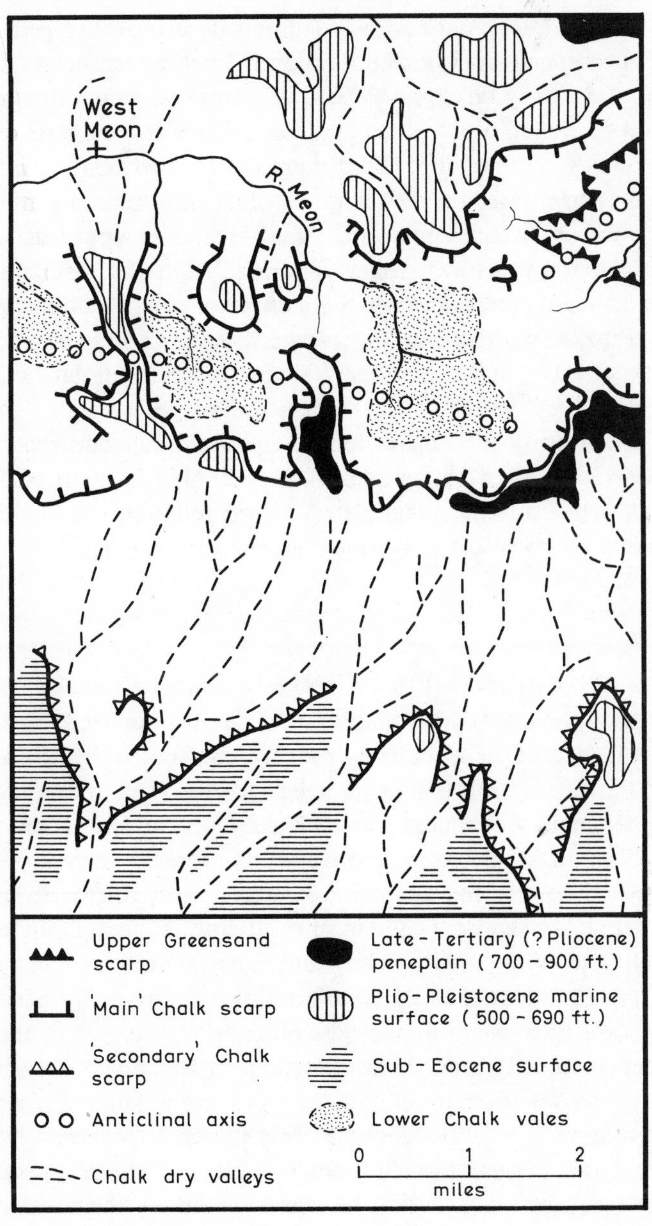

1 **Genetic geomorphological map of part of eastern Hampshire**

evolution of the landscape or as shedding light on other geomorphological phenomena (such as the form and development of the drainage pattern, which may be closely related to former surfaces of erosion) are alone recorded, and much of the landscape is ignored altogether. Recent geomorphological mapping, based more on detailed observation in the field, has concentrated on the recording of as many features as possible, whether or not their significance is immediately apparent, and on obtaining thereby a much fuller 'coverage'. On the resultant maps, breaks and changes of slope (whether convex or concave), directions and angles of maximum slope, terraces (regardless of their mode of origin), the forms of valley floors (whether V-shaped, rounded or flat-bottomed), breaks of valley gradient and other similar information is shown (Fig. 2). Obviously such maps depend upon accurate and reliable field survey, and are often so complex that great skill is required in their interpretation—indeed, 'complete' interpretation, involving the explanation of every break of slope and every slope angle, is rarely if ever possible.

THE CLASSIFICATION OF LANDFORMS

Any geomorphologist who undertakes a detailed investigation of the physical landscape of an area is quickly confronted by the need to classify the forms and/or processes he has observed. For example, a student of slopes will almost inevitably subdivide his measured profiles on the basis of form (convex, concave, rectilinear, convexo-concave, complex and so on) and probably according to gradient (it has been noted that slopes tend to occur most frequently at certain angles, such as 45°, 42°, 36–38°, 32–33°, 26–28° and so on, which are referred to as 'characteristic angles'). Such classifications are essentially descriptive, for they take no account of the possible modes of origin of the forms under consideration. Indeed, classifications of this kind are normally a prelude to the development of hypotheses of origin, and really represent an organisation of the evidence on which such hypotheses are to be founded. Other similar classifications include that of stream order and drainage basin order, and that of ocean waves on the basis of their height, length and periodicity.

By way of contrast there is the genetic classification, in which landforms are subdivided not according to shape and dimensions, but with reference to their manner of origin (though it must always be borne in mind that different origins may produce different forms, so that in

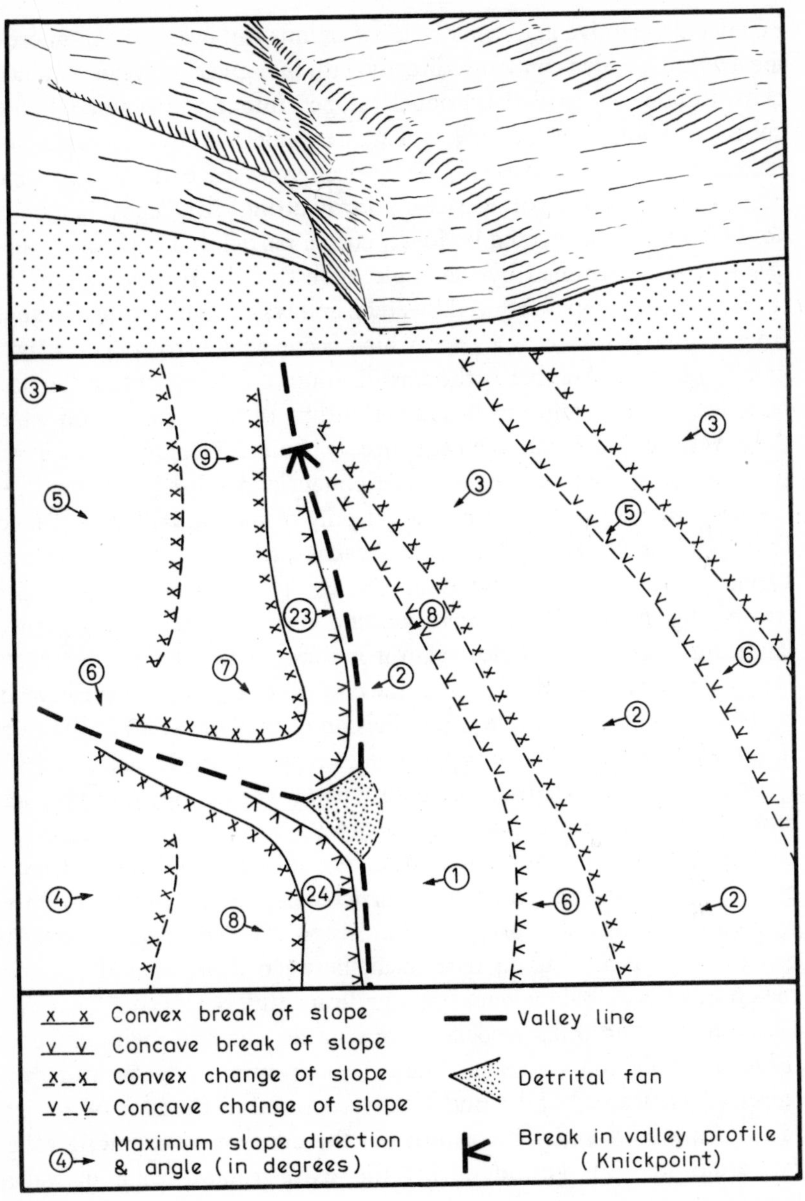

2 Descriptive geomorphological map

practice descriptive and genetic classifications can overlap). An interesting example, which compares directly with that given above, is Strahler's classification of 'graded erosional' slopes into three types: (i) 'high cohesion slopes', at 40–50°, which are underlain by cohesive, fine-textured materials such as clay, or by strong massive bedrock, and which occur in regions of vigorous stream downcutting; (ii) 'repose slopes', at 30–35°, which are covered by loose, coarse particles of rock, and whose gradient is determined by the angle of rest of the fragments lying on them; and (iii) 'slopes reduced by creep and wash', which range in angle from 20° to as little as 1–2°, and which occur where stream corrasion is greatly reduced. Another is the classification of streams not on the basis of order, but according to their relationship with the surface on which they were initiated and the sequence of their development (as in the primary division of streams into consequent and subsequent types).

Classification can be attempted not only of individual kinds of landform (such as slopes, cliffs, terraces, beaches, lakes, volcanoes, planation surfaces and so on) or types of process (mechanical and chemical weathering, mass wasting, corrosion and corrasion, and so on), but of landform assemblages. Once again it is relevant to refer to the genetic classification of landscapes, within the context of the erosion cycle, made by Davis. Another example is the division of the land areas of the earth into what are known as 'morphogenetic regions'. Each major climatic type tends to produce its own conditions of weathering, transport and erosion, for the reason that these processes are greatly influenced by climatic factors such as temperature, precipitation and wind. Consequently, it is argued by many geomorphologists, each climate produces its own distinctive landforms and landscape. The contrast, not only in general appearance but in true form, between a desert and a glacial region is a very obvious one, but significant differences may also exist between the landforms associated with other climatic regimes. Penck made a distinction between 'humid', 'sub-humid', 'semi-arid' and 'glacial' landscapes, whilst Budel has recognised (i) the zone of glaciers (as in Antarctica and Greenland), (ii) the zone of pronounced valley formation in high latitudes, (iii) the extra-tropical zone of valley formation, (iv) the sub-tropical zone of pediment and valley formation, and (v) the tropical zone of planation surface formation. A more explicit scheme is that of Peltier (1950), who has recognised nine basic types of morphogenetic region: glacial, periglacial, boreal, maritime, selva, moderate (= Davis's 'normal'), savanna, semi-arid, and arid. There are also 'morphogenetic systems' not determined primarily by climate.

These include the cycle of pediplanation, which is believed by its originator L. C. King to explain landform evolution in all non-glacial regions, the karst cycle of erosion (determined by a particular lithology and structure), and the cycle of marine erosion (which involves the geomorphological process least susceptible to climatic influence).

THE EXPLANATION OF LANDFORMS

Few would deny that the ultimate aim of geomorphology must be to explain how individual landforms and, more particularly, landform assemblages have originated. However, most landforms and landscapes are so complex and pose such a variety of problems that several genetic approaches exist.

(i) Geomorphologists may be concerned with establishing fundamental relationships which exist between certain types of landform and particular types of climate, structure or rock. For example, rock pediments are often regarded as a form peculiar to arid and semi-arid climates, whilst convexo-concave slopes are believed to be more characteristic of humid conditions. Again, attempts have been made to determine the landforms which are usually developed in, say, granite, limestone or chalk areas. Alternatively, the influence of rock-type and geological structure on more general landforms, such as valley-side slopes, might be studied. In this context, the aim might be to investigate whether slope form is affected directly or indirectly by lithological factors (for instance, it has been stated that impermeable clays favour the development of concave slopes, whilst permeable limestone and chalk give rise to dominantly convex slopes) or by the angle of dip of the underlying strata.

(ii) Other geomorphologists may elect to approach landform study from an essentially 'historical' point of view, and try to demonstrate the various stages of evolution which the landscape has passed through before attaining its present form. This method of study, which forms the basis of what is known as 'denudation chronology', stems to a large extent from the approach to geomorphology of Davis and his followers. The principal objective is to identify, date and interpret planation surfaces developed in past cycles and subcycles of erosion (see chapter 8). In work of this kind there is naturally much overlap with geology, for many of the surfaces preserved, albeit fragmentarily, in the present landscape may date back to Pliocene, Miocene or even early-Tertiary times—indeed, it is believed by some that existing surfaces may in some instances have been eroded early in the Mesozoic era. Another important aim of denudation chronology is to study the manner in which the drainage system of an area has gradually evolved. Through the identification of subsequent streams and the interpretation and dating of river captures, the original consequent pattern is reconstructed, and the possibility of superimposition, antecedence or glacial diversion investigated.

The denudation chronology approach has undoubtedly been very popular with British geomorphologists. In this we may perhaps detect the strong influence of a major work, *Structure, surface and drainage in south-east England*, published in 1939 by Wooldridge and Linton. However, denudation chronology is now being superseded by different approaches, which are associated with the recent emphasis on the quantification of landforms referred to above. An important criticism which has been levelled against the denudation chronology approach is that it succeeds in explaining directly only very small parts of the existing landscape, namely the fragments of former surfaces which have been dissected and almost totally destroyed in some cases by more recent erosion. These fragments may constitute less than 10% by area of the whole landscape, and furthermore may be visible in the field only to the very experienced observer. In denudation chronology so much remains unaccounted for, such as the precise form of the incised valleys and valley-side slopes which are far more obvious and important components of the physical landscape than planation surface remnants. Another weakness—though it is by no means confined to denudation chronology—is that it is often, even by geomorphological standards, highly speculative and controversial. The reasons are twofold: (i) the old surfaces are commonly so modified by subsequent wasting that their original form and height cannot be easily interpreted, and (ii) the geological evidence needed to date the surfaces is often missing. As a result, widely differing ages and different modes of origin may be postulated for a single surface (see the discussion of the Welsh tableland in chapter 8).

(iii) The approach which is favoured by many modern geomorphologists, notably in the U.S.A., is that concerned with the investigation of the relationship between process and form. This clearly involves in the first instance a careful analysis of weathering, transport, erosion and deposition, both as regards their mechanism and rates of operation. Secondly, an attempt must be made to relate, in a causal way, individual processes and groups of processes and particular forms. For example, in the case of valley-side slopes one basic element may be attributed to the action of one process (a convexity of profile may result from the action of soil creep), and another basic element to another process (concavity to rainwash).

It must be stated that not all geomorphologists regard the study of process as a valid task of geomorphology. Thus Wooldridge has written: '. . . I regard it as quite fundamental that geomorphology is primarily concerned with the interpretation of forms, not the study of processes'. Wooldridge argued that geomorphologists should aim at recognising 'developmental series of forms'. 'Sketch . . . a large number of meanders and you will readily convince yourself of the reality of downstream shift or sweep, as well as the stages leading to cut-off. No esoteric researches in fluid dynamics seem likely to add much to our comprehension of what is in essence a simple process in this, and like cases, of developing landscapes.' This was of course basically the approach to the study of slopes adopted by Davis, who placed individual slope profiles in a developmental series in formulating his hypothesis of slope decline. Unfortunately, it is difficult, or even impossible, always to decide whether the slopes (or meander forms) have been arranged in the correct sequence. There are undoubtedly pitfalls in this particular method, and many would now oppose the rather extreme views put forward by Wooldridge. On the other hand they might be more readily persuaded by Strahler's argument that geographically-trained geomorphologists are not well qualified to work in the field of process. The student of slopes must, in his view, 'be aware of the principles of fluid and plastic mechanics, thermodynamics, meteorology and hydrology'—all fields in which the geographer does not normally excel.

There are certainly many difficulties in the way of the process-form approach. Many of the processes being studied act very slowly (for example, chemical weathering or soil creep) or even intermittently (rainwash and certain types of mass movement), and therefore patient and precise measurement is needed. Some landforms, such as cliffs in very hard rock, may undergo no detectable change in a man's lifetime. Furthermore, it must always be remembered that the present landforms of an area may be inherited wholly or in part from the past, when climatic conditions and active processes were different from those of today. In short, many landscapes contain 'fossil' or 'relict' features, and to explain these satisfactorily one must attempt to reconstruct past processes. Another fundamental problem is the sheer difficulty of *proving* a causal relationship between process and form. How can it be demonstrated conclusively that a particular process results in a particular form? Rarely can *one* process be isolated for study, since in nature it is usual for several processes (mechanical and chemical weathering, soil creep and other mass movements, concentrated and unconcentrated rainwash and so on) to operate simultaneously and to be, to a greater or lesser extent, mutually interdependent. Even if one could relate one process to one form (say, soil creep to convexity of slope profile), it is not clear whether the process results in the form or *vice versa*. For example, in southern England the present-day slope profiles are largely relict features dating from the periglacial episodes of the Quaternary, and the distribution over these slopes of soil creep and wash may be intimately related to the forms and gradients inherited from the past.

In the study of process, the geomorphologist is in some instances able to undertake experiments, both in the laboratory and the field. It is often said that laboratory experiments are rendered largely invalid by problems of scale. Thus one can construct in the laboratory an accurate facsimile of a real river, as regards its long-profile and channel form, but in doing so one cannot at the same time scale down the fluidity of the water or the calibre of the load (sand, silt or clay) which the model river can move. This is undoubtedly true, yet the study of experimental models has been used to assist research into such phenomena as meander formation, terrace development, sedimentation, beach-profile changes and shore-bar formation. An actual example is the work carried out, with the aid of stream troughs, by Lewis (1944). Lewis experimented to determine the effects of the doubling and halving of discharge and/or load on the gradients of streams. He also demonstrated the formation of river terraces, unrelated to base-level changes, with the aid of model streams. These were fed at first with large supplies of sediment, much of which was deposited to provide the steep gradients needed for the movement of this heavy load. Later the sediment supply was restricted, and the stream by erosion produced an appreciably gentler gradient. In doing so it cut into its former deposits, leaving them upstanding as miniature terraces.

It must be admitted, however, that where practicable field experiments are usually more satisfactory than those attempted in the laboratory. The opportunities are, however, strictly limited, if only because of the large scale or extremely slow speed of most geomorphological processes. Those that act with comparative rapidity, and within well-defined areal limits, are most amenable to experimentation. For example, the movement of shingle along the beach by longshore drift can be studied with the aid of radioactive, fluorescent or dyed pebbles, and even offshore currents can be revealed by plotting the movements of weighted sea-bed drifters. Processes such as soil creep or glacier movement, on the other hand, are best measured in their natural state rather than 're-created' either in the laboratory or the field.

(iv) The approaches which have so far been discussed can be 'fused' together when a particular geomorphological problem is being tackled. An appropriate example is that posed by the tors of Dartmoor and similar areas (pp. 133–9). In the first place it is evident that these are much affected by the rock (granite) from which they are fashioned, and in particular by the strong local joint pattern, which largely determines their 'castellated' appearance. However, a full investigation of tor formation must also take account of the denudation chronology of the area, for many individual tors stand above well-preserved planation surfaces which may date from early-Tertiary times. Moreover, the climatic history must be reconstructed, since the tors may have been weathered either under Arctic conditions in the Quaternary or under wet and very warm climatic conditions in the late Tertiary. Finally, the actual processes responsible for the weathering of the granite and for the transportation downslope of the resultant detritus must be studied.

BASIC DIFFICULTIES IN GEOMORPHOLOGY

Any serious student of geomorphology will quickly realise that what is actually known with certainty about landforms and their origin is surprisingly small, despite the vast amount of research, testified to by innumerable books, articles and reports, which has been done during the last fifty or so years. He will observe that most individual landforms have been and still are explained in a variety of ways; that theories rarely remain unchallenged for more than a few years and sometimes for only a few months; that such fundamental questions as whether or not each climatic type produces its own distinctive landscape are still unresolved; and so on. Indeed geomorphology emerges as one of the most controversial of all subjects.

This lack of conclusiveness in so much geomorphological work is in part due to the comparative youthfulness of the subject. Much basic information about the landforms of many parts of the earth (such as the periglacial and tropical zones) has simply not been available—and as it does come to light existing theories have naturally to be revised. However, there are other reasons, equally important, which stem from the very nature of landforms themselves.

Firstly, geomorphology has adopted from geology the basic thesis of the uniformitarian creed, that the 'present is the key to the past', or in other words that landforms can be explained only in terms of processes that are observable today in some part of the earth. There is no real alternative to this assumption, yet great difficulties arise because of the recent climatic changes that most areas have experienced. As stated already, we now know that the landscape of southern England has been largely fashioned under the periglacial conditions of the Quaternary. Periglacial processes can today be observed at work in Arctic lands; but are the climatic conditions there precisely the same as those once existing in southern England? Again, where is there today an area of chalk being weathered in a frost climate? In the absence of such an area, our reconstruction of the denudational processes formerly at work in, say, the South Downs can only be tentative.

Secondly, the processes which shape landforms are themselves essentially destructive. The rocks and structures from which the landscape is carved are gradually worn away by weathering and erosion, and the forms so produced are in turn destroyed and replaced by other forms. It may be important to know what a present-day slope was once like, if one is concerned with formulating a theory of slope development, but unfortunately the old slope has gone forever. It is true that information about previous erosional activity may remain in the form of deposits, but these too are easily destroyed in later denudational episodes. Most planation surfaces must once have been covered by or associated with deposits which could tell all about the mode of origin and date of the surfaces. However, all too often such deposits have been completely effaced. Geomorphology, in short, suffers badly from a shortage of evidence—and that which remains is often so fragmentary as to be highly equivocal.

2

WEATHERING, TRANSPORT AND EROSION

INTRODUCTION

The breakdown of rocks is arguably the most fundamental of all geomorphological processes. In its absence landforms created by structural movements, such as anticlines, fault-scarps and rift valleys, would undergo little subsequent modification. The development of all erosional landforms would in fact be gravely inhibited, for the reason that stream corrasion, glacial abrasion and wind blasting all depend on the availability of coarse particles of rock such as only weathering can supply in sufficient quantities. However, it is as well to point out that weathering itself is to a large extent dependent on the operation of other processes. If there is no transportation away of the products of weathering, and no continual re-exposure of the fresh rock to the elements, weathering will necessarily be slowed down and perhaps in the long run halted altogether by a protective mantle of detritus. Again, erosive processes have the important effect of creating expanses of bare rock which can be attacked by weathering processes. It will be clear, therefore, that the processes of weathering, transport and erosion are mutually interdependent.

Weathering itself may be defined quite simply as the breakdown or decay of the rock *in situ*. In the process, the rock particles produced are not subjected to more than the very slightest displacement, such as may be involved in the loosening action. Two main types of weathering are usually recognised, mechanical (or physical) and chemical. However, the distinction is a somewhat arbitrary one, for it is very rare to find, say, mechanical weathering operating by itself, even under geological or climatic conditions which on the face of it overwhelmingly favour its action.

Transport is the process by which weathered material is moved off slopes, down valley bottoms, along and at right angles to the shore, or across the land surface in general. Numerous agents are involved; among them gravity-controlled movements, running water, glaciers, ice-sheets, waves, tidal currents and winds, but in this chapter only those associated with mass transportation on slopes and normal fluvial activity will be considered. The remainder are dealt with in the chapters on arid, glacial and coastal landforms.

Geomorphologists have coined various terms to embrace the main transportational processes, particularly those at work on valley-side slopes. Penck used the term 'denudation' to describe the stripping of weathered material from the underlying rock by falls, slumps, soil creep, rainwash and so on. However, to most geomorphologists denudation has come to have a much looser meaning, and is taken to include *all* the processes (of weathering, transport and erosion) which are responsible for the lowering and moulding of the landscape. 'Weathering-removal' is a useful and graphic term to describe the production and transport of slope detritus. 'Mass wasting' has been used frequently to describe the movement of rock debris *en masse* by landslides, avalanches, earth-flows, soil creep and solifluxion, all of which are directly controlled by gravitational pull.

Erosion is often taken to mean the sculpturing effects of running water, moving ice, waves and winds armed with rock fragments, produced in the first instance by rock weathering. In this sense it is synonymous with the terms 'corrasion' or 'abrasion'. Erosion is therefore readily distinguishable from weathering, in that a first essential is the movement of rock particles. However, the distinction between erosion and transport is not always so clear. For example, on a slope thickly mantled by weathered material a small rivulet may, by picking up some of the loose particles, form a quite deep and well-defined gully whose floor does not expose the solid rock beneath. In one sense the rivulet is performing merely the role of removal (that is, transport); in another, it is engaged in the sculpturing of a channel, through the picking up of fragments which will constitute a moving load whose impact on the sides and bottoms of the channel will displace other loosened fragments (or, in other words, cause erosion).

The term 'erosion' is also used commonly in a less strict sense, to describe the chemical breakdown of rock by moving water. Such chemical erosion may occur on a limestone coast, where the rock may be attacked not merely by corrasion but by the process of carbonation-solution. It is true that the latter is normally regarded as a weathering process, but both corrosion and corrasion are the result of wave action, and they combine to produce a single landform, a pitted 'wave-cut' platform. Other processes too, such as the breaking-off of rock fragments under hydraulic pressure or by glacial plucking (a process very much akin to frost weathering), can also be legitimately regarded as erosional.

WEATHERING

As stated above, two main types of weathering are recognisable. Mechanical or physical weathering leads to the disintegration of the rock into blocks and boulders ('block weathering' or 'block disintegration') and smaller fragments down to the calibre of small grains ('granular weathering' or 'granular disintegration'). In general, the products of mechanical weathering are comparatively coarse and often angular to a greater or lesser degree. Thus a granite may be effectively broken down by prolonged temperature changes into a 'sand' composed mainly of individual quartz and felspar crystals (p. 133). Mechanical weathering is induced (i) by heating and cooling (resulting from insolation and radiation) which, by causing expansion and contraction of rock minerals, set up powerful stresses within the rock, and (ii) by the formation of ice bodies in the joints and pores. Previous chemical weathering can be a very important aid, particularly in well-fissured, porous or crystalline rocks which have been 'opened up' by selective chemical attack.

Chemical weathering results in the decomposition of rock minerals by such agents as water, oxygen, carbon dioxide and various organic acids. The end-products of chemical decay include (i) 'residual' decomposition products, of which the most important are certain types of clay (for instance, kaolinite), and (ii) 'soluble' decomposition products (such as calcium bicarbonate), which are removable in solution by percolating waters and ultimately may find their way into rivers. In general, the products of chemical weathering are 'finer' than those of mechanical weathering—though it must be added that chemical weathering, when concentrated on certain susceptible rock minerals or along well-defined lines of weakness such as joints, can often lead to the *physical* breakdown of the rock and the production of detritus in many ways similar to that resulting from mechanical processes.

The fact that the residues of chemical weathering are normally clay or soluble minerals can have important implications so far as erosive processes are concerned. Thus, in an area where chemical weathering is dominant, the rivers may contain only suspension or solution loads, and corrasion may be non-existent. On the other hand, where mechanical weathering is most effective, the streams will move large traction loads, and corrasion can be a dominant process. It has even been suggested that existing valleys in temperate western Europe were formed wholly during the Quaternary, not directly by glacial erosion, but because refrigeration of the climate initiated a phase of intense mechanical weathering and so produced an abundance of coarse rock fragments which could be utilised by rivers in vertical incision (p. 349). However, this seems to be rather an extreme view, and does not take into account certain important facts. Firstly, as we have seen, chemical weathering *can* result in the production of coarse debris, as in the case of a quartz-conglomerate which is broken down into its constituent quartz pebbles and no further, since quartz is a 'stable' mineral that is not amenable to additional decay. Secondly, in many parts of Europe late-Tertiary gravels can be identified. In Belgium such deposits are associated with high-level terraces of the river Meuse near Liège, and have in fact been rotted *in situ* by the warm and moist late-Tertiary climate. Thirdly, it is incorrect to assume that all valleys are initiated by stream corrasion alone. Chemical weathering attacks rocks selectively, penetrating deeply at some points to give a thick mantle of decomposed material. Removal of this by streams, acting as transporting agents and effecting no corrasion at all, can result in the development of valleys and interfluves, much as in a normally dissected landscape (p. 176).

Factors controlling the type and rate of weathering

A *Rock hardness*. Rocks vary a great deal in hardness, depending on their constituent minerals, the nature of their cementation, the degree of their compression (which may in part be a reflection of their age, hence the common use of the term 'old hard rocks') and so on. The hardness of rock minerals is classified in Mohs Scale of Hardness, with grades ranging from 1 (very soft) to 10 (extremely hard). Some common rock minerals, with their grading, are gypsum (2), calcite (3), orthoclase felspar (6) and quartz (7). Most igneous rocks are hard, both as a result of their constituent minerals (which often include felspar and quartz) and because these minerals, in the process of cooling and crystallisation, are very tightly bonded together. Sedimentary rocks are for the most part softer, though often comprising very hard minerals. A sandstone, for example, may be largely made up of quartz grains, but will be quite weak because these are bonded by a soft cement, such as iron oxide or calcium carbonate. On the other hand, if the cement happens to be durable, the rock may be extremely hard. Quartzite, which comprises quartz particles cemented by silica, is one of the hardest and most enduring of all rocks—indeed it may be as much as 150 times as hard as a typical limestone.

It is a common misconception among inexperienced students of geomorphology that hardness is the most important single factor determining the resistance of a rock to weathering processes. It is true that some very hard rocks are little affected, even over a long period, by mechanical or chemical attack. Thus in parts of the English Chalk country there are spreads of large boulders, known as 'sarsens' and composed of silicified sands and flint gravels, which have apparently endured since early-Tertiary times (Fig. 138). Again, within the British Isles as a whole the older pre-Cambrian and Palaeozoic rocks of the west and north tend to be much harder than the younger Mesozoic and Tertiary sediments of the south and east. However, the marked differences of relief between 'Highland' and 'Lowland' Britain have arisen not only as a result of resistance to weathering, but also (i) because the rocks of the west and north are less prone to *erosion*, and (ii) because the west and north have been more affected by localised uplifts during the Alpine period of the Tertiary.

In fact, it can be stated categorically that rock hardness is usually one of the less important determinants of resistance to weathering (though not erosion). It is of virtually no consequence in combating chemical weathering, and only slows down insolation and frost weathering. The important point is that nearly all hard rocks possess weaknesses, either structural or of chemical composition, which can be taken advantage of by chemical processes.

B *Chemical composition.* This is obviously of prime importance in influencing the resistance of a rock to chemical decay. However, it can also have some effect on mechanical weathering. For instance, a rock composed of variously coloured minerals is subjected to strain because the capacity of these minerals to absorb the sun's heat is not uniform, and thus differential expansion is caused. Again, a dark rock such as basalt, gabbro or serpentine will heat up more rapidly, and so experience greater strain, than a light-coloured rock like chalk or limestone, which will reflect most of the sun's rays. In areas where insolation weathering is able to operate, such as the continental deserts, rock colour may help to determine the amount of block and granular disintegration, but in less arid climatic régimes it will be subservient to other factors.

Over the earth as a whole, chemical weathering is more active in the breakdown of rocks than is mechanical disintegration, and in some regions its dominance is very pronounced. It is especially effective in the presence of water (which allows the formation of certain acids) and high temperatures (which speed up the rates of chemical reactions). Chemical weathering takes many forms, but the following are the most common processes.

(i) Hydration results from the capacity of certain minerals to take up (adsorb) water and, in so doing, to expand and set up stresses within the rock. Perhaps the best-known example of this process is the conversion of unhydrated calcium sulphate (anhydrite) to hydrated calcium sulphate (gypsum). In rock weathering the process is most important when some decay of the minerals has already taken place, for some decomposition products (resulting particularly from the partial breakdown of felspars and micas) are very susceptible to hydration.

(ii) Hydrolysis involves a chemical reaction between rock minerals and water. Felspars are attacked in this way, and break down to give aluminosilicic acid and potassium hydroxide. Since water in nature is not pure, but contains carbon dioxide, a further reaction will affect the potassium hydroxide, which is converted into potassium carbonate and removed in solution. The aluminosilicic acid is itself 'unstable', and will break down into clay minerals (which at first are 'colloidal', occurring as minute particles dispersed in water, but later coagulate into an amorphous clay mineral) and silicic acid, which is also removed in solution. The net result of this complex process is therefore the weathering of felspars to leave residual clay minerals, notably kaolinite.

(iii) Oxidation is particularly effective in attacking iron compounds, which frequently act as cements in sedimentary rocks or are common constituent minerals in clay deposits. The process is only really effective above the water-table, where the rock can be penetrated via joints and pores by atmospheric oxygen. Oxidation can be detected by the changes of colour it induces, as in the case of blue, grey and green ferrous compounds which are oxidised to give red or brown ferric compounds. A good example is the greensand of south-east England, which in an unweathered state derives its 'grass greenness' from the mineral glauconite (a silicate of iron and potassium), but in a weathered state is light brown, yellow or bright red, owing to the oxidation of the glauconite.

(iv) Carbonation is one of the most simple and widespread of all chemical weathering processes, and involves the alteration of carbonate compounds by water containing carbon dioxide (weak carbonic acid solution). Chalk, limestone and dolomite are greatly affected in this way, the carbonation producing calcium bicarbonate which is soluble in water.

(v) Solution is of little direct importance in chemical weathering (except in the rare case of rock-salt exposures), but plays a very active role in the removal of certain residual products of other weathering processes, notably hydrolysis and carbonation.

The main achievements of these chemical processes are as follows. Firstly, they weaken the 'structure' of a rock, by selectively attacking the constituent minerals in igneous and metamorphic rocks, by more general attack on calcareous rocks, and by attacking the cements of some sedimentary rocks. Secondly, they set up stresses within a rock by causing the expansion of certain minerals. Thirdly, they produce compounds which can be removed in solution, leaving behind residual deposits. These include residual decomposition products (clay minerals), plus materials within the rock which are not amenable to chemical decay. Thus granite weathers not only to kaolin clay, produced by the hydrolysis of felspars, but also 'stable' quartz particles. Some residual deposits left by prolonged chemical weathering are composed almost entirely of 'rock impurities' which are released but little changed by chemical processes. Good examples are the clay-with-flints of the English Chalk country, which in part is a residue left by carbonation-solution acting over long periods in the Tertiary, and the reddish clay soils of many limestone areas.

It will be apparent that the resistance of rocks to chemical decay will depend in the first instance on their composition, though other factors such as jointing and porosity (which permit the entry of water, acids and oxygen) and the prevailing climatic conditions must also be taken into account. It is possible to arrange the common rock minerals in order of 'stability' (that is, their ability to withstand chemical alteration). The most stable minerals are quartz, muscovite (mica) and orthoclase felspar, and the least stable are olivine, augite and plagioclase felspars. It is noteworthy that the three first-named are light-coloured, whereas two of the second group (olivine and augite) are dark-coloured. In fact, as a general rule light-coloured minerals are more stable than dark-coloured, and it follows that acidic rocks, which contain a high percentage of quartz (for example, granite and rhyolite), are inherently more resistant to chemical weathering than basic rocks, which comprise a high percentage of plagioclase (for example, diorite and andesite). However, this statement must be qualified by stating that other factors may prove overriding. Thus in a tropical climate, with abundant moisture and high temperatures, a well-jointed granite will be rapidly and deeply rotted, whereas in a temperate climate gabbro, a basic intrusive rock, will be extremely resistant (as is shown by the Black Cuillins of Skye).

C *Rock jointing*. This is a factor of the utmost importance in all types of weathering, for jointing has the effect of increasing greatly the area of rock surface which is available to attack by chemical processes, of allowing the ingress of water and oxygen, and of providing lines of weakness which can be utilised by mechanical agents such as wedges of ice. Joints in rocks are minute cracks, often arranged in sedimentary rocks at right angles to the bedding-planes (which themselves perform, in terms of weathering, much the same role as joints) and in igneous rocks at right angles to the margins of the intrusion or extrusion. There is no relative displacement of the rock on either side of the joint, such as occurs in faulting. Joints may be relatively discontinuous minor features, or they may be well-defined and persistent over quite long distances, giving a 'master' (or 'major') joint pattern.

Joints are developed in three main ways. Firstly, in igneous rocks tensile stresses are set up as a result of contraction during cooling. The pattern of joints so produced is usually broadly rectangular, but it may be more complex, as in the basaltic lava extrusions of the Giant's Causeway, Antrim, where strong vertical jointing on a hexagonal plan has been formed. Secondly, in sedimentary rocks joints are produced by shearing and tensional forces set up during earth-movements. In areas of simple structure, the joint pattern may again be rectangular, with dip- and strike-aligned elements, but rocks of considerable age, such as the Carboniferous Limestone, may be characterised by a 'palimpsest' of joint patterns, reflecting the complex stresses to which the rock has been subjected at various times in its long history (Fig. 3). Thirdly, in crystalline rocks such as granite or gneiss, 'dilation' occurs after a heavy overburden of rocks has been removed by denudation. As 'pressure-release' operates, the rock mass tends to recoil upwards, and joints running parallel to the surface ('sheet jointing') are formed. The result may resemble the stratification of sedimentary rocks, and such joints are sometimes accordingly referred to as 'pseudo-bedding planes' (Fig. 4).

Weathering processes are aided by joints in a variety of ways, of which the following are only a sample. Firstly, chemical weathering by acidulated rainwater is concentrated along joints and bedding-planes. Limestone features such as dolines, dry valleys, sinks and underground passages are often closely related to master joints. Subsurface caverns

3 **The joint pattern of the chalk on a wave-cut bench at Telscombe Cliff, Sussex. Note the person ringed for scale, and the major joints running from bottom right to top left. [Eric Kay]**

may develop in zones where the joint pattern is especially close-spaced, and carbonation-solution thus highly effective. In granitic areas, the rock may be divided into cuboidal blocks as a result of chemical attack along vertical contraction joints and horizontal pseudo-bedding planes. In areas of close jointing the granite may be totally decayed because of the comparative instability of the felspar and ferro-magnesian minerals, and a kaolin residue containing particles of quartz will be formed. Where the jointing is wider and the decomposition less advanced, rounded subsurface boulders may be produced by the reduction of the joint-bounded rectangular blocks of granite (Fig. 5). Secondly, joints aid the process of frost and ice wedging in periglacial and glacial climates. Water percolates freely into the joints during the day and freezes at night. In cooling from 4 °C to 0 °C, the water undergoes a 10% increase in volume, and enormous pressure (in the order of 2000 lbs per square inch) is exerted on the rock on either side of the joint. In time, this will be broken into angular blocks, forming 'felsenmeer' or 'block-spreads'.

Indeed, in a frost climate, the presence or absence of joints is probably the overriding factor in determining the resistance of a rock to weathering (pp. 338–9). One of the principal reasons why the Chalk of southern England was so much affected by the periglacial conditions of the Quaternary, leading to accelerated retreat of slopes and the accumulation

4 **Pseudo-bedding of granite, Dartmoor. [R. J. Small]**

of extensive deposits of coombe rock, lay in its possession of a pattern of close jointing, allied to its lack of mechanical strength. Even today the action of frost on exposed chalk, in railway cuttings and quarries, can be very striking. Thirdly, the important process of exfoliation (pp. 292–3) almost certainly depends entirely on the existence of sheet jointing, resulting from dilatation.

D *Climate*. This again has a great influence on weathering processes, for many of these are totally dependent on the existence of certain climatically induced conditions. Thus, frost weathering can occur only where there are atmospheric freeze–thaw cycles; insolation weathering requires considerable diurnal fluctuations of temperature between very hot during the daytime and cool at night; all types of chemical weathering operate most effectively in very warm climates, for the intensity of chemical reactions is approximately doubled for every 10 °C rise in

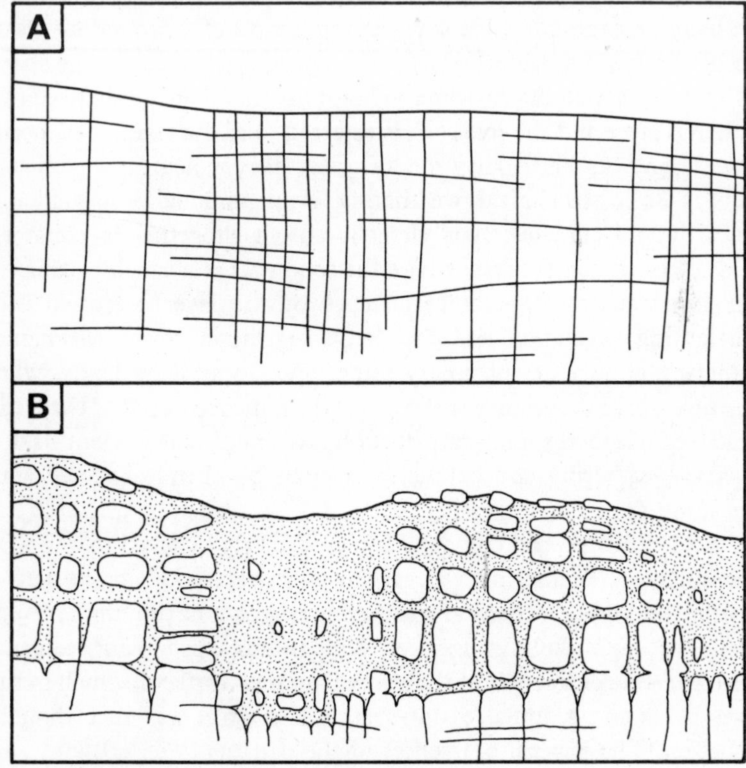

5 Deep weathering of differentially jointed rock

temperature; and chemical weathering is most active in wet climates, for water is essential to processes such as hydration, hydrolysis and carbonation. From the weathering point of view, it is convenient—if oversimple (p. 8)—to make a very broad division of the earth's climates into four main types.

(i) Tropical humid climates. The high temperatures and excessive rainfall (either seasonal or throughout most of the year) promote very active chemical decay. Except on the steeper slopes, where mud-flows and landslips occur, much of the weathered material remains *in situ*, particularly in tropical forest areas where the root mat impedes surface creep. Even rocks which are usually classified as inherently resistant, such as acidic igneous rocks, are decomposed to a great depth, sometimes exceeding 100 ft and perhaps reaching in exceptional circumstances to 300–400 ft (p. 175). In the process the rock may be little disturbed, and the various weathered and unweathered minerals maintain their positions within the rotted rock. The weathered layer frequently contains unweathered blocks ('corestones' or 'woolsacks'), which may eventually be exposed at the surface by removal of surrounding waste. One of the most notable features of tropical weathering is the sharpness of its lower limits in some areas. This junction-line between the weathered layer and the live rock is referred to as the 'basal weathering surface', and its form may be of great morphological significance (pp. 175–80). Mechanical weathering is of minimal importance in tropical humid climates. It is virtually absent altogether in equatorial and tropical rainforest areas, where the solid rock is normally cushioned from atmospheric temperature changes by the layer of rotted rock, the decaying vegetation and the dense vegetation cover. Mechanical weathering is only of subsidiary importance in savanna lands, where insolation weathering may have a little influence on the bare rock surfaces of inselbergs and scarp-faces but overall is subservient to deep chemical weathering and exfoliation caused by chemical attack along sheet joints (p. 175).

(ii) Arid and semi-arid climates. It was formerly believed that in these areas chemical weathering was largely inactive, owing to the lack of moisture, and that mechanical weathering was at a maximum. These are regions of high daily ranges of temperature, producing alternate expansion and contraction of the many bare rock surfaces, which in turn causes block and granular disintegration and (so it was once thought) exfoliation. The general coarseness of the detritus in desert lands, and the relative lack of decomposition products such as clay, certainly points to the importance of *physical* breakdown of the rock here. However, for the reasons given on p. 294, it is now believed that chemical processes, which can act in the presence of minute quantities of moisture, assist greatly in exfoliation and the 'basal weathering' of steep slopes and isolated rock masses in deserts. None the less, perhaps it would be unwise to infer too much from such evidence. There are many rocks that are normally very susceptible to chemical attack, such as limestone and dolomite, but appear to be highly resistant under arid conditions.

(iii) Temperate climates. Most of the main weathering processes are encountered here. For instance, frost weathering occurs on exposed jointed rocks in mountain areas and on cliff faces; oxidation affects rocks containing iron minerals; carbonation is active in chalk and limestone country; and hydrolysis and associated processes of carbonation and solution operate on igneous rocks. However, the important point is that none of these processes acts very rapidly in temperate conditions. Chemical weathering is, in sum total, more active than mechanical weathering, for it is aided by the more or less continous soil and vegetation cover, which favours infiltration of rainwater and the generation of humic acids. But the moderate temperatures mean that the rates of chemical reaction are not fast, and one never encounters deep rotting of the rock, such as characterises tropical humid climates, operating today. Only in limited circumstances is chemical weathering today very active in temperate areas. The obvious case, already referred to, is that of limestone, in which analyses of spring water have shown that carbonation-solution is operating effectively. In general, however, the landscapes of humid temperate areas are, in terms of weathering (and probably erosion too), in a state of comparative stagnation. Even the great scree accumulations of temperate mountains, apparently pointing to the power of present-day frost weathering, are largely relict features of the Quaternary cold periods.

(iv) Arctic climates. The dominant weathering process here is frost action ('congelifraction'), which produces spreads of blocky debris together with large quantities of finer material resulting from the granular disintegration of the larger boulders. The weathering occurs in two main ways. Firstly, on steep bare rock faces water can penetrate into cracks, freeze and cause ice wedging. Secondly, in the active zone above the permafrost (p. 322), freeze–thaw cycles can lead to the rapid breakdown of the solid rock or weathered material being moved by solifluxion. The part of chemical weathering in glacial and periglacial climates is usually considered to be insignificant, for in spite of the abundance of water during the thaw season the rates of chemical reactions are slowed down by the low prevailing temperatures. However, as in deserts, chemical weathering may be far from totally inactive. The bad drainage of low-lying or flat areas, owing to the presence of impermeable permafrost, favours peaty accumulations and the formation of organic acids. Furthermore, water emerging from beneath snowpatches has been found to contain calcium bicarbonate in solution, thus proving that carbonation is taking place. This could be due to the fact that the solubility of carbon dioxide in water increases as the temperature is lowered towards 0° C, so that comparatively high concentrations of carbonic acid can form. Measurements by the French geomorphologist Corbel (1959) have indicated that in 'cold snowy climates' (marked by the development of seasonally frozen ground and heavy winter snowfall), rates of limestone solution both at the surface and underground are abnormally high. However, it seems certain that in the drier and colder periglacial areas deep and continuous permafrost completely inhibits subsurface chemical weathering.

E *Relief*. This is a factor not generally taken into account in a consideration of weathering, but it can exert a good deal of influence. Renewal of exposure of the live rock is essential to the continuation of mechanical weathering. Hence in areas of high relief and steep slopes, which favour mass transportation processes such as landslides, slumping, soil creep and solifluxion and the constant laying bare of rock surfaces, frost and/or insolation weathering can operate effectively. Conversely, water will run quickly off the slopes, particularly if they have no mantle of soil or peat, and chemical weathering may be reduced. In areas of gentle relief, on the other hand, a thick layer of soil and other weathered material is usually present. This protects the rock from further mechanical breakdown and, by holding water, enhances chemical decomposition. Therefore, other things being equal, the cycle of erosion is marked by a gradual change from dominantly mechanical to dominantly chemical weathering. This is one of the explanations of the development, on the pediplained surfaces of South America, Africa and Australia, of very hard weathering crusts ('duricrusts'). In hot and seasonally dry climates, salt solutions are drawn up through the rock by capillarity and, after evaporation, are deposited as hard nodules just below the ground-surface. 'Ferricrete', or lateritic ironstone, may be formed to a depth of several feet in this way. It comprises innumerable small concretions of limonite (hydrous ferric oxide). Other deposits of similar origin, usually associated with low relief surfaces, are 'silcrete' (in which the cementing mineral is silica) and 'calcrete' (bonded by calcium carbonate).

TRANSPORT ON SLOPES

The removal of weathered material from slopes, one of the most important of all geomorphological processes, is effected in two ways. Firstly, the waste mantle may undergo downslope movement *en masse*, under the influence of gravitational pull and without the aid of agencies such as running water, moving ice or wind. This process is referred to as 'mass movement' or 'mass wasting'. Secondly, the surface layers of the regolith, particularly where composed of fine particles, may be transported by running water. The latter often takes the form of a very thin sheet, perhaps less than one-tenth of an inch in depth, but there may also be some concentration into rills (small threads of running water) especially on the middle and lower parts of the slope.

Mass movements

These are in turn divisible into two types. Firstly, when the weathered mantle is quite thick, and is composed of fine material capable of retaining moisture, a slow but more or less continuous 'flowage' will take place, even over comparatively gentle gradients. In movements of this type—as in streams—the maximum velocity will be at or just below the surface, and the rate of flow will diminish sharply with depth to zero at the junction between the regolith and the solid rock. This is due to the excessive friction which movement would generate towards the base of the weathered layer, where the rock fragments are normally larger, often more angular and less mobile than the overlying fines. It follows that flowing movements are wholly transportational, and that there can be no corrasion of the solid rock by the moving debris. None the less, it is possible that mass flowage may indirectly affect the profile of the underlying rock, simply because this will be protected from mechanical weathering if there is an excessive build-up of slope debris and subjected to more effective disintegration if the soil cover becomes thin. Secondly, where the regolith consists of coarse and angular debris, as in deserts and in mountain areas where frost weathering is rife, there will be little mass movement on gentle slopes. However, in favourable circumstances —for instance, in a deep gully between two steep rock faces—such material may accumulate in large quantities. Any subsequent disturbance, perhaps by stream undercutting or earth tremors, may precipitate rapid sliding of the weathered material. Such 'slides' differ from 'flows' in attaining much higher velocities, in affecting rock debris which is often dry, and in not being characterised by reduced movement at depth.

A *Mass movements of the flow type.* The various flowing movements observable on slopes grade, in terms of mechanism and velocity, imperceptibly into each other, but it is convenient to make a broad distinction between those which act comparatively rapidly and those in which mass transportation is very slow.

(i) Rapid flows. So-called 'earth-flows' are a frequent occurrence in humid regions, where the rock is deeply weathered and mobile. Masses of earth on hill-side slopes may slip quite suddenly, leaving a crescent-shaped scar above and giving rise to a bulging 'toe' beneath. In cross profile the scar tends to be concave and the slipped mass convex. Common causes of earth-flows are the undercutting of steep soil-mantled slopes by stream corrasion, the existence of spring- and seepage-lines, and the occurrence of very heavy rainfall. In the last two causes, great quantities of water may be absorbed by the earth-layer, and its weight thereby increased. At the same time its inertia will be reduced because of the lubricating effect of the constituent water, and a sudden collapse may occur, as in the case of the disastrous slip of coal-tips overlying springs at Aberfan in 1966. Movement is very rapid indeed on steep slopes. An extreme type of earth-flow is known as a 'mud-flow', and is developed where the soil is particularly highly charged with water and comprises a large percentage of clay minerals. Mud-flows occur in mountain areas after heavy rainfall, in periglacial areas during the thaw season, on the slopes of erupting volcanoes and even in deserts, when a heavily loaded stream-flood is gradually transformed as it loses its water by evaporation and percolation. Mud-flows are sometimes associated with the transportation of large weathered blocks, which tend to find their way to the snout of the 'mud-lobe', forming a dam which in time will be burst as a result of pressure from above. Subsequently the fine material may be washed from such mud-flows to leave 'streams' of rocky debris.

(ii) Slow flows. The overwhelmingly dominant process here is soil creep, which is encountered in a wide variety of climatic régimes, including that of humid-temperate regions such as the British Isles.

Soil creep may be defined as the slow and almost imperceptible movement of rock and soil particles downhill under the pull of gravity. The French geomorphologist H. Baulig has compared creep with the movement of lumps of coal at a steady speed down an inclined chute that is mechanically shaken. The evidence of creep may be seen on almost any moderately steep soil-covered slope. Soil is built up against walls, hedges and fences, which are often noticeably displaced in plan; posts and trees tend to lean over in a downslope direction, owing to the fact that movement in the upper soil layers is greater than at the base; the upper edges of weak strata such as shale and slate are sometimes bent over; rock fragments are displaced downslope from their parent outcrop; and the grass cover is commonly ruptured, as in the development of terracettes or 'sheep-walks', which are accentuated by the trampling of animals but are in the first instance undoubtedly a phenomenon of soil creep.

There are several factors which promote or hinder soil creep. Anything which disturbs the soil particles (such as heating and cooling, freezing and thawing, and the burrowing of earthworms and animals) inevitably causes some displacement in a downhill direction. The presence of water is important, not only for its lubricating effects, but also because it causes certain clay minerals to swell and move adjacent particles. The gradient of the slope is obviously significant. Penck has suggested that the minimum angle for effective soil creep is 5°, and observation shows that as the slope becomes progressively steeper instability of the soil cover becomes marked. On slopes in southern England, when an angle of 35° is exceeded soil creep is so rapid that it is difficult to sustain a continuous grass mat. However, it is curious that field observations by Schumm (1956) have revealed that erosion by creep (that is, the lowering by removal of the soil surface) can be greatest at or near the crest of the slope (1–1·5 ins per year) and less at the base (0·4–0·8 in) despite the steeper angle there. Vegetation is undoubtedly a very important influence on the rate of creep. Tree roots tend to stabilise the soil, though the bending of tree trunks shows that creep is not halted altogether, but probably proceeds by the filtration of the soil particles through the root mat. A grass cover is possibly more effective in retarding creep, for measurements show that in England today soil movement can be much slower than on poorly vegetated slopes of comparable angle in semi-arid areas of the U.S.A.

A more rapid form of flowage, intermediate between soil creep and earthflows, is solifluxion, which occurs mainly in periglacial climates.

Here saturation of the weathered mantle is caused by periods of rapid thawing of ground ice and surface snow, and by the presence of an underlying layer of impermeable permafrost. The loosening and disturbance of the soil by the formation of ice crystals and needles is a major contributory factor (pp. 323–4).

B *Mass movements of the sliding type.* The most noteworthy of these are landslides and rock avalanches. These involve the very rapid sliding of accumulated masses of rock debris, with little or no flowage. Landslides occur on very steep slopes, usually in mountain areas which have had their relief sharply accentuated by valley glaciation, in areas which have been deeply rejuvenated, and on precipitous scarp-faces. The transported material usually collects at the foot of the slide, forming talus cones at an angle of approximately 35° (the common angle of repose of scree). The largest fragments, because of the momentum they gain during the sliding movement, accumulate at the base of the talus, whilst finer material, travelling at a lower velocity, lodges near the apex of the cone. Thus a crude grading of the constituents of the talue slope is achieved. A more advanced form of the sliding process is 'free fall', which occurs on cliffs and exceptionally steep slopes. As the rock is weathered by processes such as frost wedging, the loosened fragments slide or fall immediately to the base of the slope. There is thus no build-up of a weathered mantle, and the underlying rock is constantly exposed to attack by mechanical weathering.

One type of mass movement which involves a form of sliding, but which in most cases is also associated with some flowage, is 'rotational slipping'. A mass of weathered material often builds up until the weight of the deposit exceeds its resistance to shearing (which in turn may be suddenly reduced by an accession of soil water). The upper part of the mass then slips bodily downwards over a shear plane which is curvilinear in profile. The movement is such that the displaced mass is tilted backwards, with the result that the toe actually rises. Detailed examination of rotational slips reveals usually the presence of a series of curved shear planes, so that the upper part of the slipped mass develops a stepped profile. Rotational slips occur not only in weathered material, but in suitable lithological conditions affect the solid rock. Where mechanically strong and permeable sandstone, chalk or limestone overlies weak and impermeable clay or shale, the latter is likely to founder, particularly at points where underground water is held up in large quantities at the junction between the two rock-types. The slip plane itself will be well lubricated, and the toe of the slipped mass becomes so sodden that flowage, often involving earth- and mud-flows, occurs. Rotational slipping is at the present day affecting the coastal cliffs of southern England (Fig. 6), especially in localities where Chalk and/or Upper Greensand overlie Gault Clay, as at Ventnor in the Isle of Wight and at the Warren, Folkestone. There is much evidence to show that it affected inland scarp-faces in the same area during the Quaternary, when movement was promoted by ground ice disturbance and the abundant underground water derived from the melting of snow.

Running water

The transport of weathered material on slopes by running water is a vital process in all but the most arid climates. By contrast with soil creep, water action is intermittent, and is really effective only during and after prolonged or heavy rainfall or when thawing of a snow cover takes place. When rain falls on a slope, unless the weathered mantle is already saturated, the water will sink quickly into the ground and there will be no run-off. As the soil gradually becomes saturated, however, the rate of infiltration will slow down and in time become constant. This constant rate is determined by 'the infiltration capacity of the soil'. Any excess of rainfall over the infiltration capacity will naturally collect on the surface, forming individual puddles in small depressions. In time these will overflow to give a layer of water covering the ground surface, and this will begin to flow as 'sheet-wash' or 'rainwash' in a downslope direction. If the water layer is very shallow and the ground smooth, the flow may be of the 'laminar' type, but if the water is deeper and the surface rough

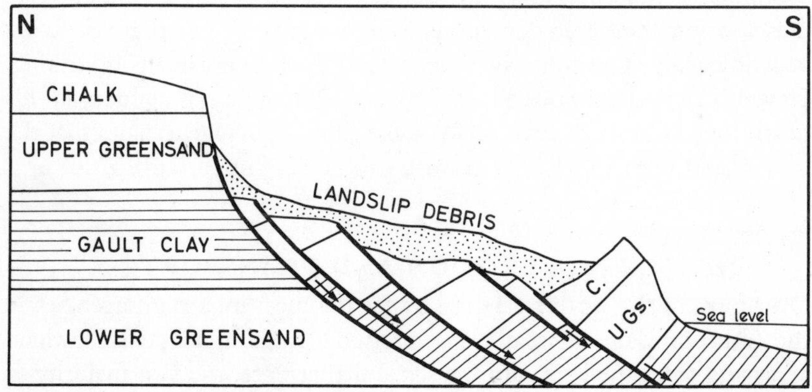

6 Rotational land-slip, southern Isle of Wight

it will become 'turbulent', with much eddying and a considerably increased capacity to pick up and transport fine particles. Irregularities of the surface may also cause the concentration of the rainwash into rills, which subdivide and rejoin to give an anastomosing pattern. Downslope the amount of surface water will increase, and the rills may become sufficiently large and powerful to initiate gullies. The impact of the raindrops nourishing the wash may be important, both in directly dislodging soil particles and in compacting the surface, so reducing the infiltration capacity of the soil.

The amount of rainwash generated on a slope will depend on various factors, including the intensity of the rainfall (fine drizzle is unlikely to exceed the infiltration capacity of the soil), the steepness of the slope (run-off is generally greater on a steep slope, because the increased velocity reduces the time available for loss due to percolation), the infiltration capacity (a clay soil will promote greater run-off than coarse sand and gravel) and the nature of the vegetation cover (grass reduces the impact of raindrops, impedes surface flow, and aids infiltration via the root passages).

The transport of fine weathered material by rainwash is believed to be most effective away from the crest of the slope, where the depth of the flow is zero. Horton (1945) has suggested that on every divide there is a belt of no erosion, whose width varies according to the infiltration capacity of the slope mantle, the intensity of the rainfall, and the rate at which the slope steepens away from the divide. The last point is important, because observations show that transport by rainwash increases steadily with a steepening of gradient up to about 40°, after which there is a steady decrease to nil erosion on a vertical face. It is common to find in highly dissected 'badland' areas rounded slope crests, where the action of running water is apparently negligible, giving way downslope to a zone where trenching by individual rills is striking. However, it must be added that there is no unanimity of opinion on this downslope increase of erosion, for some geomorphologists claim that the lowering of crest slopes by rainwash can be very important (p. 208).

FLUVIAL EROSION AND TRANSPORT

Over most of the earth rivers are by far the most important agencies in the transportation of weathering products from highland to lowland areas and from the land into the sea. Furthermore, the vast majority of valleys, although in many instances modified by glaciation and periglaciation, owe their origin to vertical and lateral erosion by streams. In order to understand something of the complex processes of fluvial transport and erosion it is necessary to consider briefly some aspects of the dynamics of stream flow. This is a highly technical subject, and only some of the simpler and more basic concepts will be touched upon here.

Stream energy

Rivers are able to do geomorphological work because they possess energy. The total energy, or 'power', of any stream is conditioned by a number of factors.

(i) The pull of gravity is the cause of water movement through a channel, and anything which effectively increases this force will promote greater stream energy. The amount of water in the stream (its mass) is a case in point, for the total gravitational pull on a large body of water will be greater than on a small one. Therefore, a stream in spate, simply because of its increased volume, will possess greater energy and so have the potential ability to achieve more in the way of transportation and erosion than a dry season flow.

(ii) The height of the stream above the base-level of erosion largely conditions the supply of 'potential energy' (which is the energy stored up in the water simply because it occupies, if only temporarily, a position well above that to which gravity is impelling it).

(iii) The steepness of the stream channel is extremely important, for it is one of the main controls over the velocity of water flow. The greater the velocity, the greater will be the amount of 'free' or 'kinetic' energy (that is, the energy which is actually being used up in the movement of the water and its load). This is the reason why, in mountainous areas, the rivers—flowing over steep gradients and augmented in discharge by very heavy rainfall or snow and ice meltwater—are able at times to transport phenomenal quantities of very large boulders with apparent ease.

The potential and kinetic energy possessed by a stream is dissipated through the development of friction (involving conversion into heat energy), both between the moving water and the stream bed ('external friction') and within the body of the stream ('internal friction'). The losses of energy due to friction with the bed may be partially 'expended' on the process of corrasion, since this results from the impact of rock particles moved by the stream against the channel.

The various factors that influence external frictional losses include the following.

(i) Most streams move a 'bed' or 'traction' load, and the greater and/or coarser this load the greater will be the energy loss, for the reason that the movement of water alone over rock generates less heat than the movement of solid particles over rock.

(ii) The configuration of the stream channel is important, in that it determines the length of the line of contact between the moving water and the bed (the so-called 'wetted perimeter'). For a given discharge, a very wide and shallow channel will give a comparatively large wetted perimeter, whereas a narrower and deeper channel will produce a much reduced wetted perimeter (Fig. 7). The efficiency of a stream channel in this respect is defined by its 'hydraulic radius', which is the ratio between the cross-sectional area of the stream and the length of the wetted perimeter. A simple calculation will show that the greater the hydraulic radius (resulting from a reduced wetted perimeter), the more efficient the stream will be. The ideal cross-section of a stream, leading to the smallest losses of energy from external friction, is semicircular. An examination of stream channels in the field, however, will quickly reveal that streams rarely attain or even approach the ideal. Most are comparatively broad, with flattish bottoms and steep, even undercut banks. The reasons for this are numerous. During periods of flood, streams tend to attack their banks and so widen their channels, but when the water-level subsides the eroded material cannot be restored, for the

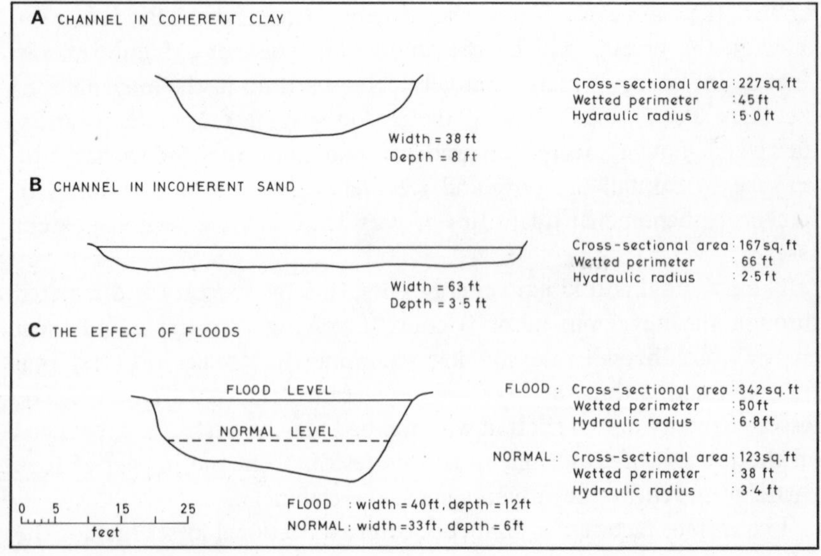

7 **Stream channel cross-sections**

streams are no longer at the 'bankfull stage'. Instead, deposition occurs on the channel floor, forming banks and shoals. In areas where the channel banks are composed of loose friable materials, such as incoherent sand or alluvium, stream widening is again very marked, and is accompanied by wholesale shoaling and braiding. On a small scale this can be observed on a sandy sea-shore, drained by rivulets as the tide ebbs. Continual bank failure leads to the formation of absurdly large channels, the floors of which are in part abandoned as the flow cuts into one or both banks. On a much larger scale, similar broad and complex channels characterise the extensive sandy outwash deposits beyond many glacier snouts. Stream channels with a comparatively efficient hydraulic radius can in fact only form where the banks comprise coherent materials, such as coarse gravels or clay.

(iii) The size of the stream channel is another important factor in determining its efficiency in terms of external frictional losses of energy. Obviously, as the discharge of a stream increases, so the wetted perimeter must increase in length. However, the hydraulic radius will also tend to increase, in some instances very markedly. As a result, the loss of stream energy owing to external friction will be proportionately reduced (Fig. 7). A very large stream would therefore seem to possess more available energy for transportation than a small one—an inference which is strongly supported by the vast amounts of detritus moved under flood conditions (Fig. 8). However, one must be careful not to oversimplify the analysis. An increase of stream discharge will normally be accompanied by the acquirement of a much larger bed load, in which case external friction (and also river corrasion) will be enhanced. This is an illustration of the dangers of trying to deal separately with aspects of fluvial activity that are in fact inextricably linked. Some writers have attempted to calculate the percentage of stream energy used in (i) the movement of the water, (ii) the movement of the load, and (iii) corrasion. However, this is unrealistic in that movement of load cannot take place without movement of water, and corrasion cannot take place except in the presence of a moving bed load. Returning to the question of the efficacy of large streams, it must be admitted that they are undoubtedly very powerful agents both of transport and erosion, but this may be due in the main to the possession of great total energy and a very marked reduction in the effect of 'channel roughness' (p. 38). Finally, it needs to be pointed out that sometimes an exceptional increase in discharge can lead to a marked diminution of efficiency. When a lowland stream overtops its banks and spreads widely over the adjoining flood-plain,

8 Rock debris transported and deposited under flood conditions, Glen Nevis, Scotland. [Eric Kay]

there is in effect a very considerable increase in the wetted perimeter (and a corresponding reduction in the hydraulic radius), together with a marked diminution in velocities of flow. Stream energy is thus much reduced, and deposition of the stream load necessarily follows. This is broadly the mechanism by which flood-plains are steadily built upwards by successive increments of alluvium.

(iv) The 'roughness' of the stream bed will greatly affect external frictional losses. Where there are numerous boulders and large stones, deep scour pools and rocky outcrops on the channel floor—as in a youthful stream in a highland area—these losses will be much greater than in a stream smoothly floored by sand and mud. Indeed, the overall velocity of mountain streams may be so restricted by the effect of roughness that, though flowing over steep gradients, they may flow no more quickly than lowland rivers passing over much gentler slopes. Observers have noted a tendency for roughness actually to decrease in a downstream direction, and thus for the stream to become more efficient as it grows larger and approaches the sea. Certainly there must be a relative decrease in the effect of roughness away from the source, since as the stream becomes deeper the configuration of the bed is less able to impede movement in the main body of flowing water.

The loss of energy in a stream as a result of 'internal friction' is likewise affected by a number of factors. The degree of turbulence (eddying or confused flow) is important, for when it is well developed neighbouring threads of water travel at different speeds, setting up what is known as 'viscous shear'. Turbulence itself is promoted by such causes as increased stream velocity, extreme roughness of the stream bed, and sudden changes of direction of the stream channel. The internal losses of energy are also larger when the suspended load is great, thus increasing the viscosity of the stream, as during periods of heavy rainfall and flooding when much fine material is washed in from adjacent valley slopes or released by bank erosion.

Stream transport

It is well known that streams transport material in four ways.

(i) Traction is the process whereby particles of various sizes are rolled and bumped along the channel bed (hence the terms 'traction load' and 'bed load'). During periods of large discharge and rapid velocity, the bed load of the stream may be much increased in volume, since large rock fragments that cannot be moved at all under conditions of low flow will now be set in motion. However, at the same time the finer particles which are normally moved along the stream bed may begin to saltate, or may be carried in suspension, so that it is at least theoretically possible that an increase of stream energy could be accompanied by a decrease of bed load.

(ii) Saltation is the 'jumping' of comparatively small grains along the stream bed. Any particle resting on the bed will impede water movement to some extent, and hydraulic pressure will be built up behind it. In time the inertia of the particle will be overcome, and it may be rolled (as in traction) or actually be thrust up into the body of moving water. However, unless it is very small and light, gravity will impel it again to the stream bed, and the whole process will begin anew.

(iii) Most of the finest material moved by a stream is carried in suspension. In the first instance this is lifted from the bed by hydraulic upthrust or washed in from the banks, and enters that part of the stream where turbulence is most marked. This is usually some distance above the stream bed, where the flow may be more of a laminar type. The complex eddying movements buoy up the fine particles, which thus remain suspended as they are moved downstream. It should be added that turbulence is not the only cause of suspension, for some clay minerals have such a slow settling rate that very little stream movement is needed to keep them up almost indefinitely.

(iv) In areas where, for climatic and/or lithological reasons, chemical weathering is very active, streams may contain much of their load in solution, and traction, saltation and suspension may be correspondingly small or even virtually non-existent.

It will be readily apparent that the total load of a stream will vary greatly, both from time to time and from place to place. In the first place, there is the obvious point that the energy of a stream is not constant, but may occasionally be greatly increased, as during times of flood, when volume and velocity grow rapidly. It has been suggested that the 'capacity' of a stream (that is, its ability to move *total* load) varies according to the third power of its velocity; in other words, if stream velocity is doubled, capacity is increased eight times. This can, of course, only be regarded as an approximation, for much will depend on the calibre of the load. In general, it is much easier for a stream to move fine sand, silt and mud, by saltation and in suspension, than large stones and boulders by traction.

In nature there are numerous reasons why a stream rarely attains its full capacity, or in other words becomes 'fully loaded'. For instance, in a deeply dissected mountainous area, the bed of a small stream may be littered with pebbles, rock fragments and even boulders. These obviously constitute the potential stream load, yet under normal flow conditions they are not moved at all (Fig. 8). Indeed, in the absence of fine weathered material, the stream may for much of the time be performing no transportational work, and therefore be underloaded. The reason is that all streams, at any given moment, are 'competent' to move objects only of a certain maximum weight, depending on the actual velocity and power of the stream. If the critical velocity is not attained, no displacement of the objects can occur, and the free energy of the stream will not be expended on load movement unless other small objects are available. It is only when a substantial increase of velocity or discharge is experienced, as in a spate, that the competence of the stream will be sufficient for transportation of the largest boulders. It is often said that the competence of a stream varies according to the sixth power of its velocity. When in a moderate flood the velocity of a stream is increased four times, the weight of individual boulders it can move is increased something like 2048 times. However, this again can only be regarded as a rough-and-ready rule, for angular boulders of a given weight will become wedged against each other on the stream bed and be much more difficult to dislodge than rounded boulders of the same weight.

The availability of load is a factor that needs emphasis. In the Chalk country of southern England streams such as the Test and Itchen, though flowing strongly throughout the year and exhibiting a fair degree of turbulence, are quite remarkable for the clarity of their waters. The suspended load is evidently very small, and the solution load (of calcium bicarbonate released by chemical weathering of the Chalk) comparatively large. There is little tractional movement of coarse debris, such as flints, though there is evidence in the form of river terraces and spreads of gravel that such transportation was important in the past, when periglacial solifluxion fed abundant quantities of such detritus into the streams and valley bottoms. By way of contrast, in an area such as the Weald, where clay is the dominant rock outcrop, rivers such as the Arun and Ouse carry considerable suspension loads, particularly during and after prolonged rain, when the waters become brown and opaque.

The nature of the stream load may also be affected by distance from the stream source. In the upper reaches, where the stream normally drains an area of steep gradients and rapid weathering, the load tends to comprise coarse material which can be moved most easily by traction. However, farther downstream the load becomes comminuted, largely through the process of attrition (involving the bumping of particles against each other, and also against the bed of the stream), and is transported more readily by saltation and in suspension. This gradual diminution in the calibre of the load downstream also reflects the comparative rapidity of movement of the finer particles, which are carried quite swiftly towards the mouth of the stream whilst the coarser debris is left behind.

Finally, 'time' or 'stage of landscape evolution' may be a determinant of both the amount and nature of stream loads. In a high relief or youthful landscape, characterised by steep slopes undergoing rapid retreat, the amount of load both coarse and fine shed into the streams will be very large. If the base-level of erosion remains stable, and relief is reduced and slopes affected by decline, to give an old age landscape, the total load entering streams will inevitably be much diminished, and it will consist of much finer material, having undergone more advanced weathering whilst moving slowly down the gentle slopes.

Stream erosion

Stream erosion is usually said to involve three distinct processes: downward corrasion, which is particularly effective in high gradient streams moving large bed loads; lateral corrasion, resulting from the uniclinal shifting of streams and the development of meanders; and headward extension, which is most obviously associated with knickpoints, rapids and waterfalls of 'cyclic' and 'structural' origin. In many streams these three processes act together, though at different times one may become dominant over the others. Thus Gilbert (1909) argued that when in an 'underloaded' condition streams cut mainly downwards, but that when 'fully loaded' they cut only in a lateral direction (p. 56). Crickmay (1933), on the other hand, has stated that old age streams, though incapable of transporting large loads because of their small velocity, continually shift their courses, develop large meanders, and produce eventually wide plains of lateral corrasion (p. 168). In streams which have been rejuvenated, however, the process of vertical erosion and knick-point recession are dominant, and deeply cut valleys and incised meanders are formed.

9 Active pot-hole drilling, Devil's Bridge, Rheidol valley, central Wales. [Eric Kay]

That river valleys are extremely commonplace features of the landscape, and that most valleys owe their initiation to river erosion, are geomorphological truisms. However, it should not be too readily inferred that at the present day all streams are effectively corrading their channels. Many rivers, especially those of lowland areas with well-developed flood-plains, are manifestly depositing agents. Even where such deposition is not apparent, erosion may be today of little significance, simply because the load of the stream is quite inadequate for corrasive purposes. As Rastall (1944) has pointed out, it is difficult to see how small particles of density 2·75–3·0, being moved gently along the river bed, can achieve any real mechanical effect, particularly since their effective weight is being reduced by displacement. There is indeed no evidence that, say, the Chalk rivers of southern England (such as those referred to above) effect other than minor erosion of the channel banks, where these are composed of unconsolidated alluvium. It seems certain that the present streams are often no more than transporting agents carrying away the weathered material fed into them by soil creep and rainwash from the valley-side slopes. It is an axiom of geology, often applied also in landform study, that the present is the key to the past. In the question of valley erosion by streams this does not always seem to be so.

A *Downward corrasion.* Vertical incision by streams results from a variety of individual processes. The best known is abrasion (otherwise true corrasion) which results from the impact of a large bed load moved rapidly over the channel floor. The reality of the process is shown by the existence of pot-holes, which have been drilled by circulatory pebble movements in high velocity streams (Figs 9 and 10), and by the deep pools below waterfalls—though processes other than corrasion are also involved here. However, abrasion has on the whole almost certainly been overstressed, whereas processes such as hydraulic action, weathering of the stream bed, and chemical erosion are often underestimated. The hydraulic force exerted by fast-running water is clearly of great importance in certain favourable circumstances, for example where the rocks are well bedded, closely jointed, possess numerous cleavage planes, or are mechanically weak (such as poorly cemented sands or crumbly shales), and at points where the sheer impact of the water is unusually powerful, as in rapids and plunge pools beneath waterfalls. Weathering of the underlying rock may sometimes pave the way for stream 'erosion', and weathering of the channel itself can occur effectively in streams

10 Dissection of resistant gneiss exposed by recent glacier recession, Val d'Hérens, Switzerland. The channels were evidently formed by sub-glacial streams under considerable hydrostatic pressure. [M. J. Clark]

which flow intermittently, as in seasonally dry climates. In both these cases the stream merely removes the loosened material, and the channel is incised without the operation of any corrasion. Chemical erosion is obviously a factor of real importance in the formation of valleys in calcareous rocks (for example, the gorges of limestone uplands and the dry valleys of the Chalk).

B *Lateral erosion.* Lateral erosion, except where uniclinal shifting of streams is induced by the dip of the rocks or by unequal weathering of the valley-side slopes (pp. 339–43), is usually achieved by meandering streams. Why streams should develop meanders at all is a very difficult geomorphological problem. Simple explanations based on the assumption of initial chance irregularities in the configuration of the land, or the occurrence of hard rock outcrops, either of which could cause a deflection of the stream that subsequently grows into a bold meander, are totally inadequate. Meanders are not 'random' either in their size or occurrence, and their often perfect symmetry cannot be explained as a mere accident. They are clearly a normal development to be related to the mechanics of river flow and transport. This can be demonstrated in laboratories, where straight channels constructed in sand of uniform consistency quickly develop meandering patterns when fed with running water. Another conclusive fact is that meanders are often most marked on flood-plains which have been built up by the rivers themselves, so that initial irregularities have long since been masked by alluvium.

Several authors have in fact noted the existence of relationships between the dimensions of meanders and stream discharge, valley slope and the size and nature of the stream load. For example, as the discharge increases, so does the overall width of the meander belt, though not in the same proportion. Bates (1939) has shown that, on flood-plains, streams that are 100 ft in width have a meander belt approximately 1600 ft wide, whereas streams that are 1000 ft in width have a meander belt some 12,000 ft wide. An increase of discharge also causes an increase in the wave-length of meanders (the distance between the crests of two adjacent meanders). The discharge factor would appear to account for the fact that meanders are in general a more obvious feature of the lower tract of a valley, whereas they may be poorly developed towards the source of the stream. It has also been demonstrated that meander growth is induced by small bed loads. This is in accord with the tendency for 'old age' streams to form meandering patterns, and with the fact that meltwater channels on glacier surfaces, moving no sediment at all, frequently develop fine meanders. It is often suggested that meanders are promoted by very gentle stream gradients (again as in old rivers and on flood-plains), but in fact in equal conditions meandering streams may have steeper gradients than non-meandering streams, simply because greater energy must be generated to overcome the extra frictional losses resulting from curvature of the channel.

Recent work carried out by Dury (in the 1950s) has focused attention on the fact that valleys, as well as streams, often exhibit meandering patterns. Such valleys are usually marked by an alluviated floor, over which the existing stream itself meanders (Fig. 11). However, these stream meanders, cut into alluvium, are very much smaller than the valley meanders, which are eroded into solid rock. Measurements have shown that the valley meanders obey similar rules to stream meanders. Thus, as the valley width at channel level increases, so the wave-length of the valley meanders becomes larger. Borings through the alluvial infill of valley meanders in southern England have revealed the existence of buried stream channels, with dimensions very much larger than those of existing streams and with courses closely related to those of the valley meanders. Dury has concluded that such meandering valleys were in many instances eroded by streams which occupied all or most of their bottoms, and possessed discharges some 80–100 times those of present-day streams. Such great discharges can only have occurred during the Quaternary, when various factors such as higher rainfall, reduced percolation, scanty vegetation, low evaporation and snow melt caused at certain periods far greater run-off than that of today. In post-glacial times, the old channels have been choked with alluvium, and the meandering valleys are now followed by 'underfit' streams, possessing insufficient power to effect further erosion.

Although the factors that underlie meander development are obviously complex and as yet not fully appreciated, the actual mechanism by which a stream erodes one bank and deposits on the other is perhaps better understood. Many authorities are agreed that some form of

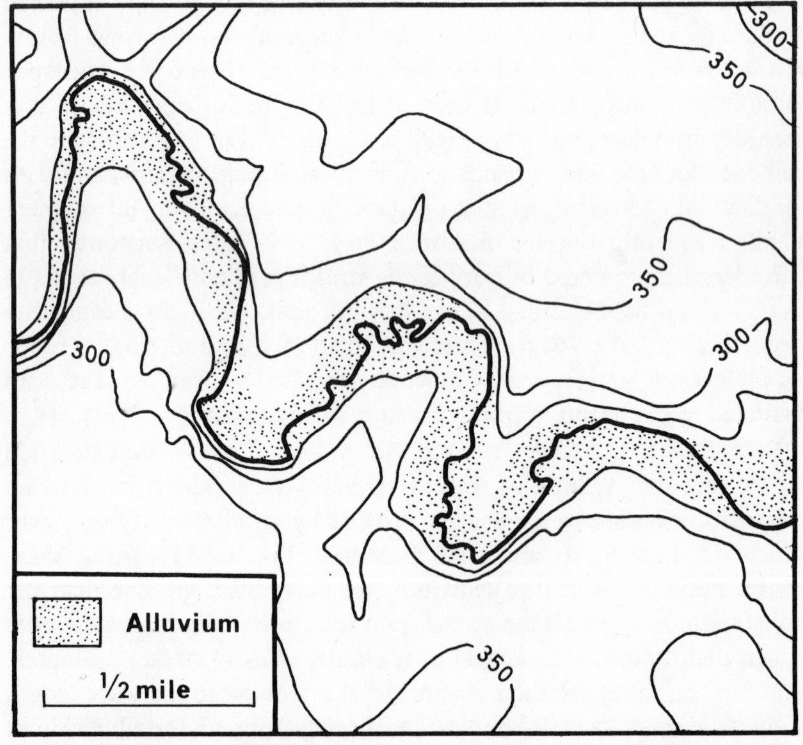

11 Meandering valley of the river Evenlode, Oxfordshire

spiral (helical) flow is involved, and that this produces a surface movement of water towards the outside of the meander bend and a return flow to the inside of the bend at the bed of the stream. A simple explanation of this phenomenon is that, once a meander has been initiated, the tendency for a stream to flow in a straight line results in a slight head of water being built up on the outside of the bend. This must produce a compensatory return flow across the channel, presumably along the stream floor, which is of sufficient velocity to transport eroded material (Fig. 12A). The actual erosion of the meander bend is due to the fact that the thread of maximum velocity in a stream, which in a straight channel follows a median line, is deflected against the outside bank so that a comparatively large hydraulic force is exerted there. This would satisfactorily explain the broadening of the meander belt as a whole (Fig. 13A), but does not account for the well-known phenomenon of downstream meander migration or the progressive narrowing of meander

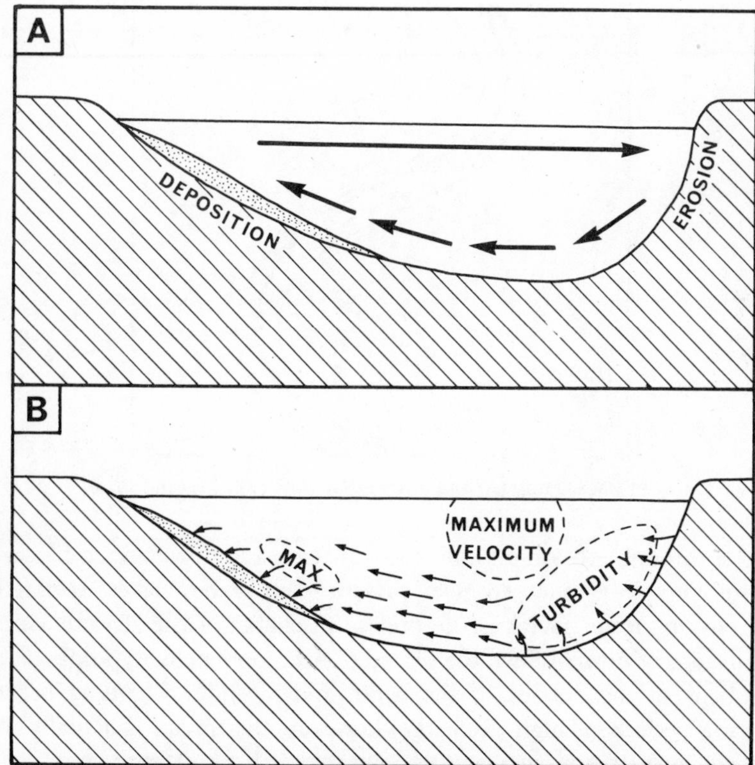

12 **Helical flow (A) and turbulence and turbidity (B) in meanders**

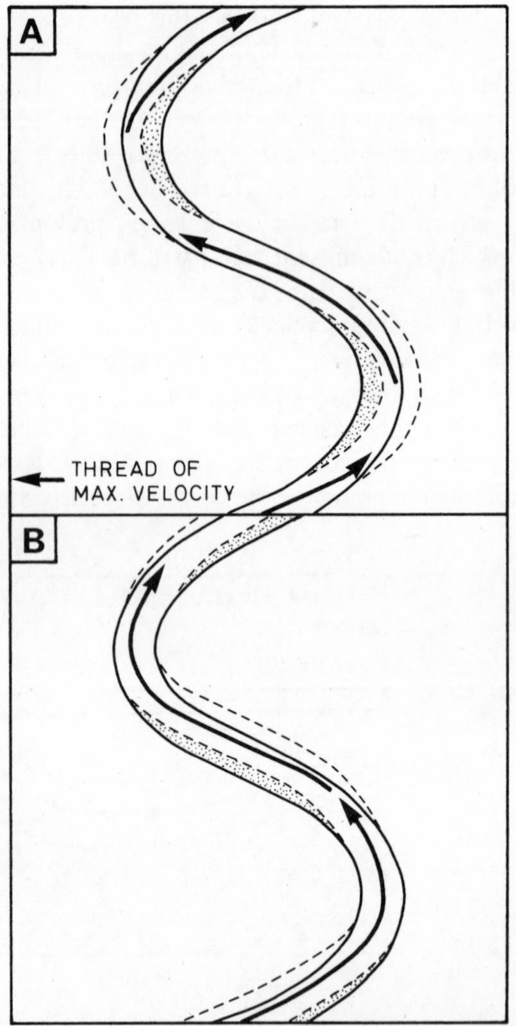

13 **Widening (A) and downvalley shift (B) of meanders**

necks, through erosion on both the upstream and downstream banks of an individual meander, to give 'cut-offs' and ox-bow lakes. It may be that an important factor is the precise discharge of the stream. Under conditions of normal flow, the thread of maximum velocity impinges on the bank at the outside of the bend, and meander widening occurs. During periods of high discharge, however, the thread is actually displaced downstream, so that the downstream bank of each meander is attacked, resulting in meander migration (Fig. 13B). The pattern of flow required to produce meander cut-offs is clearly more complex. The process cannot operate until the meanders have attained a considerable width, at which stage the mechanism depicted in Fig. 14 may operate.

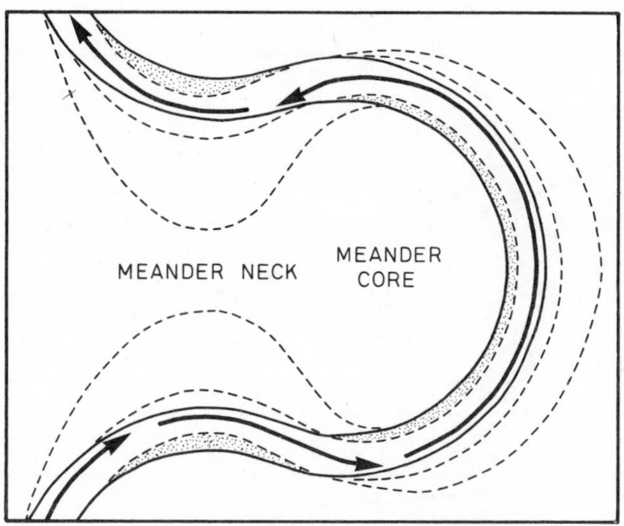

14 **The development of meander cores and necks**

A rather more sophisticated explanation of meander growth has been proposed by Leighly (1934), based on a study of the cross-sectional distribution of turbulence within a stream. Under normal conditions, on either side of the thread of maximum velocity there is a zone of maximum turbulent flow (Fig. 12B). Most of the suspended load of the stream is concentrated into these two zones, so that they are also zones of maximum turbidity (muddiness). In a meandering stream, as the thread of maximum velocity is displaced towards the outside of the bend, the 'outer' turbulent zone is much accentuated (causing increased load pick-up and thus enhanced turbidity), whereas the inner zone becomes less pronounced. In such circumstances, there may be a 'turbidity gradient' from the outside to the inside of the meander belt, and cross-sectional transference of fine material towards a depositional area on the inside of the bend may be effected.

However, it cannot be pretended that the processes associated with meanders have all been satisfactorily explained. The configuration of the channel floor in many meandering streams shows features of interest which have to be accounted for. In particular deep scour pools are often found on the outside of bends, where they are associated with active undercutting of the banks, whilst between these pools are banks of sand or gravel, sometimes running diagonally across the stream and with a steep downstream side indicative of migration. Such 'pool-and-riffle'

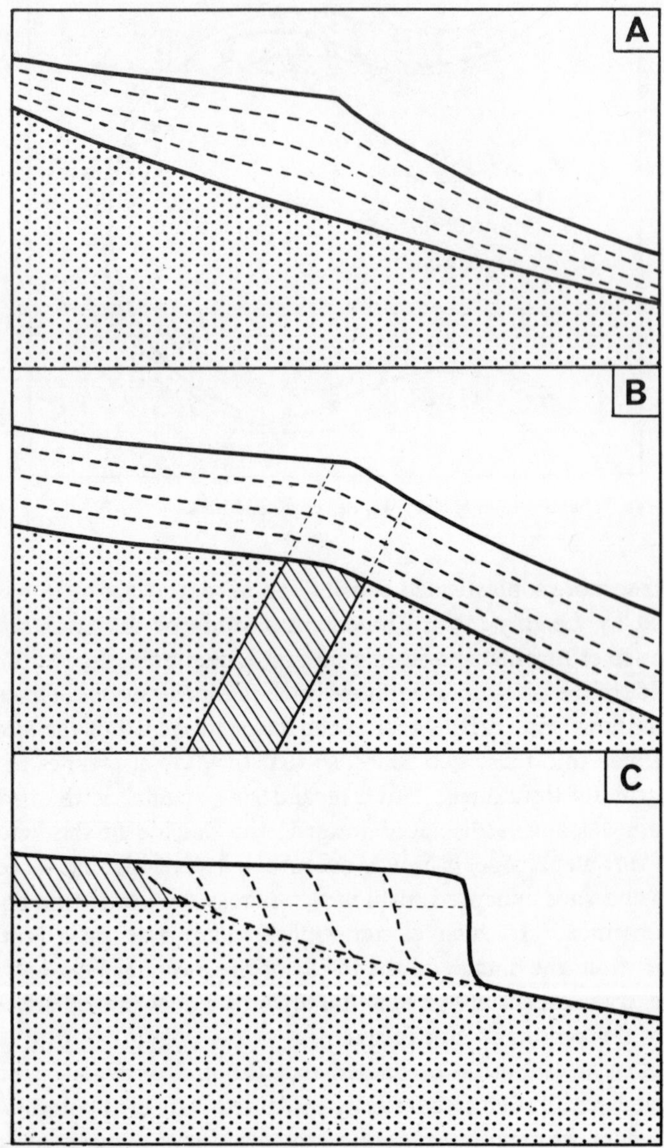

15 The behaviour of knickpoints

sequences may well hold the key to remaining problems of meander formation.

C *Headward erosion.* This process is most frequently associated with locally steepened sections of the river profile, produced either by outcrops of hard rocks or falls of base-level. Many textbooks describe such 'knickpoints' or 'headcuts' as inevitably receding upstream until the river source is reached. However, field and laboratory studies have shown that in reality, providing the stream flows across weak or incoherent sediments, knickpoints can undergo progressive flattening with little or no headward migration (Fig. 15A). Similarly, when the break of river profile is associated with a hard steeply dipping stratum it will maintain its position until it is gradually smoothed away over a long period of time (Fig. 15B). Finally, those knickpoints which do experience headward recession are often reduced in effect and eventually disappear altogether (Fig. 15C).

Despite these reservations, headward erosion must not be ignored, for it can be very effective in certain types of stream and under certain structural and lithological conditions. When gullies form on hill-sides, they are at first discontinuous, comprising several quite distinct sections. Each section will contain a headcut, immediately below which the gully is incised deeply into the slope. Lower down it becomes progressively shallower, until the eroded section is replaced by an alluvial fan. Recession of the headcuts leads in time to the coalescence of the individual gully sections and the formation of a unified channel. On a larger scale knickpoint recession is an important process in areas where the strata include alternately hard and soft bands and are gently dipping or horizontal. The outcrop of the hard rock will commonly give rise to a waterfall, the plunge pool of which is associated with the undermining of the soft rock beneath the fall (Fig. 16). An interesting example of this is found on the river Hepste, in south Wales, where the Scwd-yr-eira falls were initiated at a fault-line in the Millstone Grit, bringing hard gritstone against weak shales (Fig. 17). The falls are now dissociated from the fault, however, for shales lying within the gritstone have been attacked, mainly by 'splash' from the plunge pool, to give an impressive overhanging fall which has retreated upstream. Basal sapping is also active in permeable rocks, where water which has percolated into the stream bed emerges at the foot of the knickpoint, thus accelerating its recession. In rocks such as chalk and limestone the actual sources of streams are extended into escarpments and steep slopes by the process

16 Glenashdale Falls, Isle of Arran. The falls are developed where the Glenashdale Burn crosses quartz-dolerite sills. [Eric Kay]

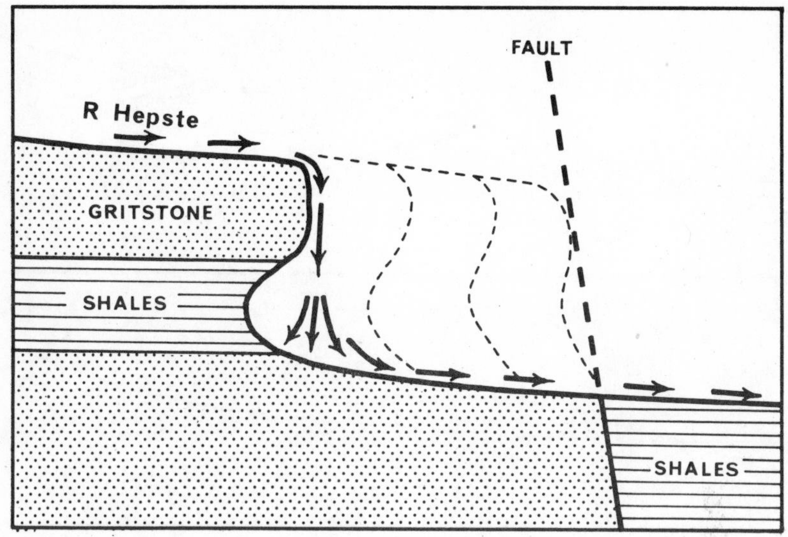

17 The Scwd-yr-eira falls, river Hepste

known as 'spring sapping'. This involves underground chemical erosion, surface stream erosion, and slumping of moistened debris around the springhead.

The long-profiles of rivers

The majority of rivers, whatever their initial form, develop in time profiles that exhibit a systematic decrease of gradient in a downvalley direction. In youthful rivers this concavity may be very approximate, for the continuity of the profile will be broken by falls and rapids (of structural and lithological origin) which alternate with comparatively smooth and gentle sections (Fig. 18A). Even in mature streams the concavity of the profile may be far from perfect, for in some circumstances a gentle section may be succeeded downstream by a steeper section. For instance, when a stream transporting a small bed load is joined by a tributary carrying a very large load, a steepening of profile below the confluence will be necessary to afford the extra energy needed to move the suddenly increased load of the main stream.

Streams which have succeeded in eroding a smoothly concave profile are said to have 'graded' their courses (Fig. 18B). In this sense (of smoothing out irregularities) the term is comparable with that used in engineering, for those engaged in constructing roads and railways in hilly country 'grade' their profiles by cut-and-fill. However, as will be

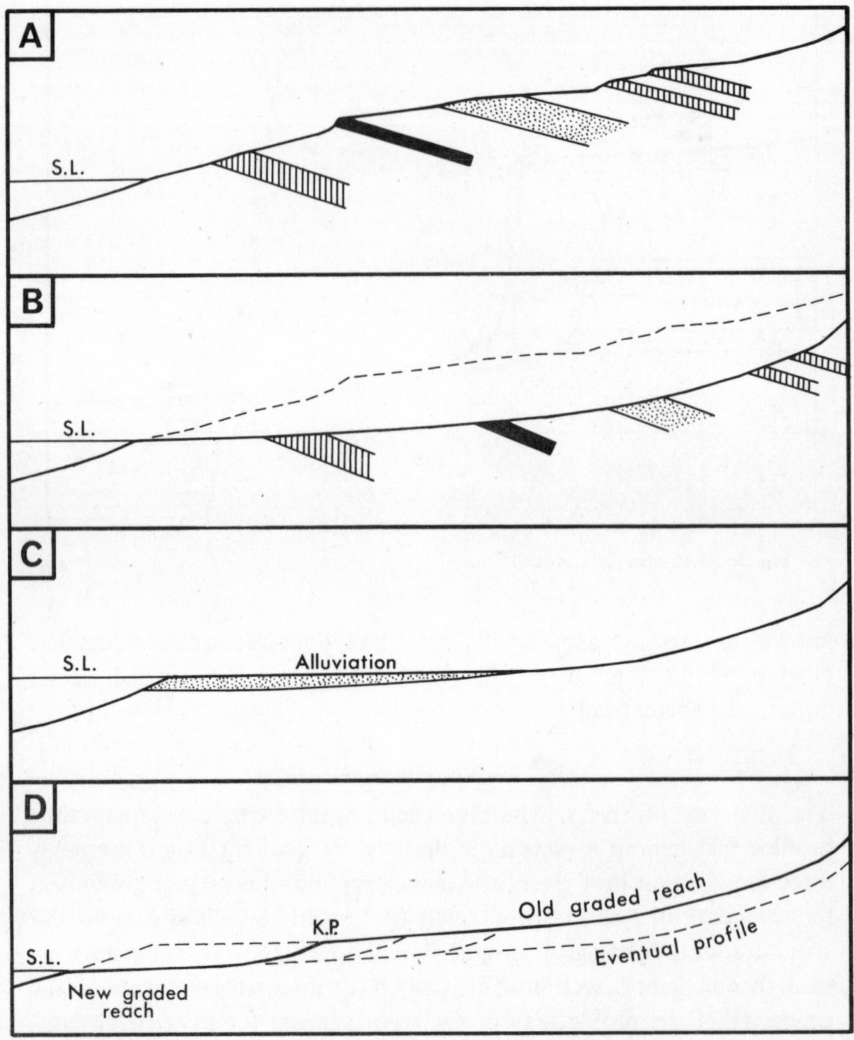

18 River long-profiles: ungraded (A); graded (B); affected by rise of base-level (C); rejuvenated (D)

shown below, geomorphologists have also attributed to the term 'grade' a slightly different meaning. It should be added that a graded river is also often said to have attained its 'profile of equilibrium'. This implies the development of a condition of balance in the stream which is intimately associated with the smoothing process.

The reasons for the concavity of river profiles are complex. It might be thought at first sight that the explanation lies in (i) the restriction of downcutting near the mouth of the stream, where it stands only a little above sea-level, and also at the head of the stream, where the discharge is small and the load as yet insufficient for corrasion, and (ii) the existence of a central section of maximum erosion, coinciding with that part of the stream where both energy and load have been much increased. However, this is altogether too crude a thesis, which could not account, among other things, for the mathematical form of the profile curve.

It seems likely, in fact, that several factors contribute to the development of the concave profile. Firstly, although the total amount of load to be transported increases in a downstream direction, the river discharge itself also increases steadily as more and more tributaries enter. The cross-sectional area of the channel therefore grows larger, and the hydraulic radius will also increase. Thus the river becomes more efficient, and has greater 'spare' energy to devote to load transportation; in other words, although the load increases it can be transported over a gentler and gentler slope towards the mouth of the stream. Secondly, the calibre of the load tends to change downstream, for the reasons given on p. 41. The coarse material of the upper tract gives way to finer material in the middle and lower course of the river, and this can be transported again over a progressively gentler slope. The actual rate of comminution may be an important control over the curve of the profile. Observation suggests that this is very rapid towards the head of the river, but tails off sharply at first downstream and then more gradually, so that in the lower part of the stream the calibre of the load remains almost constant. This would help to explain the marked flatness of many rivers for several miles above the mouth, and the suddenly increasing gradient in the upper reaches (in other words, the typical hyperbolic form of many curves).

Needless to say these arguments would be rendered invalid if stream energy itself were to undergo a continuous reduction downstream. It is true that discharge and channel efficiency both increase in that direction, but this could be completely offset if stream velocity, a major determinant of stream energy, were at the same time sharply diminished. Until quite recently it was believed that this was so, for at first sight the rushing mountain torrent contrasts vividly with the sluggish river meandering across the lowland flood-plain. Measurements made in the field by American workers have revealed the surprising fact that the *mean* velocity of streams (that is, the total discharge in cubic feet per second divided by the cross-sectional channel area) either remains constant or actually increases slightly downstream. The reason seems to be (i) that the roughness of the channel decreases downstream, and (ii) as the depth of water in the channel increases roughness of the bed is less able to impede the movement of water through the channel (p. 38). The rapid flow of upland streams is in fact illusory, for only threads of maximum velocity are observed, not the movement of the water body as a whole.

The concept of grade

From the early days of modern geomorphology students of fluvial activity have believed that streams can achieve a condition of equilibrium, whereby the total energy of a river is exactly sufficient to move the water and its load. Such streams are described as being in a 'state of grade', 'in a graded condition', or quite simply as 'graded'. Streams which have succeeded in fashioning a smoothly curving long-profile are considered to have attained the graded condition, hence the term 'profile of equilibrium' (p. 54).

The first geomorphologist to outline the concept of grade clearly was the American G. K. Gilbert, who devised the following classic definition. 'Where the load of a given degree of comminution is as great as the stream is capable of carrying, the entire energy of the descending water is consumed in the translation of the water and its load, and there is none applied to corrasion.' Thus Gilbert believed that graded streams were incapable of deepening their valleys or changing the form and gradient of their long-profiles directly, though he did consider that fully loaded streams were capable of lateral erosion, presumably on the grounds that moving water is always capable of undermining and attacking the channel banks whatever the conditions of erosion, deposition and load transport on the river bed.

Davis also adopted the concept of grade, though he modified it in certain important essentials. He did not accept that graded rivers had no power to attack their beds, and he considered that the graded condition was established relatively early in the cycle of erosion (normally by the onset of the mature stage). However, if Gilbert's definition were true, grade must be a late development, since it marks the end of vertical erosion. Davis argued that in fact maintenance of the graded state necessarily involved some erosive lowering of the stream bed. 'In virtue of continual variations of stream volume and load, through the normal cycle, the balanced condition of any stream can be maintained only by an equally continuous, though small, change of river slope.' Thus, as valley-side slopes steadily undergo wasting, and the load entering streams becomes smaller in total amount and finer in calibre, streams will possess more and more surplus energy (that is, will tend to become ungraded), unless they succeed in reducing their channel gradients by erosion and so experience a lessening of available energy. Presumably, therefore, for a stream to maintain its graded state in the long term, for short periods it must become ungraded and have the ability to attack its bed.

Although the concept of grade has been accepted by most geomorphologists, for the reason that it offers a working hypothesis of why at some times streams erode vertically (because they are underloaded), at others they erode laterally (because they are graded), and at others they deposit (when they change from a graded to an overloaded condition), some have been sceptical. The main difficulties are to prove the existence of this delicately balanced equilibrium, and to recognise a graded stream in the field. Davis himself believed that streams which lacked waterfalls and rapids, the typical features of a youthful profile, were graded, but obviously this does not constitute final proof of a condition of balance between energy and transportation. It might be thought that the state of grade could be demonstrated by the finding of a stream which is neither eroding its bed nor depositing alluvium—in other words, is poised between an underloaded and an overloaded state. The problem here is that a stream may be attacking its bed in order to maintain or restore the balanced condition; such a stream is merely 'degrading' its course. Similarly, deposition of alluvium might indicate only 'aggradation', or deposition to maintain grade. It has been argued that in many streams there is an upper section where vertical erosion is pronounced, as a result of underloading of the stream, and a lower section where deposition is dominant, owing to overloading. Between the two there must logically be a section where the stream is fully loaded or at grade. However, no one has been able to prove that the change from underloading to overloading occurs over a section of the stream rather than at a point.

One of the foremost antagonists of the concept of grade has been the American geomorphologist Kesseli (1941). Kesseli has brought forward many persuasive arguments to show that, at any given moment of time, there cannot be a perfect balance between stream energy and transport. The fact that the discharge of most streams varies greatly, sometimes over very short periods of time, between dry weather and flood conditions means that the total energy of a stream is itself exceedingly variable. It is hard to visualise the stream load being changed in perfect harmony with these energy fluctuations, without there being any time lag in load pick-up or deposition. Kesseli believes, in fact, that streams can never become fully loaded, but are always underloaded. The reason is that it requires more energy to move particles from rest than to keep them moving. Therefore, unless a stream possesses sufficient energy in the first instance it will not overcome the inertia of a particle (in which case its energy will be surplus). If, on the other hand, it does have enough energy to start the particle moving, then immediately part of that energy becomes redundant. The availability of load is another factor in the permanent underloading of streams (p. 41).

In spite of these difficulties, grade can still be a useful concept, at least in the view of Mackin (1948), providing it is not applied too rigidly and is defined with care. Mackin's own definition of grade is as follows: 'A graded stream is one in which, over a period of years, slope is delicately adjusted to provide, with available discharge and the prevailing channel characteristics, just the velocity required for the transportation of all the load supplied from above.' It is clear that Mackin, like Davis, regards grade as an average condition, which from time to time and for short periods may be departed from, but which is always re-established by an adjustment of the channel slope, either by erosion or deposition. There are certainly some reasons for accepting Mackin's view. It is noticeable that many river valleys possess terraces, planed across the solid rock and veneered by alluvium or gravel, whose down-valley gradients are virtually the same as that of the present stream channel. How can this be explained, except by the inference that the channel slope is 'delicately adjusted' to conditions of discharge and load, and that so long as these are unchanged, even though there are interruptions resulting from an uplift of the land or a fall of sea-level, this precise slope will always be restored? Mackin has also argued that the graded stream is a 'system in a state of equilibrium'. Any change in one of the controlling factors (discharge, load, climate and so on) will cause a 'displacement of the equilibrium in a direction that will tend to absorb the effect of the change'. Mackin's meaning may be illustrated by reference to a simple example. The load of a stream may be suddenly increased, perhaps as a result of climatic change inducing more rapid weathering, and the graded condition will be replaced by overloading. Some of the extra load will therefore be deposited immediately below the point of influx, thus causing a steepening of the channel slope. The energy of the stream will thereby be increased, and in time it will be able to transport the increased load. Thus the overloaded condition will be overcome by an increase of channel slope, involving aggradation, and grade will be restored.

Today students of fluvial processes regard the term 'grade' as somewhat outmoded, though accepting that streams are able to achieve a state of apparent equilibrium ('quasi-equilibrium'). Furthermore, though Mackin's views are considered to be a useful starting-point, it is thought that he places too much emphasis on channel slope. In the words of Leopold, Wolman and Miller (1964), 'there are eight interrelated variables involved in the downstream changes in river slope and channel form; width, depth, velocity, slope, sediment load, size of sediment debris, hydraulic roughness and discharge'. Any state of stream equlibrium must involve all of these, and a change in one may have repercussions on all or some of the others. For instance, an increase in sediment load may be compensated not just by a change of slope (as in the example cited above), but by an alteration in the depth and width of the channel, the velocity of the stream, and so on.

The influence of base-level on stream action

It is one of the most basic facts of geomorphology that fluvial activity is strongly influenced by the base-level of erosion. The vast majority of rivers flow ultimately into the sea, the surface of which provides a limit below which erosion cannot proceed. Lakes can act as temporary base-levels for streams flowing into them, but as they are susceptible to draining (as a result of vertical erosion by streams draining from them) or infilling by alluvium they cannot exert a long-term effect. Indeed, one of the features of the development of graded river profiles, under conditions of stable sea-level, is the obliteration not only of rapids and falls but also lakes. The influence of base-level on the profile of equilibrium is obvious, for in their lower parts such profiles are very approximately tangential to the sea-surface. To be more precise, river profiles are asymptotic; in other words, their gradients become gentler and gentler downstream, but never become quite horizontal, for the reason that some slope, however slight, is necessary to produce water flow.

The effect of a rise of sea-level (sometimes rather confusingly referred to as a 'positive' movement of base-level) will be to flood the lower part of a river valley, producing an estuary in which extensive alluviation will eventually occur. The rivers of southern England have been affected by such a rise in post-glacial times, and many of the flood-plains of the area have been produced by sedimentation of the resultant inlets (Fig. 18C).

19 The valley of the upper Taff, Brecon Beacons, south Wales. The 'rejuvenated' valley form is due not to a fall of base-level, but to penetration of a resistant stratum in the Old Red Sandstone. [Eric Kay]

The effects of a fall of sea-level (a 'negative' movement) are rather more complex. The immediate result will be an extension of the river course from the old to the new shoreline (Fig. 18D). Since the offshore gradient is normally greater than that of the lower course of a river, the added section will be characterised by greater stream energy and a renewed capacity for downward erosion, more appropriate to a youthful stream. Thus rejuvenation will affect the mouth of the stream, and in favourable circumstances (p. 51) the steep new section will undergo headward recession as a knickpoint. In time the original river profile will thereby be regraded to the new sea-level. This process may operate many times, a series of intermittent base-level falls resulting in a succession of knickpoints. Because of the numerous changes of base-level in the Quaternary (pp. 421–7), most British rivers exhibit several knickpoints, each separated by a gentle reach graded to a sea-level higher than at present (Fig. 19).

The existence of these polycyclic profiles has led geomorphologists to investigate the possibility of determining precisely former high sea-levels (also denoted by raised beaches and old marine abrasion platforms backed by degraded cliffs) by extrapolation of the preserved graded reaches (Fig. 20). This can be done very approximately by eye, but a more scientific method is that of 'mathematical extrapolation'. It has been shown that graded river profiles closely resemble mathematical curves, usually of simple logarithmic form. It follows that, if a mathematical formula can be devised to fit a preserved graded reach, the whole of the former profile can be accurately reconstructed and the contemporary height of the sea discovered. Green (1934), in his study of the river Mole, used the formula

$$y = a - k \log (p - x)$$

in which y is the height of the stream above sea-level, a and k are constants to be determined for each river, p is the length of the stream, and x is the distance from the stream mouth.

However, it must be admitted that, in practice, certain problems arise when mathematical extrapolation is attempted. Firstly, it must be realised that the graded reaches which have not yet been overtaken by knickpoint recession are often of very limited length, and it may be possible to fit several curves, using slightly different formulae, which

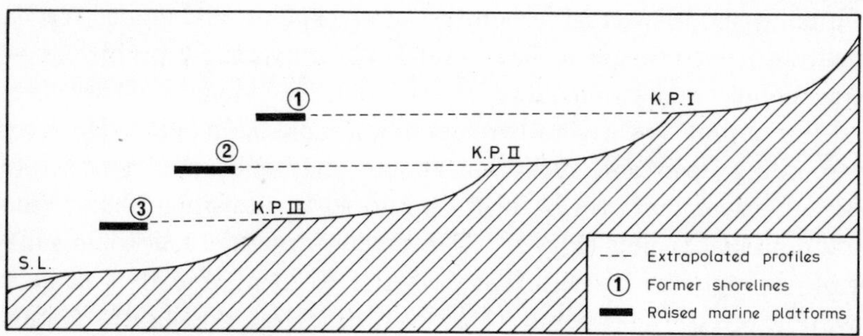

20 **Poly-cyclic river long-profile**

give several distinct sea-levels. Some means must be found of deciding which formula, if any, is correct. Sometimes river terraces can be used as a control in mathematical extrapolation (Fig. 21), as was shown by Brown (1952) in his study of the river Ystwyth. In Fig. 21, three possible former profiles are shown, together with terrace remnants. Reconstruction A is clearly too high, C too low, and B probably correct. Secondly, for complete accuracy the line of the former shoreline (or, in other words, the exact length of the old stream) must be known. This is because of the asymptotic nature of river profiles. Ideally, therefore, the use of extrapolation methods should be combined with the study of remnants of raised beaches and marine platforms.

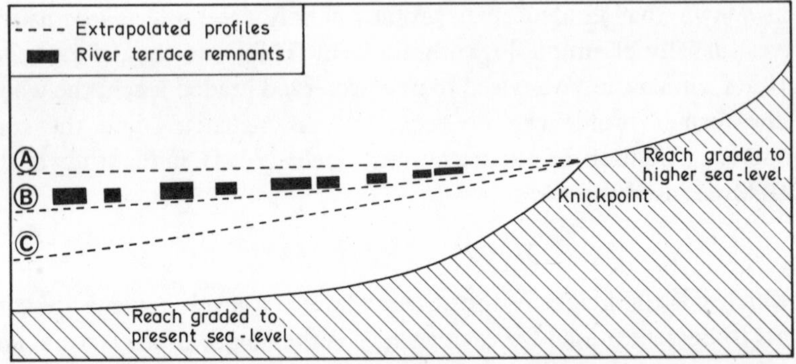

21 **The extrapolation of river profiles with the aid of terrace remnants**

Other difficulties confronting extrapolation are perhaps more fundamental. In using a so-called 'graded reach' to determine a former sea-level, one is making the tacit assumption that the reach has not subsequently been altered by erosion and/or deposition. Needless to say, this is hardly realistic, since flowing water transporting a load will inevitably produce some modification of its channel over a long period of time. Again, the actual recognition of knickpoints and graded reaches resulting from changes of base-level is often extremely problematical. Knickpoints are of two main types, 'cyclic' (related to falls of base-level) and 'structural' (related to hard rock outcrops, faults, etc.). However, in receding upstream, cyclic knickpoints may encounter a structural barrier, where they will be held up and so be indistinguishable from structural knickpoints (Fig. 22). Conversely, structural knickpoints may break through the barriers that gave rise to them, and thus come to resemble cyclic knickpoints. However, this last possibility is more doubtful, for in the view of some, breaks of channel slope will remain tied to structural barriers until destroyed by the grading process (Fig. 15B). It is possible that confusion may be avoided if in one area several rivers, either within one basin or separate basins, are surveyed. If similar graded reaches, bounded downstream by knickpoints, are identifiable in all or most of these rivers, and can be extrapolated to approximately the same sea-levels, then it is highly unlikely that they can be random features.

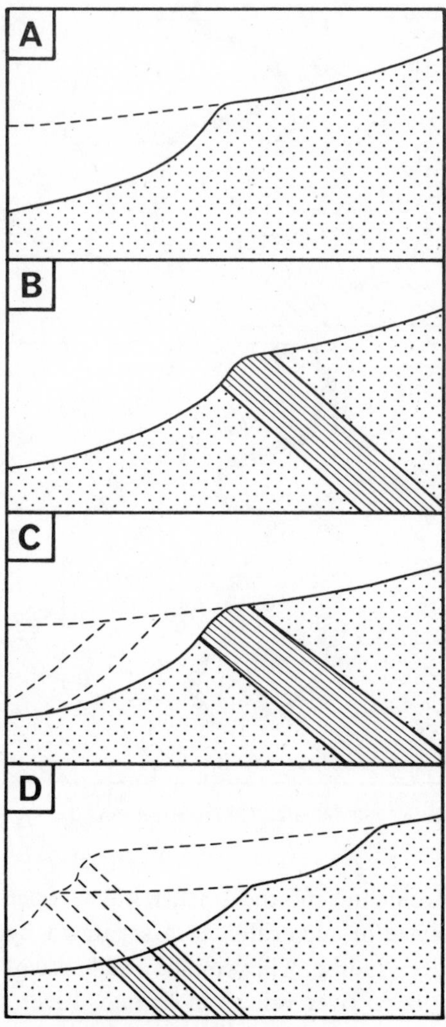

22 Cyclic (A) and structural (B) knickpoints; complex knickpoints (C, D)

The effects of base-level falls on the cross-profiles of river valleys are most striking when the stream has previously begun to meander, planing off the valley floor and spreading gravel and alluvium over the resultant plain. A sudden lowering of sea-level will cause incision of the

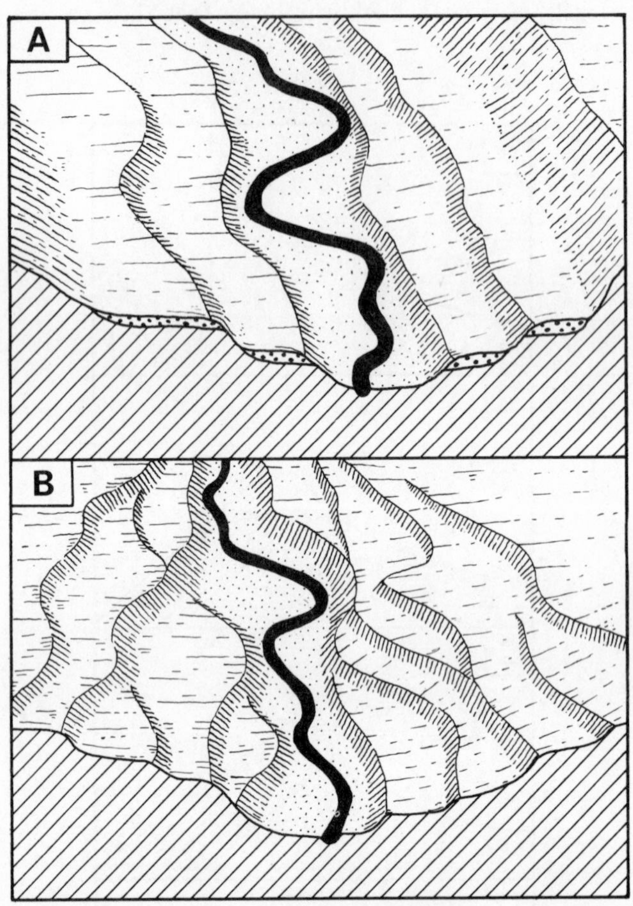

23 **Paired (A) and unpaired (B) river terraces**

stream, leaving remnants of the former valley floor standing up as terraces. A sequence of such falls, each separated by a 'stillstand' of sea-level and valley grading, will produce a staircase of terraces, such as is found in most large river valleys in Britain. River terraces resulting from rejuvenation of this type will be paired across the valley (Fig. 23A). Unpaired terraces (or meander terraces) are formed when laterally shifting streams are cutting down steadily (probably because base-level is also falling continuously), as depicted in Fig. 23B. It must be added that river terraces do not result only from intermittent or continuous falls of base-level. Certain climatic conditions, especially

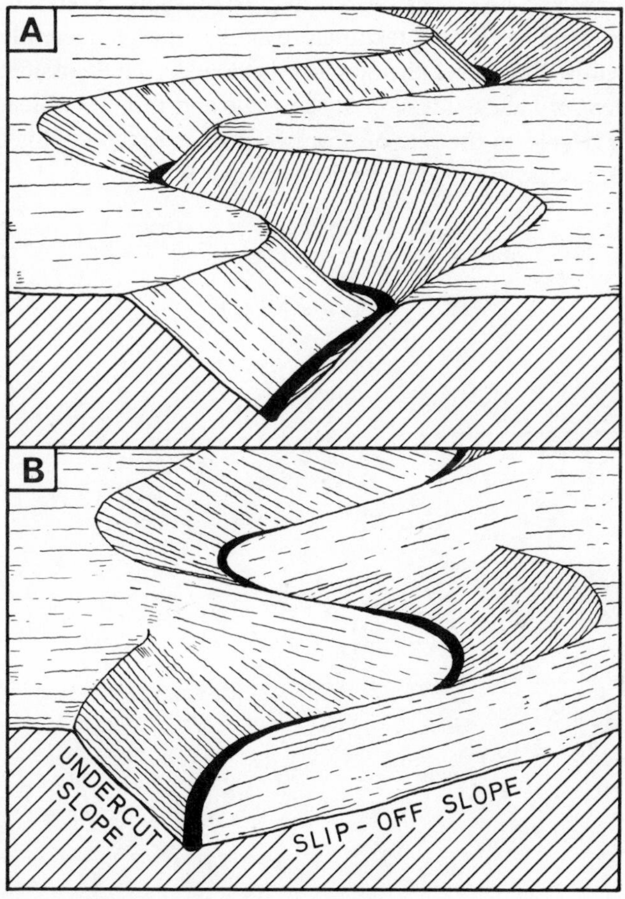

24 **Incised (A) and ingrown (B) meanders**

glacial and periglacial, promote the movement of very large quantities of weathered and eroded material into streams. These will become overloaded and considerable aggradation will result. A change of climate, leading to a reduction in the debris supply and a condition of stream underloading, will be accompanied by incision of the rivers into these deposits, which are left upstanding as terraces (see also Fig. 165).

Other important results of vertical erosion induced by a falling base-level are incised meanders (which are formed by the rapid incision of an already meandering river) and ingrown meanders (which develop where the stream is cutting down more gradually and at the same time enlarging its meanders) (Fig. 24). Characteristic features of ingrown meanders are slip-off slopes, which may show signs of terracing if the fall of base-level has been interrupted by stillstands, and sharply undercut river cliffs. Base-level changes also have wider repercussions on the development of valley-side slopes, the angle and form of which may be intimately related to the rate of downward river erosion. This complex question is discussed in some detail in chapter 6.

3

GEOLOGICAL STRUCTURE AND LANDFORMS

INTRODUCTION

Geological structure—in its broader sense the nature of the rocks of the earth's crust and the manner of their disposition—is one of the major determinants of surface form. It hardly needs saying that the landforms of a limestone area usually bear little resemblance to those of a boulder-clay plain; that the relief features of a strongly folded area are normally more complex than those developed in horizontally bedded rocks; or that faulting is sometimes associated with spectacular scarps and troughs, as in the Rift Valley of East Africa. Even so, geological structure is sometimes referred to as a passive element in land sculpture. This is not strictly true, in the sense that some landforms of a purely structural origin may be found, but as a broad generalisation it is acceptable. The great majority of landforms, even those exhibiting a strong structural influence, are not the immediate result of crustal movements and dislocations. Rather they result from differential weathering and erosion, which etch out unresistant rocks and lines of weakness such as faults and joints into valleys or lowlands and leave the more resistant strata upstanding as plateaus, ridges, cuestas and so on. Under certain circumstances, denudational processes may in fact completely dominate geological structure, resulting in the development of surfaces of planation on which both hard and soft rocks alike have little or no topographical expression.

In this chapter, the emphasis will be on structure in the narrow sense of the term. It will, in fact, be taken to mean the arrangement of rocks brought about by earth-movements, in particular those involving uplift, tilting, folding and faulting. The influence of rock-type as such will be considered more fully in the next chapter. The object here will be to illustrate, with the aid of examples taken from various parts of Britain, how the most basic types of structure affect landform development. At the same time full account will have to be taken of the other relevant factors, both lithological and denudational. One of the primary aims of this book will be to emphasise continually that landforms cannot be ascribed to one cause, but are the result of a complex interplay between several factors and processes, both endogenetic (originating within the earth's crust, and including structure and rock-type) and exogenetic (originating from without, and including weathering, transportation and erosion). As E. H. Brown has aptly said: 'There is a tendency to regard structure as the dominant control of surface form and doubtless this is true in many instances. But structure is not always the principal control and never the only one.'

HORIZONTAL STRUCTURES

The most simple structures in existence are those comprising sequences of sedimentary strata, igneous intrusions of the sill type, and basic lava extrusions which are either totally undisturbed or have been slowly uplifted by epeirogenetic movements involving no tilting or dislocation. In reality, perfectly horizontal structures are extremely rare, and are largely confined to regions where a very rigid basal complex, of great antiquity, has been covered by more recent marine or continental deposits and then affected by bodily uplift. For example, a wide extent of the pre-Cambrian and Palaeozoic rocks of the Sahara and Egyptian deserts are overlain by undisturbed Cretaceous formations, including the limestones of southern Algeria and the Nubian sandstones of southern Egypt (p. 296). In the British Isles, however, such stability has not obtained. The Palaeozoic and older rocks of the west and north have been complexly folded and faulted by the Caledonian and Hercynian orogenies, and more recently have been uplifted, tilted and dislocated by movements associated with the Alpine orogeny. The latter has also been responsible for the widespread if generally rather gentle folding, sometimes allied to faulting, of the Mesozoic and early-Tertiary sedimentary rocks which occupy most of lowland England. Nevertheless, there are still to be found areas within Britain where the structure shows a 'generalised' horizontality, and where the landforms typically formed by the dissection of truly horizontal strata are well in evidence.

Salisbury Plain

Over most of this area the underlying Chalk is virtually horizontal or dipping at a very gentle angle (1–2°), and is little affected by the strong anticlinal flexuring associated with the Vale of Pewsey to the north and the Vale of Warminster axis to the south. The Upper Chalk itself

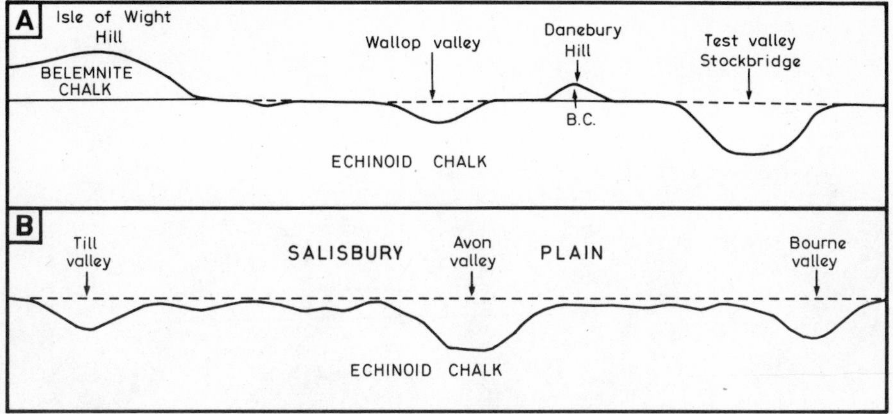

25 Cross-sections of eastern (A) and western (B) Salisbury Plain

contains slight but significant lithological variations (largely of flint and marl content), and as a result is by no means uniformly resistant to erosion (p. 123). It is possible to make a fundamental distinction between an upper layer, the Belemnite Chalk, which is weak towards its base, and a lower layer, the Echinoid Chalk, which because of its possession of abundant hard flints and its high degree of permeability is the most durable horizon within the whole of the Chalk. The streams of Salisbury Plain and the area immediately to the east (the Chitterne Brook, Till, Avon, Bourne, Nine Mile River, Wallop Brook and Test) run generally southwards and have incised their courses quite deeply into the Chalk, though never sufficiently to expose the Middle Chalk. Over a long period retreat of the valley-side slopes has led to the partial or complete stripping of the Belemnite Chalk from the underlying Echinoid Chalk on the broad plateau-like interfluves (Fig. 25). The topmost surface of the Echinoid Chalk is sometimes almost perfectly preserved, affording a striking example of a structural surface or plain. The feature is especially evident to the north-west of Stockbridge, on either side of the Wallop Brook valley, and is terminated to the west by a discontinuous line of hills, with steep scarp-edges, formed by remnants of the Belemnite cover (Fig. 26). Farther still to the west, beyond the valley of the Bourne, the Echinoid surface seems to have been exposed at an earlier date, and as a result is dissected to a certain extent by networks of shallow dry valleys. Even so, when viewed from a distant vantage point this part of Salisbury Plain gives an impression of remarkable overall flatness.

26 The structural surface of the Echinoid Chalk near Stockbridge, Hampshire. The line of hills in the background forms part of the 'secondary' (Belemnite) escarpment of the Chalk (see p. 123). [Eric Kay]

South-east Devon and south-west Dorset

In this area, although gentle folding and some important faulting of pre-Upper Cretaceous and mid-Tertiary date is observable, the structure as a whole is broadly horizontal. The dominant feature is a formerly continuous, but now fragmented, cover of resistant Upper Greensand and Chalk (the latter greatly reduced in thickness here by a long period of erosion, probably in early-Tertiary times) resting unconformably on older and for the most part appreciably weaker rocks of Triassic and Jurassic age. This Upper Cretaceous cover, though locally downfaulted to sea-level at Beer Head and broadly domed in the Marshwood Vale area, is elsewhere approximately constant in elevation, and gives to much of the region a remarkable plateau-like character. It should be added that in some localities there is little divergence of dip between the Cretaceous cap-rock and the underlying strata, and since the latter include some stronger elements, such as the Bridport and Thorncombe Sands, plateaus are formed even where the Greensand and Chalk have been removed.

Deep dissection of this part of Devon and Dorset has been effected by a system of south-flowing streams (the Sid, Axe, Char and Brit) and their many tributaries. Once the resistant Chalk and Greensand have been penetrated, the impermeable and soft clays (particularly those of the Lias and Keuper) have been rapidly attacked and broad-floored, steep-walled valleys have been formed. An early stage in the process can be seen to the east of Sidmouth (Fig. 27). Here the plateau has been divided into a series of flat-topped interfluves, each about a mile in width and at a height of 400–600 ft O.D., trending southwards to the coast at Salcombe Hill Cliff, Dunscombe Cliff and Coxe's Cliff. On the margins of these plateau-remnants the ground falls steeply away at an angle of 20–30°, but at the junction between the Greensand and the Triassic marls there is a sudden break of slope, accentuated by a change from woodland, scrubland and rough pasture to cultivated fields, below which there are gentle slopes at 5–10°. To the east of Lyme Regis, the removal of the cap-rock is at a more advanced stage, and very broad valleys are separated by interfluves occasionally surmounted by butte-like hills (for example, Black Ven, Stonebarrow, Golden Cap and

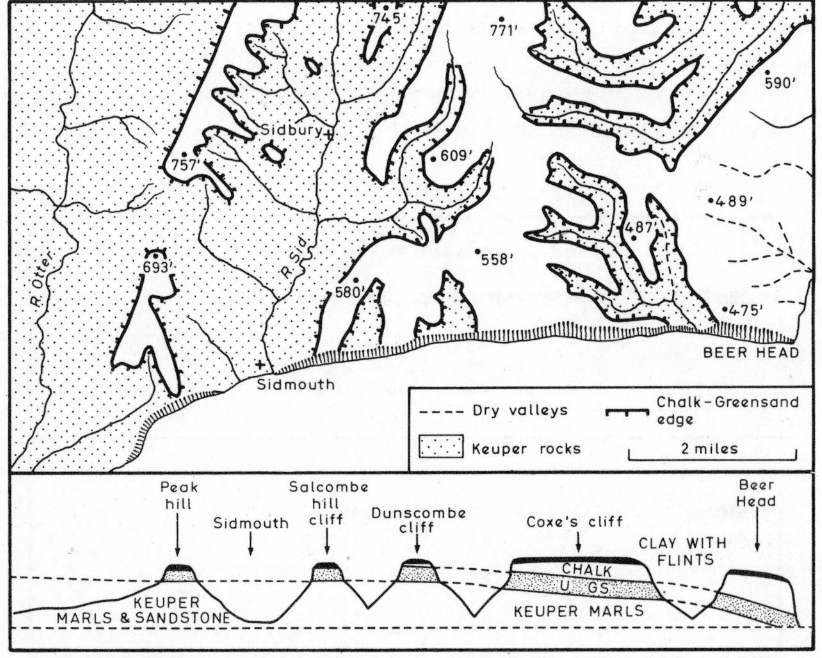

27 The dissected plateaus of east Devon

Thorncombe Beacon, all protected by small outliers of Upper Greensand) and mesa-like plateaus developed in the near-horizontal Upper Liassic sandstones (as around Bridport). Farther inland, however, the Chalk, Upper Greensand and resistant Liassic formations have been completely removed, and the Marshwood Vale has been excavated by the upper Char and its tributaries from the unprotected Lower Lias clay.

The New Forest of Hampshire

This region is underlain by unresistant Eocene and Oligocene sands and clays dipping gently southwards towards the main axis of the Hampshire Basin syncline. However, over much of the Forest these weak rocks are protected by spreads of plateau-gravel, which is composed of up to 20 ft of hard subangular flints set in a matrix of sand and sandy clay. The gravel, of Quaternary age, has been deposited by a variety of agencies (marine, fluviatile and periglacial) on a series of marine and sub-aerial surfaces developed in relation to an intermittently falling sea-level (pp. 254–6). Although as a whole the plateau-gravel possesses a slope towards the Solent, in any particular locality it is virtually horizontal. Since its accumulation, the gravel has been breached by small streams, but has remained intact on the interfluves, where it has acted as a cap-rock preventing the wasting of the underlying sands and clays.

There are notable geomorphological contrasts between the northern and southern parts of the Forest (Fig. 28). In the south, where the

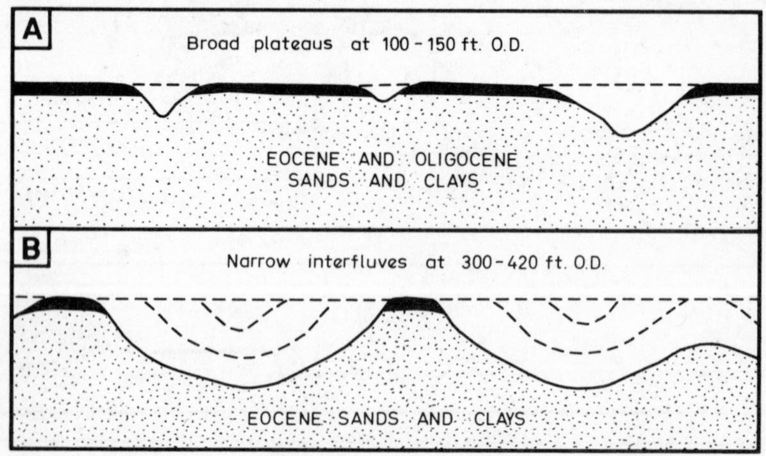

28 Cross-sections of southern (A) and northern (B) New Forest valleys

gravels are of mid- and late-Quaternary age, the valleys are usually constricted and steep-sided, and the gravel remains as extensive heathy plateaus, up to 150 ft above sea-level, which are still the dominant element in the landscape. To the north, on the other hand, the gravel dates from earlier in the Quaternary and is appreciably higher (300–420 ft O.D.). Its dissection is accordingly more advanced. The valleys here are unusually broad, and often possess a U-shaped cross-section, the result of lateral extension by springs and seepages issuing (i) at the margins of the permeable gravel and (ii) at junctions between sand and clay in the exposed Tertiary rocks. The gravel itself has been reduced to narrow tongue-like cappings, or has been locally consumed altogether, so that the Eocene rocks, deprived of their protection, have been rapidly attacked and the relief greatly reduced.

The Tertiary volcanic district of western Scotland

The islands and peninsulas of western Scotland were in the Eocene period the scene of a vast outburst of igneous activity. Large plutonic masses were injected into the crust, and later exposed to form the granite mountains of Arran and southern Skye (the Red Hills) and the jagged gabbro peaks and aretes of the Cuillins. Literally thousands of dolerite and basaltic dykes were intruded, either in the form of ring-dykes and cone-sheets (as in Mull and Ardnamurchan) or as swarms of subparallel dykes trending generally from north-north-west to south-south-east (as in southern Arran). Of particular interest in the present context are the near-horizontal intrusions developed in some places, mainly where relatively undisturbed Mesozoic sediments have been preserved. In south Arran, for example, the Permian and Triassic sandstones and marls are interrupted by many sills, which are substantially harder than the country rock. Among the more important of these are the so-called 'crinanite' (a form of dolerite) and quartz-dolerite sills of the areas inland from Dippin Head and in upper Glenashdale, together with the magnificent quartz-porphyry sill exposed at Drumadoon. Also of great interest in western Scotland are the huge spreads of 'plateau-lava', which cover some 400 square miles of northern Skye and a further 300 square miles of Mull (where the lavas, although reduced by denudation still reach the amazing thickness of 6000 ft). These great lava flows, which spread over and submerged the previous landscape, comprise layers of hard, often columnar-jointed basalt alternating with softer 'slags'.

Both the sills and plateau-lavas of western Scotland clearly fall into the category of horizontal structures, and the landscape where they occur is characterised by extensive plateaus, dissected by stream erosion and glaciation into numerous broad flat-topped hills, which contrast vividly with the high and sometimes spectacular mountain masses formed by the larger intrusive outcrops. On coastal cliffs and valley sides it is common to find stepped profiles, composed of steep rocky scarps about 50–100 ft in height and associated with individual basalt flows, and pronounced steps, in effect small structural terraces, developed where weak slag has been eroded from the surface of underlying basalt. In southern Arran, the sills give rise to scarp-edges and plateaus, and in Glenashdale form valley-side terraces and a stepped long-profile, the Burn plunging in a series of falls over the outcrops of the hard quartz-dolerite (Fig. 16).

It will be seen from these four examples that the characteristic landforms derived from horizontal structures are (i) plateaus, (ii) flat-topped isolated hills of the mesa and butte type, (iii) structural surfaces of wide extent, and (iv) 'benched' or terraced hill-sides and valley slopes. It should be added that landforms similar in appearance to these may also be found in areas of complexly folded and faulted structure, where these have been affected by past episodes of planation or partial planation followed by uplift and rejuvenation. A good example is the Forest of Dean, a faulted synclinorium incorporating Old Red Sandstone and Carboniferous rocks, which was base-levelled by erosion at 600–1000 ft O.D. either in the Triassic or late-Tertiary periods and subsequently incised by the Wye and other streams. Even in the areas described above planation has sometimes played an important role. In Devon and Dorset it produced the unconformity on which the Greensand-Chalk cap-rock rests, and has also been responsible for removal of the Chalk, either wholly or in part, the deposition of an extensive layer of plateau-drift (mainly broken flint and chert), and accentuation of the plateau-like appearance of the region.

UNICLINAL STRUCTURES

Uniclinal structures (sometimes referred to as 'homoclinal') are those in which a general or regional tilt has been given by gentle earth-movements to the constituent rocks. Usually such dipping strata are found on the margins of broad dome-like uplifts or major synclinal depressions (such as the Paris Basin), or occupy the flanks of ancient massifs which have been affected by successive bodily uplifts. For instance, the Mesozoic sediments of lowland England, which show a broad overall tilt towards the east and south-east, largely owe their present structural form to a series of vertical movements which have raised the Palaeozoic 'oldlands' of the Pennines, Wales and the South-west Peninsula in mid-Cretaceous, pre-Tertiary and mid-Tertiary times.

One of the most important features of uniclinal structures is that they commonly affect a considerable thickness of rocks of differing lithological composition (limestones, sandstones and clays) and resistance to denudation. Such structures are therefore ideal for the process of differential erosion, which is effected by streams developing their courses along lines of weakness, such as those afforded by clays (which are mechanically weak and because of their impermeability favour a high degree of surface run-off) and incoherent sands. In this way, strike vales are formed, and intervening harder or more permeable strata (well-cemented sandstones, limestones and chalk) are left upstanding by default as cuestas or ridges. The drainage pattern associated with a uniclinal structure is usually of the trellised type (p. 227), comprising large dip-slope streams which traverse the cuestas and ridges at right angles and subsequents which have grown by headward erosion along the strike vales.

The most important landforms of such scarp-and-vale scenery are, of course, the cuestas themselves. These vary greatly in their morphology, for reasons to be discussed below, but in their simplest form they comprise a steep scarp-face, often exceeding 30° in angle and sometimes displaying bare rock faces, and a long and gentle dip-slope (occasionally referred to as a 'back-slope' when the gradient of the surface does not exactly coincide with the angle of dip) (Fig. 29).

By far the greater part of lowland Britain is made up of scarp-and-vale scenery, and excellent examples are also to be found in Palaeozoic rocks of upland Britain. Between Brecon and Aberdare, in south Wales, rocks of Old Red Sandstone and Carboniferous age dip generally southwards, at an average angle of 5–10°, towards the synclinorium of the coalfield. Structurally the region is complicated by faults running north-north-west to south-south-east and east-north-east to west-south-west (notably the great Vale of Neath 'disturbance-zone'), but the conditions have remained ideal for cuesta development (Fig. 30). The individual cuestas vary greatly in scale and form, and their continuity is often broken by streams running southwards (such as the Nedd Fechan and Afon Mellte). Continuous strike vales have not been formed, with the

29 The South Downs escarpment west of Fulking, Sussex. Note the rounded hollows in the foreground; these may be due to nivation in the Quaternary (see p. 343) [Eric Kay]

exception of the Neath valley, which has been extended north-eastwards to give a deep trench and in the process some disruption of the south-flowing drainage has been effected.

From north to south the cuestas of the area are as follows.

A *Old Red Sandstone*. The great escarpment of the Brecon Beacons, reaching a height of 2906 ft O.D. at Pen-y-fan, is formed mainly by the Brownstones division of the Old Red Sandstones (Fig. 31). The highest summits of the Beacons, usually of table-like form, are capped by the tough and resistant topmost grits and conglomerates of the Old Red Sandstone (for example, Fan Llia and Fan Fawr). The escarpment face, rising a thousand feet or more above the scarp-foot plateau developed in the Red Marls division of the Old Red Sandstone, is precipitous and deeply indented by large cirques and nivation hollows; the latter are separated by spurs showing a marked horizontality at 1600–1800 ft O.D. The dip-slope of the Beacons is deeply intrenched by streams such as the Afon Llia, Afon Dringarth, Taf Fawr and Taf Fechan (Fig. 19). At the heads of these valleys, deep cols may indent the escarpment crestline to a depth of a thousand feet or more.

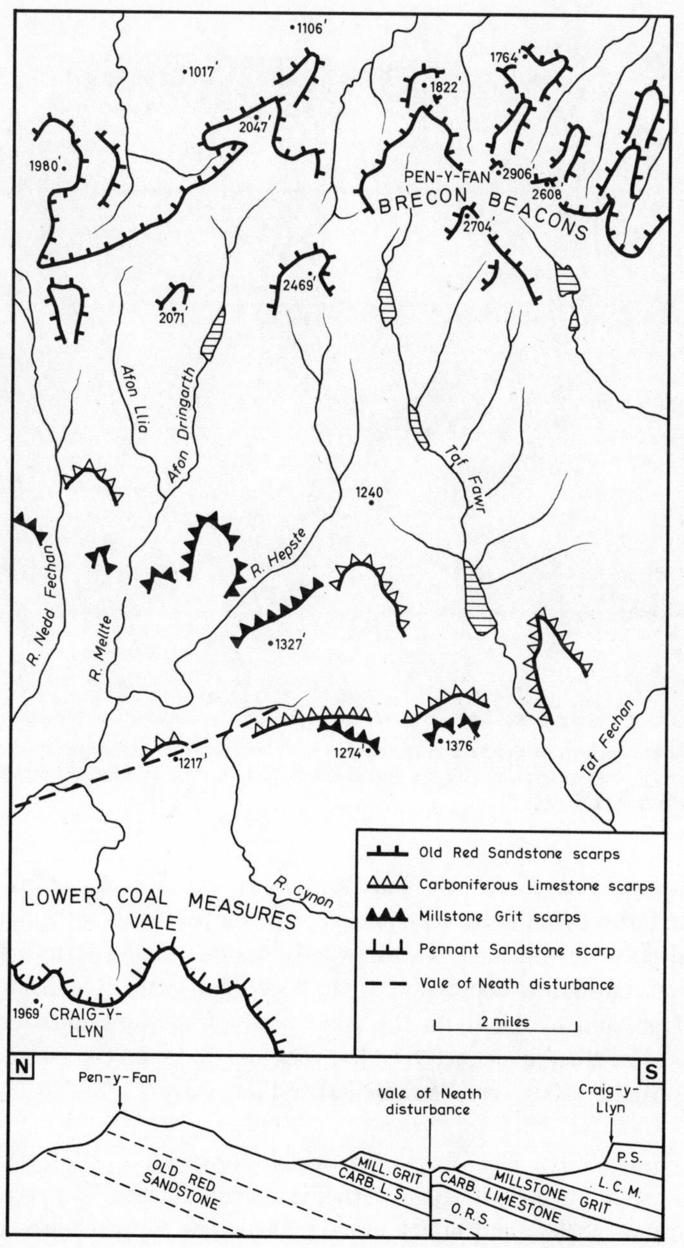

30 **The scarp-and-vale scenery of part of the northern margin of the south Wales coalfield**

31 The escarpment of the Brecon Beacons at Corn du. Note the resistant cap-rock, the well-developed recession-col on the near side of Corn du, and the glacial steepening of the scarp face [Eric Kay]

B *Carboniferous Limestone*. This formation, which in other parts of Britain forms magnificent escarpments, has comparatively little morphological effect in this area. The Lower Limestone Shales form a gentle but not continuous depression at the foot of the Old Red Sandstone dip-slope, and to the south the main body of the limestone forms a plateau-like area, diversified by a few rocky edges (as on the eastern side of the Mellte valley near Ystradfellte) and low ridges.

C *Millstone Grit*. This formation, which comprises in its lower part tough conglomerate (the Basal Grit), in its middle part comparatively unresistant shales, and at its top the sandstone layer known as the Farewell Rock, has a generally greater influence on surface form. The Basal Grit in particular forms a broken but nevertheless distinct escarpment, characterised by small free faces, which reaches to between 1500 and 1800 ft O.D. The scarp may be seen at Gwaen Cefn-y-Gareg, near Ystradfellte, where in places it assumes something of a 'flat-iron' form (p. 83).

D *Lower Coal Measures.* These comprise a considerable thickness of weak shales (though some bands of sandstone are present) which have been eroded into a major depression trending east-north-east to west-south-west from Brynmawr through Merthyr Tydfil and beyond. The vale is by no means continous in the sense that it is broken by deeply cut streams traversing it from the north; indeed it comprises rather a series of large saddles, reaching to 1000–1250 ft O.D., which occupy the interfluves between these streams.

E *Pennant Series.* These are composed of often massive and very resistant sandstones and grits, and form a major scarp-face bounding the Upper Coal Measures depression on the south side. In places (as to the east of Aberdare) the scarp is steplike in profile, owing to the effects of individual bands of sandstone, but to the west of Hirwaun there is one major scarp, reaching 1969 ft O.D. at Craig-y-Llyn and descending by way of near-vertical cliffs into the cirques of Llyn Fach and Llyn Fawr. The Pennant scarp as a whole is broken by large water-gaps, such as those of the Cynon (at Aberdare) and Taff (at Merthyr Tydfil), the entrances to which are broadly funnel-shaped (p. 83). In the Quaternary these gaps, together with the Neath valley, provided outlets for ice moving south-wards from the Brecon Beacon summits, and their form must have been modified to some extent by glacial erosion.

The profiles of cuestas

The height and cross-sectional form of cuestas is controlled by the following factors.

A *The thickness of the cap-rock.* Other things being equal, the thicker the resistant stratum the higher will be the resultant escarpment. This is admirably illustrated by the Hythe Beds (a division of the Lower Greensand) of the western and southern Weald. Around Hindhead, Surrey, these are composed of resistant sandstones and hard chert, and reach a thickness locally of 250–300 ft. They accordingly give rise to a very prominent cuesta, exceeding 700 ft at most points and culminating at Blackdown, near Fernhurst, at 918 ft O.D. However, from Petersfield eastwards the beds thin steadily, and the escarpment declines in height from 676 ft at Telegraph Hill to 495 ft and less east of Petworth, Sussex.

Beyond the Arun valley the Hythe Beds are attenuated very rapidly and also undergo lithological changes, and the cuesta virtually dies out altogether.

B *The durability of the cap-rock.* This will clearly be a major determinant of cuesta height. For example, in Dorset the Upper Greensand, consisting of well-cemented sandstone containing hard cherty concretions, forms a major escarpment on the southern side of the Vale of Wardour, reaching a height of over 800 ft O.D. immediately north of Shaftesbury. A few miles to the south-west, however, the chert beds disappear and the Greensand is represented by soft sands which have no morphological effects at all.

C *The angle of dip of the cap-rock.* It is axiomatic that the gentler the dip of the resistant stratum the higher will be the resultant cuesta (Fig. 32A). This is clearly brought out by the central Chalk uplands of the Isle of Wight. These reach a maximum height of 702 ft O.D. at Brighstone Down, in an area where the dip is 4° or less. To the west the Chalk rapidly decreases in elevation to 500 ft in the vicinity of Freshwater Gate, where the dip is as high as 70–80°. There is little contrast in slope between the scarp-face and back-slope here—indeed the Chalk forms a narrow ridge of hog's-back form rather than a true cuesta. Again, to the east of Newport, as the dip increases to 60–85°, the Chalk declines to a little over 400 ft, forming a fine hog's-back reaching the coast at Culver Cliff.

D *Erosional history.* Episodes of past planation which have affected a scarp-and-vale area may be recorded in the form of horizontal bevellings. These may either truncate the crests of the escarpments or interrupt the scarp- and dip-slopes (Fig. 32B). Bevelling of the cuesta as a whole is evident in the North Downs near Folkestone and Dover, where an extensive 'summit-plateau' at 500–600 ft O.D., the result of marine planation (pp. 278–9), separates the steep scarp-face on the south and the long and gentle dip-slope to the north-east. Farther to the west in the North Downs, between the Darent and Mole gaps, marine erosion during the same episode was confined wholly to the dip-slope, forming a bench at 500–650 ft above which the escarpment crest rises beyond the old degraded cliff-line to a maximum of 882 ft O.D. Scarp benches, of more limited dimensions, may be seen in the eastern South Downs between Lewes and Eastbourne, at heights of 200 and 400 ft. On an altogether

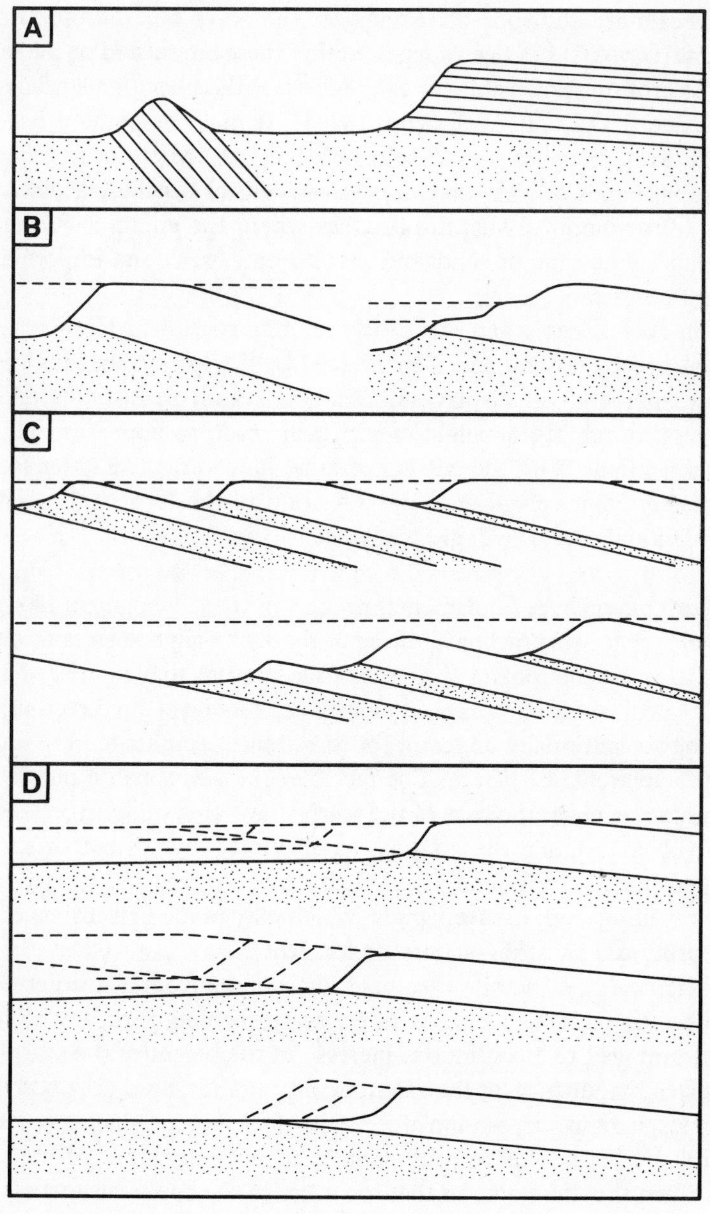

32 **Cross-sections of cuestas: the influence of dip (A); escarpment bevels (B); cuestas in a peneplained area (C); the recession of cuestas (D)**

larger scale are the spur-flattenings on the scarp-face of the Brecon Beacons (p. 76). It has been suggested that these are related to the major cycle of erosion responsible for the highest of the planation surfaces (at 1700–2000 ft O.D.) in Wales (pp. 281–6). It must be pointed out that not all scarp benches are of 'erosional' origin. Many are related to individual bands of rock, and thus are structural benches, whilst some result from rotational slipping in areas where the cuesta is formed of permeable limestone or sandstone resting on a weak and impermeable foundation of clay (p. 33).

In an area of heterogeneous gently dipping rocks which has recently been planed by erosion and then affected by limited stream incision, all the escarpments will display summit bevels and, irrespective of rock thickness, durability or angle of dip, will reach to approximately the same elevations. With the passing of time, however, these latter factors will reassert themselves, and diversification in the form and height of the individual cuestas will gradually occur (Fig. 32C).

E *Scarp recession.* A fundamental process in the development of scarp-and-vale scenery is uniclinal shifting of the strike-aligned streams. In an effort to maintain contact with the weak stratum that nourished their early growth, these streams tend to migrate in a down-dip direction, and in doing so continually undercut (or to be more accurate tend to undercut) the adjacent scarp-face. The fact that such streams do not always flow immediately at the base of the scarp shows that other processes too are involved. Among these are (i) basal erosion by scarp-foot springs and associated streams, and (ii) accelerated weathering and retreat of the scarp resulting from its steep angle (which may in the first instance have been produced by strike-stream undercutting).

In the case of a 'new' escarpment, developing from a 'feather-edge' produced by previous planation of the scarp-forming rock, scarp retreat will at first lead to a continuous increase in the height of the scarp-face and a steady reduction in the extent of any summit bevel (Fig. 32D). At a later stage, however, erosion of the scarp-foot vale, producing recession beyond the bevel, will be counterbalanced by 'migration' of the scarp crest down the dip-slope, so that the relief of the cuesta as a whole will tend to remain constant. Later still, under conditions of base-level stability which prevent any further downcutting of the scarp-foot vale, escarpment retreat will necessarily continue and be accompanied by a reduction in the height of the scarp-face and a decline in the importance of the cuesta.

The plans of cuestas

Other notable morphological features of escarpments, more apparent in their plans than their cross-profiles, are as follows.

A *Water-gaps.* The continuity of most cuestas is broken at some points by large through valleys occupied by permanent streams. The latter are usually, though not always, major dip-slope consequents which for one reason or another have resisted encroachment by neighbouring consequent systems or have actually enlarged their own catchments, ahead of the escarpment, through subsequent stream growth and river capture (pp. 233–4). Notable examples of water-gaps are those of the Usk (cutting through the Old Red Sandstone escarpment between Brecon and Abergavenny), the Humber (crossing the Chalk between the Lincolnshire and Yorkshire Wolds), the Witham (through the Jurassic limestone scarp at Lincoln) and the Bristol Avon (which cuts, unusually, against the dip through the southern Cotswolds at Bath).

In areas where water-gaps are small and numerous (as in the case of a 'young' escarpment where little modification of the initial stream pattern has as yet been achieved), the escarpment may take on a highly individual form. Each stream crossing will be associated with a V-shaped re-entrant, and between the gaps the scarp will project forward as a prow-like hill or 'flat-iron'. The reason for this is simply that, within each re-entrant, the weak stratum lying beneath the cap-rock of the cuesta rises higher and higher above the stream in an 'up-dip' direction. It therefore occupies an increasing proportion of the valley-side slopes, and so promotes more rapid slope retreat and opening-out of the entrance to the gap. It has been suggested that, in time, these funnel-shaped re-entrants will lose their significance, the intervening spurs being driven back by wasting and the escarpment as a whole becoming straighter in plan. However, much will depend on (i) the degree to which the initial drainage pattern has been changed by capture, since the growth of dominant subsequent streams and a reduction in the through-flowing drainage will lead to the disappearance of many gaps, and (ii) whether a stable base-level of erosion prevents further incision by the gap streams. If it does, the re-entrants can no longer be extended in a down-dip direction, while at the same time the flat-irons must be slowly worn back by normal scarp retreat, thus giving in plan a rectilinear escarpment.

An interesting example of an escarpment with well-developed flat-irons is that formed by the Lower Calcareous Grit (a division of the

Corallian) on the southern slopes of the North York Moors (Fig. 33). The scarp is broken by a series of streams, spaced some 2–3 miles apart, flowing towards the Vale of Pickering. On its northern side the scarp-face rises some 200 ft above an incipient strike vale in the weak Oxford Clay. A similar pattern of gaps and flat-irons can be seen on the dip-slope of Wenlock Edge in Shropshire. Wenlock Edge itself is a bold and, in plan, remarkably straight escarpment formed by the eastward-dipping Wenlock Limestone of Silurian age. A very short dip-slope leads down to a vale developed in the Lower Ludlow Shales, beyond which there is a much broken 'secondary' escarpment, formed by the Aymestry Limestone and traversed by small brooks rising in the Ludlow Shales depression. A final example is found in the South Downs and Chalk of eastern Hampshire, where the *Gonioteuthis quadrata* fossil-zone of the Upper Chalk gives rise to numerous flat-irons projecting up the dip-slope towards the 'main' Chalk escarpment (usually capped by the *Micraster* fossil-zones). The gaps here, however, are no longer occupied by streams, for these have vanished with the lowering of the Chalk water-table—indeed, the secondary escarpment may be described as being in a state of arrested development.

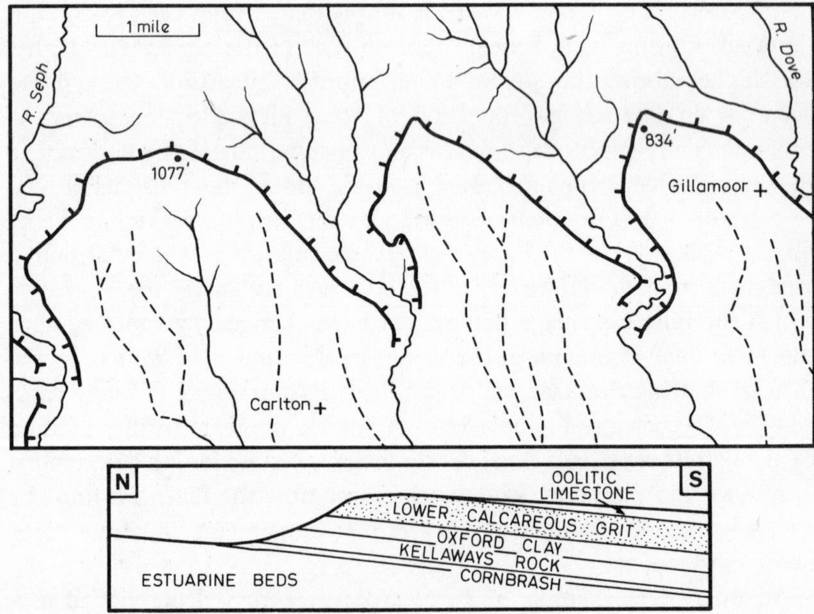

33 The Lower Calcareous Grit escarpment, North York Moors

B *Wind-gaps.* In some cases these mark important through valleys, once occupied by sizeable consequent streams which have been since beheaded by subsequent streams developing ahead of the escarpment. They are now dry in their uppermost parts, but may contain small misfit streams in their lower courses. An interesting sequence of such wind-gaps, together with intervening water-gaps, can be seen in the South Downs. From west to east, in a distance of some 40 miles, the following gaps are encountered: the Cocking wind-gap; the Arun water-gap; the Washington wind-gap; the Adur water-gap; the Poynings wind-gap; the Ouse water-gap; the Cuckmere water-gap; and the Jevington wind-gap. It has been argued from this evidence that the Downs were once crossed by eight south-flowing consequent streams, on average some 5–6 miles apart, and that the Arun, Adur, Ouse and Cuckmere, by enlarging their catchments in the Weald Clay vale both westwards and eastwards, took over the corresponding catchments of the former Cocking, Washington, Pyecombe and Jevington streams.

In addition to the larger consequent streams, with sources well in advance of the line of the escarpment, smaller streams—usually in considerable numbers—actually rise on the upper part of the dip-slope itself. Frequently scarp recession has been so great that these dip-slope valleys *sensu stricto* have been truncated, and as a result the escarpment crestline is broken by numerous cols at the heads of existing valleys. Such recession-cols are normally smaller and less deeply cut than the gaps of the through-flowing streams, but a clear distinction between the two is in practice not always easy to make. A fine series of high-level recession-cols may be seen in the North Downs around Caterham, Surrey, where the close spacing of the gaps and their obvious relationship with small dip-slope tributary valleys indicate clearly that the streams responsible for their formation rose only a little way to the south of the present escarpment (that is, within the area of the dip-slope at a rather earlier stage).

A more unusual case is found in the Chalk Downs of Dorset, west of the Stour water-gap (Fig. 34). Here the Upper Chalk cap-rock has been greatly reduced in thickness, and past and present dip-slope streams have cut down in their upper courses to reveal inliers of Middle and Lower Chalk. These divisions—and especially the latter—are much less resistant than the Upper Chalk, and where exposed have been hollowed out into large, elongated and sometimes amphitheatre-like 'basins' (for example, the valleys north and west of Milton Abbas and the remarkable Lyscombe Bottom, north-east of Piddletrenthide). In some instances, recession of the escarpment from the north has subsequently overtaken the heads of such valleys, giving cols on a scale suggestive of formation by large streams once rising far to the north. Not all are, strictly speaking, wind-gaps. That at the head of the Devil's Brook, at Bingham Melcombe, is still occupied by a small stream (actually fed by scarp-foot springs) with the appearance of a typical misfit. However, there is little doubt that its catchment has undergone little or no reduction, and there is no question of the gap being comparable, in terms of origin, with the main wind-gaps of the South Downs, North Downs or Chilterns.

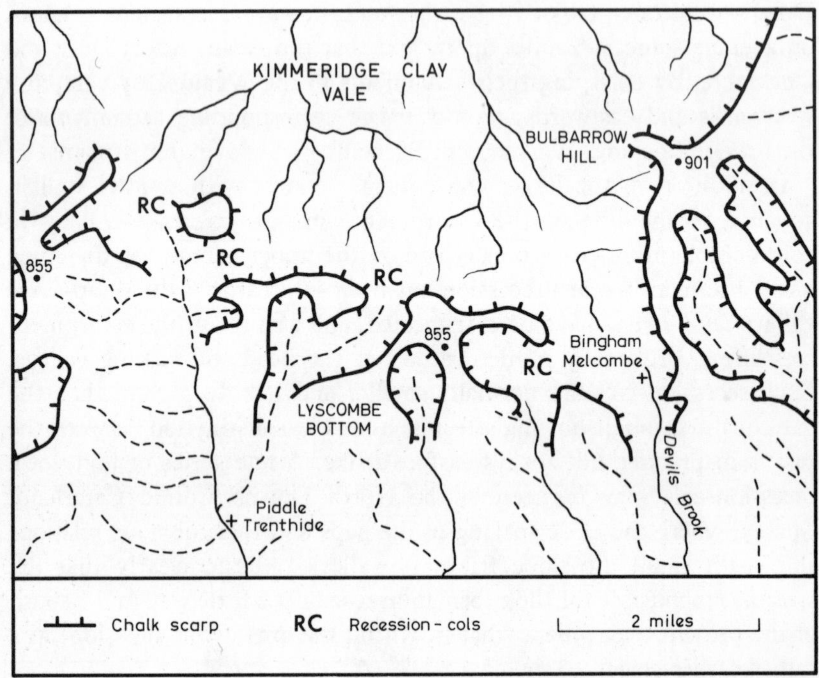

34 **The Chalk escarpment west of the Stour valley, Dorset**

C *Escarpment valleys.* In plan, escarpments show an almost infinite variety. As we have seen, some are virtually straight, with little or no evidence of stream dissection except in the vicinity of large water-gaps where deep indentations may occur. Others are rendered highly complex by the occurrence of numerous deeply penetrating and ramifying escarpment valley systems. Nowhere in the British Isles can this be seen to better effect than in the Cotswolds, a magnificent cuesta capped around Stroud and to the north-east by the Inferior Oolite limestones and to the south, nearer Bath, by the Great Oolite (Fig. 35). The upper part of the scarp-face is formed by Upper Lias sands (the Cotteswold Sands) and clay, and overlooks a fine structural bench (that of the Marlstone Rock Bed, the uppermost division of the Middle Lias). The lower part of the scarp is developed in Middle Lias sands and clay, and the important Lower Lias clay.

Dissection of the Cotswold escarpment by 'anti-dip' streams, though a feature of the whole length of the scarp, is at a maximum around Stroud. Here the gorge-like Golden Valley penetrates several miles eastwards into the upland, and is, in its lower part, joined by the impressive strike-aligned Painswick and Nailsworth valleys, together with many smaller but still striking tributary valleys (Fig. 36). To the east of Cheltenham scarp dissection is again pronounced, and the Winchcombe valley, though broader and more mature than those at Stroud, forms a major re-entrant. The valley pattern at Winchcombe is

35 The Cotswold escarpment near Birdlip, Gloucestershire. Note the development of a free face, in Inferior Oolite limestones, at the crest of the scarp. [Eric Kay]

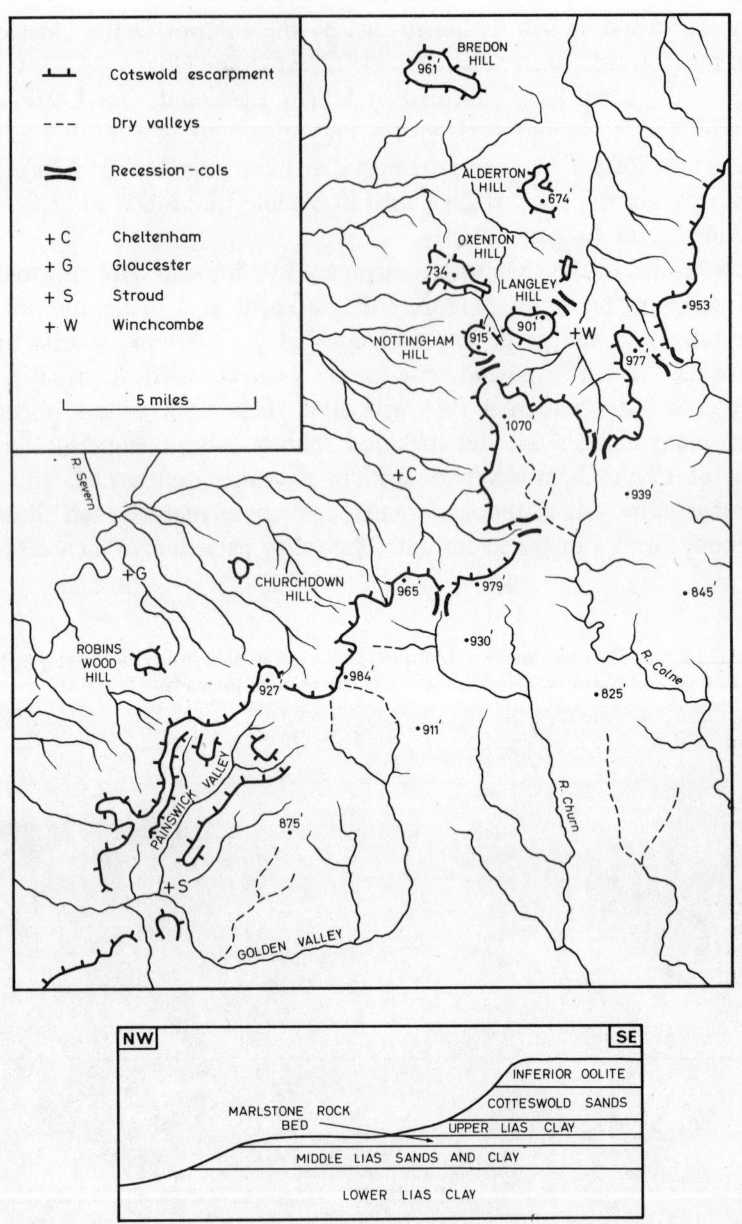

36 **The Cotswold escarpment between Stroud and Winchcombe**

especially interesting in that the western tributary streams have, by headward erosion, almost succeeded in isolating the upland masses of Langley and Nottingham Hills. In form and size these two hills in fact closely resemble the completely detached Oxenton Hill, lying a little to the north. Such outlying hills are a notable feature of the Cotswold escarpment, though they are generally rare elsewhere in Britain. Other local examples include Bredon and Alderton Hills, north of Cheltenham, and Robins Wood and Churchdown Hills, south and east of Gloucester. Outliers of this type can be developed only (i) where the angle of dip is exceptionally low (that of the central Cotswolds is in the order of 1°), and (ii) where scarp dissection is abnormally advanced (for the outliers are, in effect, the unconsumed remnants of interfluves between adjacent escarpment valley systems). Conversely they are virtually never present when the angle of dip of the cuesta forming rock exceeds 5°, or where the scarp-face undergoes more uniform retreat by mass wasting processes.

The unusual morphology of the Cotswold escarpment poses some difficult problems. What is apparent is that certain individual valleys (for example, that at Winchcombe and the Chelt valley east of Cheltenham) have been eroded back along the lines of former dip-slope valleys, and that the agencies responsible have been able to take advantage of important recession-cols incised well below the general summit-level of the cuesta. Scarp valleys in other permeable rocks (such as the Chalk) have been explained in terms of erosion by springs ('spring sapping'). Although the low angle of dip has probably facilitated the movement of underground water towards the Cotswold escarpment, where in turn the various clay horizons have favoured the development of springs and seepage-lines, spring sapping can hardly be invoked to explain features such as the Golden Valley at Stroud. It does seem clear, to judge from the evidence of the outliers, that the dissection of the scarp-face is not a recent phenomenon, developed since the escarpment as a whole had retreated to its present line, but that it has been a long-term process in the Cotswolds.

FOLDED STRUCTURES

Over much of the earth compressive forces in the crust, stemming mainly from orogenetic movements of diverse age, have produced folding of sedimentary strata. The complexity of such folding, and the precise scale and form of the individual structural components, differs greatly from place to place. In the main mountain chains, such as those produced by the Alpine movements of the Middle Tertiary, intense compression of vast thicknesses of geosynclinal sediments has given an assemblage of complicated overfolds, isoclinal folds, recumbent folds and nappes. Farther away from the centres of orogenesis the folding is of a more simple type. The Jura mountains, for instance, consist of a series of arcuate, roughly parallel and sharply folded anticlines and synclines, resulting from the less extreme compression of marine limestones which have slipped over a 'lubricating' layer of Triassic anhydrite. In southern England, where what are usually termed the 'outer ripples of the Alpine storm' have been felt, the folds are still smaller and gentler, and have resulted not merely from compressive forces acting on late-Mesozoic and Tertiary sediments, but also from deep-seated movements, along old axes formed by the Hercynian orogeny, in the underlying Palaeozoic basement; such revived folding is normally referred to as 'posthumous'.

In the following discussion, examples will be used to illustrate detailed relationships between the more simple types of folded structure and landform development. In areas such as the Alps or even the Scottish Highlands, where dissection of highly complicated and often imperfectly understood structures by fluvial and glacial action has reached an advanced stage, the principles underlying such relationships are far more difficult—or perhaps impossible—to demonstrate.

Normal or Jura-type relief

At the time of their formation, anticlinal and synclinal folds will have an immediate and obvious effect on relief. The landscape will simply comprise a series of anticlinal ridges, whose form will depend on the symmetry, amplitude and pitch of the upfolds, separated by synclinal depressions which will guide the main streams of the area. However, the intervention of erosion will disturb this close correspondence between structure and surface form, and may ultimately cause the anticlinal ridges to be replaced by scarp-rimmed valleys or depressions (anticlinal vales), and the original synclinal lowlands to be left upstanding as ridges (synclinal hills). It must be stressed, however, that many folded structures, which have had a long history of denudation, have not experienced such 'inversion of relief', but continue to display a 'normal' relationship between structure and relief form. This is true, in fact, of the Jura mountains themselves. It has to be borne in mind too that where inversion of relief has once been achieved, continued erosion can lead to a restoration of the original structure-relief relationship; the synclinal valleys and anticlinal ridges are then, strictly speaking, 'resequent'. Some selected examples of 'normal' relief patterns follow.

A *Eastern Hampshire.* In a transect through eastern Hampshire from Portsmouth towards Basingstoke, a number of east-west Chalk ridges (for example, Portsdown and that trending east-south-east from Winchester towards Butser Hill) and broad valleys (including the Forest of Bere depression and the upper Itchen and Dever valleys) are encountered (Fig. 37). In every case the Chalk uplands are anticlinal in structure, and the intervening vales are either shallow synclines containing some Tertiary infilling (the Forest of Bere is underlain by Reading Beds, London Clay and Bagshot Beds) or are associated with downfolds in the Chalk from which Tertiary sands and clays have been removed by fluvial action. The anticlinal ridges have themselves been affected to some extent by erosion since their formation in the Miocene period. Their coverings of weak Tertiary rocks were obviously destroyed at an early date, perhaps as the folds were being upraised. More recently much of the Upper Chalk too has been worn away; for instance, the whole of the Belemnite Chalk (p. 69) has been removed from the crest of the Winchester anticline, possibly as a result of marine erosion during the late Pliocene (pp. 278–9). Such planation may exert a very strong influence on subsequent relief development. It has been suggested that, in this particular area, the folds were planed across by wave action at a height of 540–690 ft O.D., and that on the emergent sea-floor belted outcrops of weak Tertiary rocks (coinciding with synclinal axes) and resistant Upper Chalk (along anticlinal lines) were exposed. During the Quaternary streams such as the Itchen and Dever etched out the synclines, leaving the Chalk ridges upstanding by default. Had marine planation been effected at a substantially lower level (say, close to the present base-level), the results would have been quite different, for

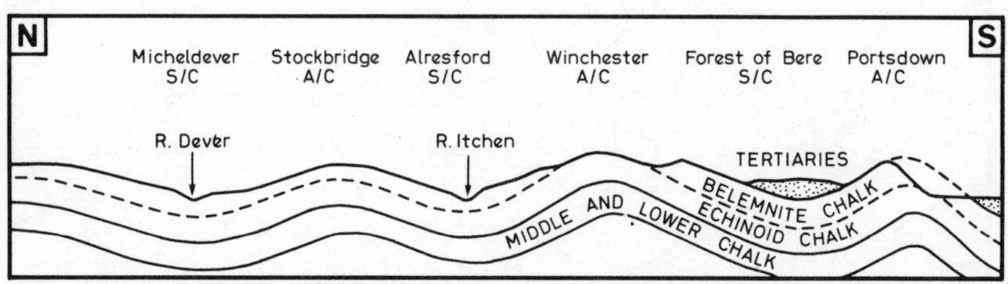

37 Cross-section of the Alpine folds of eastern Hampshire

complete removal of the Upper Chalk crests and exposure along many anticlinal axes of weaker Middle and Lower Chalk would have favoured the subsequent development of inverted rather than normal relief.

B *The Gower peninsula.* This area, lying to the south-west of Swansea, again displays admirably the normal relationship between folds and surface form, despite the fact that it has experienced a much longer and more eventful denudational history than eastern Hampshire. Structurally, Gower consists of a series of anticlines and synclines, of Hercynian age, trending west-north-west to east-south-east (Fig. 38). Notable individual folds include the Cefn Bryn and Oxwich Point anticlines, and the Oxwich Bay and Port Eynon synclines. The most important elements in the present landscape of Gower are:

(i) The anticlinal ridges and hills, composed mainly of quartz-conglomerates of the Old Red Sandstone (though Silurian shales are exposed as an inlier near the crest of Cefn Bryn) and rising to heights of 609 ft (Cefn Bryn), 609 ft (Llanmadoc Hill), 633 ft (Rhossili Down) and 500 ft (Harding's Down). These elevations may be coincidental, but also possibly indicate the former existence in Gower of the 600-ft coastal platform (supposedly of late-Pliocene age) found elsewhere in Wales.

(ii) The pronounced plateau at 200–400 ft O.D., evidently the result of planational processes affecting the steeply dipping Carboniferous Limestone and locally masked by deposits of till and glacial sands (Fig. 115). Again, it has been postulated that episodes of marine erosion during the Pliocene period have been responsible for the cutting of this plateau, parts of which have been equated with the widely developed 200-ft coastal platform of Wales.

(iii) The small 'lowland' which has been excavated between Oxwich and Penrice at a point where the Oxwich Bay syncline preserves an outcrop of weak Millstone Grit shales.

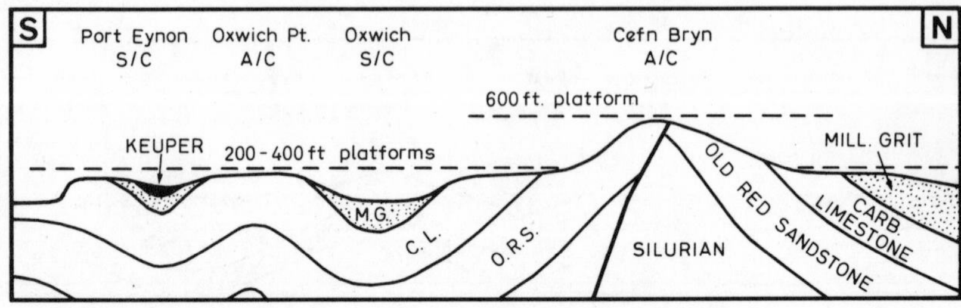

38 Cross-section of the Gower peninsula, south Wales

The relationship between structure and surface in Gower is therefore basically very simple, but even so the history of relief development is decidedly obscure. The first thing that must be appreciated is that the formation of the present landscape has been achieved at the expense of enormous erosion of the original Hercynian folds. Over most of Gower literally thousands of feet of sedimentary rock (Old Red Sandstone, Carboniferous Limestone, Millstone Grit and Coal Measures) have been removed. It is likely that most of this erosion was accomplished, under 'desert' conditions, during the immediate post-Hercynian period (Permo-Triassic times), and that the resultant landscape was fossilised beneath younger deposits of upper Triassic (Keuper) and Jurassic age (such as are widely preserved in the vale of Glamorgan to the east, where they still rest on the eroded remnants of similar Hercynian structures). Some evidence in direct support of this interpretation lies in the existence of a small outlier of Keuper conglomerate, resting on Millstone Grit at Port Eynon, together with 'gash-breccias' of Triassic age preserved in former caverns in the limestone of Gower. The removal of these Mesozoic deposits, and the exhumation of the ancient relief, may be a comparatively recent (perhaps late-Tertiary) phenomenon. If so, the so-called 'coastal platforms' may in essence be sub-Triassic surfaces which have merely been retrimmed by late-Pliocene wave erosion. Another possibility is that active chemical weathering during the Tertiary era (pp. 135–6) may have affected the Gower landscape. It is noticeable that the so-called 'planation surfaces' are largely confined to the Carboniferous Limestone (a hard rock, but one very susceptible to chemical attack) and never extend on to the Old Red Sandstone (likewise hard, but not susceptible to rotting). This is difficult to account for if the Gower plateau is the product either of 'backwearing' and pedimentation in an arid climate or of cliff recession and wave-cut platform extension under conditions of higher sea-level. It is more readily explicable in terms of differential *downwearing*, resulting from deep weathering and transportation in a 'humid tropical' climate (p. 176), though it must be added that there is no direct evidence of such an episode in Gower today.

It is interesting to make a brief comparison between Gower and the Mendips, the structure of which is in many ways similar. The Mendips form in general an anticlinal upland of Carboniferous Limestone, bounded by steep edges to the north and south where the limestone plunges steeply beneath Keuper marls and conglomerates. Although deeply dissected in parts by dry valleys and gorges (such as Cheddar and Burrington Combe), much of the Mendips is plateau-like, at a height of 800–900 ft O.D. Individual ridges of Old Red Sandstone (such as Blackdown, North Hill and Pen Hill) rise somewhat above the main plateau, though they are less bold than comparable features in Gower. These ridges mark east-west periclines which are superimposed on to the main Mendip upfold. That the eastern part of the Mendips is an exhumed landscape is shown by the occurrence on the plateau of outliers of Dolomitic Conglomerate (of Keuper age) and Rhaetic and Liassic strata, and by the 'banking up' of Keuper conglomerate against the northern and southern limestone scarps. To the west, the evidence is even more striking. In its western section the Blackdown pericline has been breached to give a small lowland, the Vale of Winscombe, framed by ridges of Carboniferous Limestone. On the floor of the anticlinal depression, resting discordantly on Old Red Sandstone, are Keuper deposits, which prove conclusively that (i) the Vale of Winscombe was first etched out under conditions of desert erosion in the Triassic period, (ii) the breached anticline was fossilised by a Keuper infilling (and probably overlying Jurassic strata), and (iii) recent exhumation of the vale has been carried out by the Lox Yeo river, which itself takes advantage of an ancient Triassic exit towards the south-west.

Inverted relief

As described already, long-continued denudation of anticlines may lead eventually to their breaching, and to the formation of continuous or discontinuous vales bounded by infacing escarpments. The factors which promote such breaching are threefold.

A *Crestline weaknesses*. It is often said that the strong tensional forces arising from uparching produce structural weaknesses (mainly in the form of a well-defined pattern of joints and fissures) at the crest of an anticline. These weaknesses can be later utilized by weathering and stream erosion, and an anticlinal valley etched out. However, it seems probable that this factor, although of some importance, has been overstressed.

B *Weak cores*. Most anticlines, like the uniclinal structures discussed above, comprise strata of varying resistance to erosion. Often they possess an outer 'shell' of durable rock (limestone, chalk, sandstone) and an inner 'core' of weaker material (clay or unconsolidated sands). This is, of course, the reverse of the present situation in Gower, where the cores are evidently composed of hard Old Red Sandstone contained within weaker Carboniferous Limestone and younger rocks (though it should be remembered that in due course, as the Cefn Bryn anticline shows, erosion will reveal weak cores of Silurian rocks). Where a soft core does exist, and providing erosive processes are in a position to take advantage of it, breaching can be initiated. The controlling factor in this respect seems to be not so much the thickness of the overlying rock layers as the relationship of the fold and its constituent strata to the prevailing base-level of erosion. If the fold (and thus its soft core) stands high above base-level, breaching is feasible; if it does not, so that the core is beyond attack by streams incising the flanks of the fold, breaching is a manifest impossibility. This is the explanation of why an individual fold may be breached at one point (where there is a periclinal culmination) but not at another (where it stands lower, but is still a sufficiently pronounced structural feature to give rise to crestline weaknesses due to tension). It also accounts for the fact that adjacent folds, involving the same rock-types, are differently eroded. An interesting example of this is to be found in the western Weald, where the Hindhead anticline (with a shell of resistant Hythe Beds sandstone and a core of weak Atherfield and Weald Clay) remains virtually unbreached, while the neighbouring Fernhurst fold is fully breached, with a fine anticlinal vale floored by Weald Clay and bounded by high infacing Hythe Beds scarps. The difference between the two stem from the fact that the Fernhurst anticline, at least in its central and eastern parts, has always stood higher—indeed the Hythe Beds must once have reached to over 1000 ft O.D. and the Weald Clay to something like 600 ft on the axis of the fold. Towards the west, however, the Fernhurst anticline pitches strongly, the scarps 'close in' and eventually coalesce, and between Hill Brow and Petersfield the Hythe Beds give rise to an unbroken anticlinal ridge.

C *Past Planation.* An episode of planation may pave the way for inversion of relief, by leading to the removal of the resistant crests of anticlines and exposing the weaker rocks beneath. With uplift and rejuvenation, the latter may be hollowed out by stream erosion and anticlinal valleys rapidly formed.

The whole process of relief inversion is vividly illustrated by the Chalk country of central southern England. This area is crossed by numerous fold-axes, developed during early-Tertiary and 'Alpine' times and trending broadly from west to east. The individual folds are usually periclinal in nature; indeed several separate periclines may be

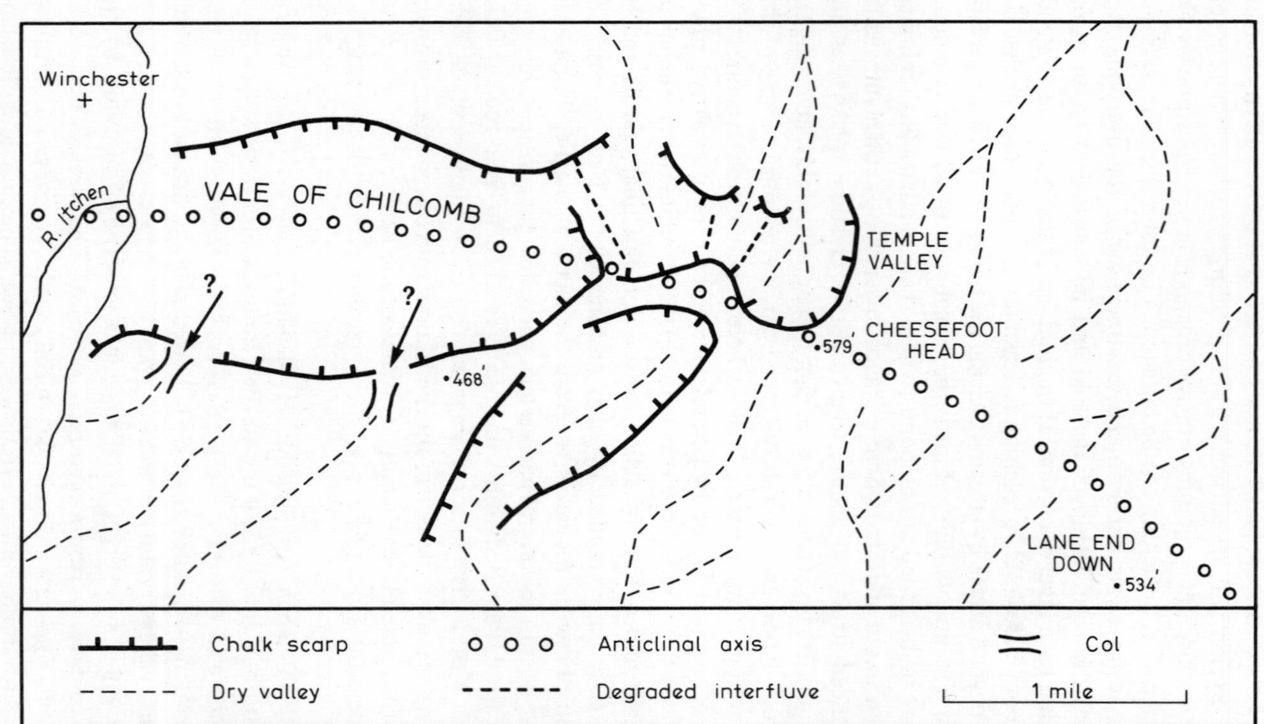

39 The breaching of the Winchester anticline at Chilcomb

developed along one extended axis. These periclines are for the most part strongly asymmetrical, with dips of from 2 to 10° on the southern flanks and from 30 to 90° to the north. The outer shell of most of the folds is formed by the resistant Echinoid Chalk (the Belemnite Chalk having been removed at some time in the past) and the core by weaker Middle and Lower Chalk. The inversion process can only begin effectively when streams are able to penetrate to these lower divisions of the Chalk. At first the breaches are very localised, occurring at periclinal culminations where the weaker Chalk stands highest above base-level. Away from these points the pitch of the folds carries the Middle and Lower Chalk below the level of possible stream incision, and only narrow and usually shallow valleys are formed in the strong Upper Chalk.

The actual process of breaching seems to have been initiated in two ways:

(i) In a few instances anticlines are crossed discordantly by main streams, which have probably had a history of superimposition (pp. 254–60). These have been able to cut deep gaps and expose in their valleys Middle and Lower Chalk. Valley widening has then proceeded, giving a small circular or elliptical vale overlooked by infacing Upper Chalk scarps (Fig. 39). This process is perhaps best seen in the Vale of Chilcomb, near Winchester, where the breaching has been extended some two miles eastwards from the discordant Itchen by former miniature tributary streams developing in the Middle Chalk and Chalk Marl (the lower division of the Lower Chalk).

(ii) More commonly the early stages in the breaching process have been achieved by small streams draining the flanks of the Chalk folds, particularly on the northern side where, owing to the steepness of the dip, comparatively little headward erosion is needed before the soft core is revealed. In their upper courses these streams, when reaching the Middle and Lower Chalk, are able to hollow out large amphitheatres on or close to the fold-axis. In time these hollows are extended laterally by such processes as periglacial mass wasting and spring sapping, and the interfluves between neighbouring valleys are partially destroyed (Fig. 40). There is thus a growing measure of integration, and a true anticlinal vale begins to emerge. The ultimate step is the development of a subsequent stream unifying the drainage of the vale.

In some cases it is apparent that these two methods of breaching have combined. For example, the breaching of the Winchester anticline immediately east of the Itchen has, as described above, been effected by tributaries of that stream. However, farther to the east individual amphitheatres, now beginning to coalesce, have been opened up by former north- and south-flowing minor streams; that at Temple Valley is an outstanding feature (Fig. 39). It appears likely that other such amphitheatres once existed nearer the Itchen, for the cols in the southern Chalk escarpment at St Catherine's and Deacon Hills could have been the outlets from these; however, their drainage has been taken over by the Itchen tributaries, which were able to work headwards along the fold-axis at a much lower level.

To the west of Salisbury a particularly fine sequence of the various stages in the inversion process may be seen (Fig. 41). In the south the

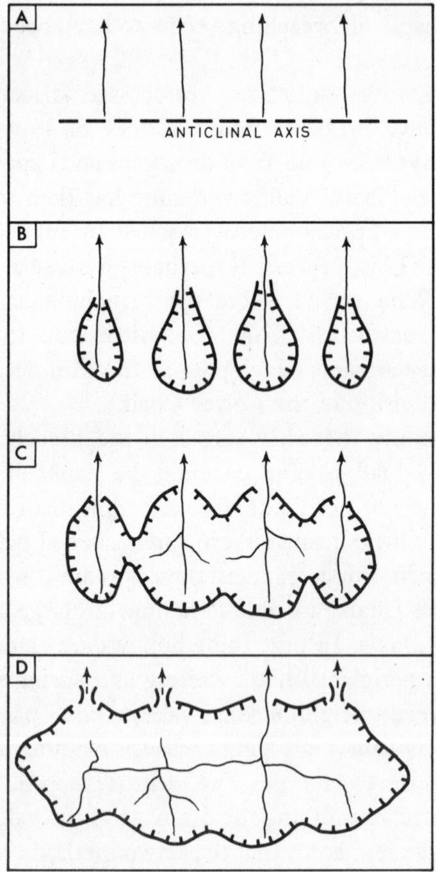

40 Stages in the breaching of an anticline by dip-slope streams

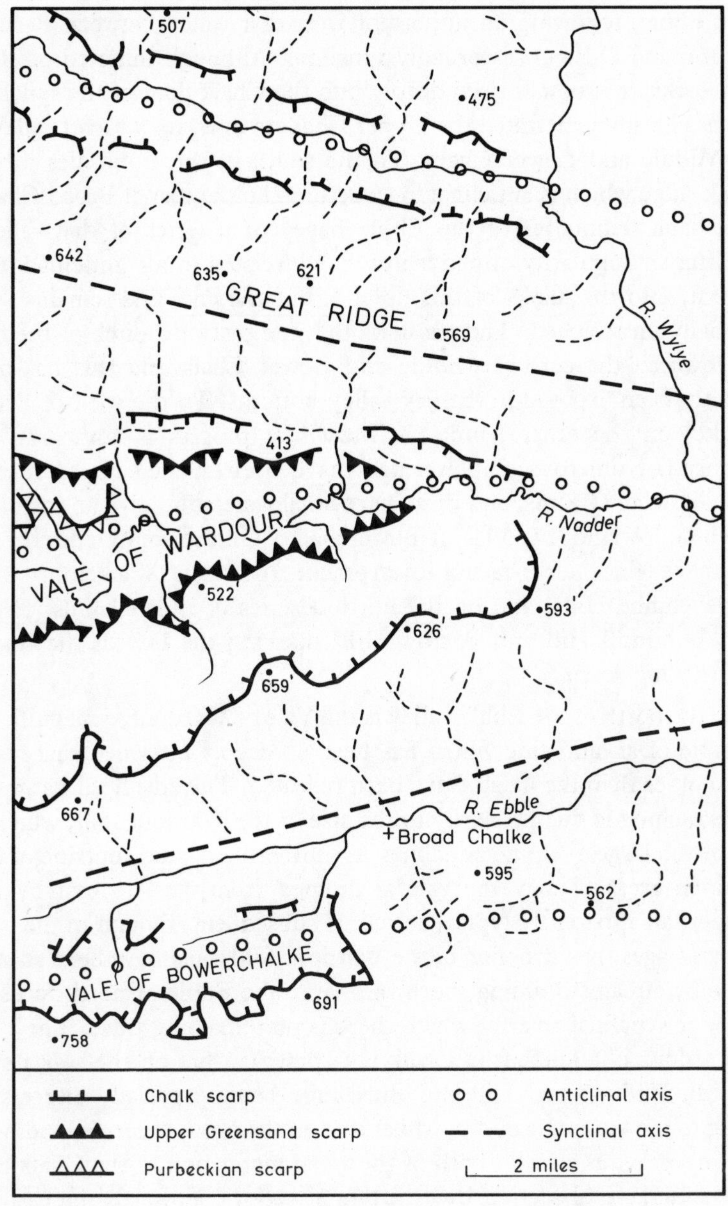

41 **The structure and relief features of the area west of Salisbury**

river Ebble, following an important west-east valley between Berwick St John and Odstock, is broadly synclinal. Although initiated on Tertiary rocks, it is now incised deeply into the Chalk floor of the syncline, and has locally penetrated the Upper Chalk to open up a broad valley in the Middle and Lower Chalk. To the south of the Ebble lies a high Chalk ridge which is anticlinal in structure. To the east of Broad Chalke right-bank tributaries of the Ebble have cut a series of dry valleys, showing an angularity of pattern probably related to the anticlinal joint pattern, into the flanks of this upland, but the anticline remains substantially unbreached. The reason is that the eastward pitch of the fold has lowered the core of Middle and Lower Chalk and this has only recently been exposed in the dry valley bottoms. To the west of Broad Chalke, on the other hand, the breaching process is comparatively advanced. Numerous amphitheatres have been opened up by former north-flowing streams, and these have amalgamated to give an incipient anticlinal lowland, the Vale of Bowerchalke. On the south side there is already a fine north-facing escarpment (between Winkelbury and Marleycombe Hills) and on the north a series of isolated hills (Horse Hill, Windmill Hill and Barrow Hill) marking the line of the future south-facing scarp.

To the north of the Ebble valley is the Vale of Wardour, a magnificent example of an anticline which has been so deeply breached that rocks much older than the Chalk have been revealed. The advanced nature of the breaching is due primarily to the size of the Wardour fold, which is one of the largest Alpine anticlines in southern England (outside of the Wealden area). Today the vale is drained from west to east by the Nadder, an apparently typical anticlinal subsequent, though in the very earliest stages the unroofing of the Wardour fold must have been started either by streams draining southwards into the Ebble or northwards to a former synclinal river of which there is no remaining trace. The Vale of Wardour is bounded by Chalk escarpments, that on the south side between White Sheet Hill and Burcombe being especially impressive owing to the low angle of dip, whilst that on the north is correspondingly less imposing, as a result both of the steep inclination of the Chalk and of considerable dissection by escarpment valleys. Towards the east, as the Wardour fold pitches, the two scarps become closer together, and would meet at Barford St John but for the Nadder, which has cut a deep gap through the Chalk to Wilton and beyond. Within the Vale of Wardour there are also escarpments formed by the Upper Greensand and the limestones of the Purbeckian and Portlandian, together with miniature 'vales' in the Chalk Marl and the Lower Cretaceous sands and clays and a much more extensive lowland, farther to the west, in the Kimmeridge Clay.

To the north again lies the major Chalk upland of the Great Ridge, which rises steadily westwards above 650 ft O.D. and exceeds 900 ft at its termination beyond Brixton Deverill. This upland has a synclinal capping of Upper Chalk, and it is clear that any stream once following the pitching synclinal floor from west to east disappeared long ago. The Great Ridge thus represents, with the neighbouring Vale of Wardour and the anticlinal valley of the Wylye farther to the north, the ultimate stage of the process of inversion. Nevertheless, it is questionable whether the process has been an uninterrupted one. For one thing a difficult problem is posed by the total disappearance of the Great Ridge synclinal stream and the rise to complete dominance of the near-by Nadder. Except where breaching is directly initiated by tributaries of a stream cutting across the fold (as at Winchester), the drainage of a newly forming anticlinal vale must be carried out by a stream or streams tributary to an adjacent synclinal river (in the case of the Vale of Wardour, either the Ebble or the Great Ridge river). The latter therefore provides what may be termed the base-level for the breaching process, and the anticlinal vale floor cannot normally be lowered below the level of the synclinal valley floor; yet this must happen if the synclinal stream is ultimately to be replaced. The problem is nicely illustrated in this same area by the upper course of the Ebble, whose southern tributaries have eroded deeply into the Bowerchalke anticline. The Bowerchalke vale naturally stands above the valley floor of the Ebble, which has begun to 'breach' the Upper Chalk flooring the syncline (p. 100). The possibility of true inversion here seems to be ruled out, for the Ebble syncline can never form a Chalk upland comparable with the Great Ridge.

The probability must hold that features such as the Vale of Wardour and the Great Ridge do not really represent the end-stage of a continuous breaching process whose early stages are exemplified in the Vale of Bowerchalke. Rather it may be necessary to postulate the existence, west of Salisbury, of a former high-level planation surface (at 700–850 ft), eroded across the Alpine structures in late-Tertiary times. The general accordance of the ridge heights certainly supports an hypothesis of planation (pp. 265–6). It has been suggested that the surface is of Mio-Pliocene or Pliocene age and of sub-aerial origin, but the evidence is equivocal and the possibility of marine trimming must be considered. It may be significant that, if the geology of the hypothetical surface is reconstructed, outcrops of weak rock (Middle and Lower Chalk and even older rocks along the Wylye, Wardour and Bowerchalke axes, and Tertiary sands and clays in the Ebble syncline) occur where the three main streams of the area now flow. Are these to be regarded as comparatively recent subsequents which have worked back headwards from the Salisbury Avon? It is worth noting that the Great Ridge, on the other hand, coincides with a line of strength—the outcrop of the Upper Chalk—at the level of planation, and would thus not have nourished such a stream.

FAULTED STRUCTURES

Faulting involves the differential movement of strata on either side of a fault-plane (involving a single plane of shearing) or fault-zone (involving a number of closely spaced fault-planes) as a result of either compressional or tensional forces in the earth's crust. The differential movement may be upwards, downwards, horizontal, oblique or even rotatory. It is the up-and-down movements which have the most obvious direct influence on relief development, and which are therefore of most immediate concern to the geomorphologist. However, it must always be remembered that the indirect effects of faulting, which may become apparent only after a lengthy period, can be equally significant. For example, where crustal movement leads to the creation of crushed or brecciated zones ('shatter-belts') which are much more easily eroded than the rock on either side, the development of river valleys may be closely guided. In south Wales the rivers Tawe and Neath have taken advantage of such lines of intense disturbance to incise deep and trench-like valleys (pp. 239–40), which have been rendered even more impressive by glacial erosion during the Quaternary.

Faulting results, either directly or indirectly, in two principal types of landform: the fault-scarp and the fault-line scarp.

Fault-scarp

Faulting may, through relative displacement of the land surface on either side of the fault-plane, immediately produce what is termed a 'fault-scarp'. However, it is important to bear in mind that the amount of vertical movement, or throw, at a fault may vary enormously between a few inches and several thousands of feet. The very large displacements cannot be accomplished by one rapid and continuous movement, but are the outcome either of a number of successive small movements during a comparatively brief period of geological time, or of fracturing which has been spread over a vast period of time. For instance, in the Massif Central of France many of the great fault-lines were initiated in Hercynian times, but a renewal of movement occurred during the mid-Tertiary period, when the massif acted as a foreland against which the Alpine mountain chain was pushed and folded. In some localities this has led to posthumous faulting in sedimentary rocks which had accumulated on top of the Hercynian basement (p. 141).

Since large faults inevitably take time to develop, the question arises as to how common true fault-scarps really are. In many cases it seems probable that erosive processes can act more rapidly than the faulting process, so that important fault-scarps cannot easily arise. Instead, the fault-lines will be marked, immediately after each small displacement, by minor 'fault-scarplets', a foot or two in height, which are ephemeral features and will be quickly erased by erosion during a succeeding period of tectonic stability. It seems certain that no fault-scarps exist in the British Isles, for the simple reason that no serious faulting has occurred since the culminating movements of the Alpine orogeny in the Miocene period, so that ample time has elapsed for any resulting scarps to be destroyed by erosion during the Pliocene and Quaternary. In other parts of the world, however, which are at present geologically more active, true fault-scarps are more likely to exist.

Fault-line scarp

Faulting may have a profound influence on landform development at a later date because of the way in which it commonly brings against each other rocks of contrasting resistance to weathering and erosion. Thus, although for the reasons given a fault may in the first instance not be associated with a scarp, subsequently a 'fault-line scarp' may result from the erosion of the weaker rocks on one side of the fault.

Three main types of fault-line scarp may be recognised:

(i) *A 'normal' or 'consequent'* fault-line scarp will be formed at an early stage after the faulting movements by the removal of unresistant rocks occurring on the downthrow side of the fault (Fig. 42A). The resultant scarp-edge will face in the same direction as the original fault-scarp (assuming this to have actually existed), and eventually the height of the scarp-face may even approximate to the amount of vertical displacement at the fault.

(ii) *An 'obsequent'* fault-line scarp faces in the opposite direction to the original fault-scarp. It is produced by erosion of weak rocks preserved on the upthrow side of the fault, and in fact denotes the faulted margins of a resistant layer which has been affected by downfaulting. As Fig. 42B shows, an obsequent fault-line scarp will normally represent a later stage of development than a consequent scarp, though this is not invariably the case. An important factor here seems to be (as it is also in the breaching of anticlines) the precise relationship of the strong and weak rocks on either side of the fault to the base-level of erosion. In the example illustrated, the 'reversal' of the fault-line scarp is possible only because a fall in base-level has exposed to denudation the weak rocks on the upthrow side of the fault.

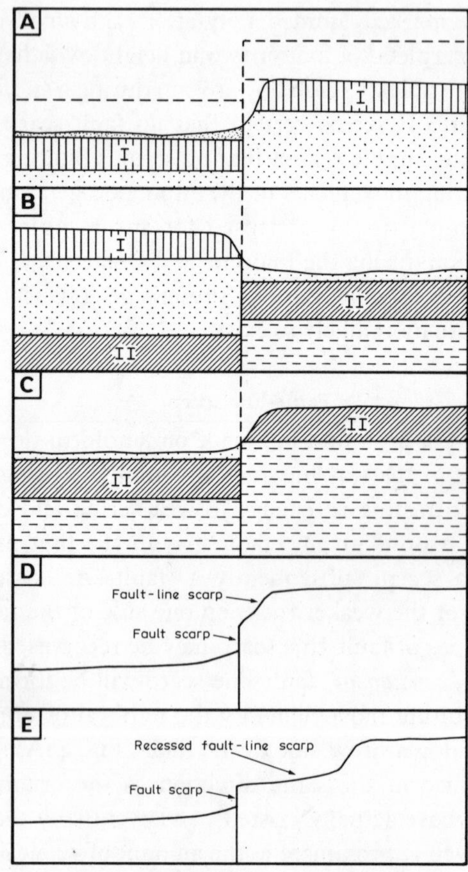

42 Fault-line scarps: normal (A); obsequent (B); resequent (C); composite (D); rejuvenated (E)

(iii) *A 'resequent'* fault-line scarp represents an even later stage of development, and results from the 'reversal' of an obsequent scarp by further downward erosion permitted by another base-level fall (Fig. 42C). It will be seen that consequent and resequent scarps are very similar, and indeed many would argue that to make a distinction between them is purely academic, on the grounds that it is invariably impossible to decide in actual cases whether or not an obsequent scarp has previously existed along a fault-line (and thus whether the scarp in question is resequent or not).

As stated already, scarps associated with faults in areas such as Britain are by definition fault-line scarps, but in geologically active areas (such as New Zealand) it is possible that both fault- and fault-line scarps exist together. Indeed, it has been suggested that what are termed 'composite' and 'rejuvenated' scarps may be developed in these circumstances. The former will arise as a result of renewed movement along a fault of earlier date, which has already been converted by erosion into a consequent fault-line scarp; this is, in effect, augmented at its base by a fresh rock face produced directly by the renewed faulting (Fig. 42D). The latter occurs where the initial fault-scarp has been weathered back some distance from the fault-line, and a more recent movement at the fault has given a new fault-scarp; thus the scarp as a whole will assume a step-like profile (Fig. 42E).

In cases where fault-scarps and fault-line scarps do co-exist, differentiation between the two may not be easy. However, the following criteria may be usefully adopted. A fault-line scarp may be inferred where the scarp faces the upthrow side of the fault, where there is a close correlation between rock resistance, structure and relief (indicating that a period of erosion has probably attacked the weaker rocks, leaving the stronger 'in charge'), where there is proof that the fault is of considerable age, and where there is superimposed drainage crossing the fault-line (revealing the existence of a rock layer not affected by the faulting movement). On the other hand, a true fault-scarp may be indicated by a poor correlation between rock resistance and relief (with a weak stratum, on the upthrow side of the fault, actually producing the scarp), by the ponding back of streams against the scarp-face (revealing recent movement at the fault sufficient to overcome the powers of down-cutting of the streams), and by the obvious displacement at the fault of recent planation surfaces (say, of Quaternary age) or recent datable deposits of alluvium, boulder clay or lava.

Finally, one of the most important types of escarpment associated with faulting is the 'resurrected' or 'exhumed' fault-scarp or fault-line scarp. This is a feature of considerable geological age which has been preserved either in part (by the accumulation of thick deposits of scree weathered from the upper face of the scarp) or wholly (as a result of the burial of the landscape in its entirety by younger rock formations, either of continental origin or laid down during a major marine transgression). At a later date these more recent protective accumulations may be removed and the ancient scarp revived as a landscape feature.

The following are selected examples to illustrate the effects of faulting on landform development.

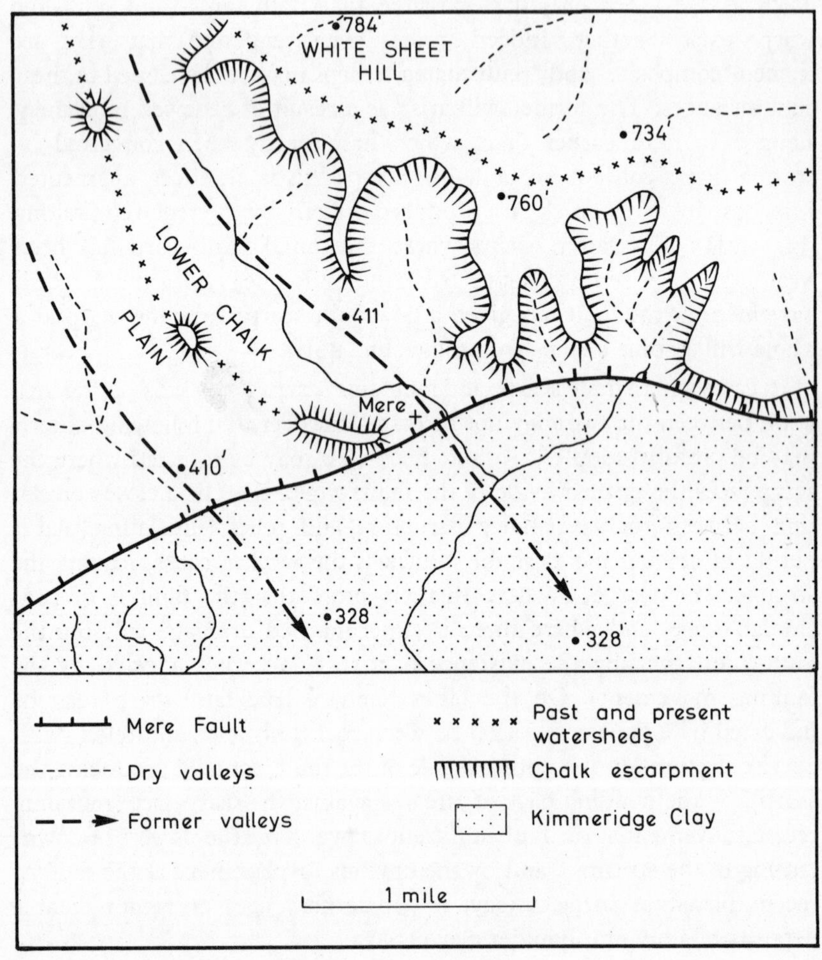

43 The geomorphology of the Mere fault, Wiltshire

The Mere area of Wiltshire

At the western end of the Vale of Wardour anticline (pp. 100–101) a large strike fault, of Tertiary age, affects the steeply dipping northern limb of the fold (Fig. 43). The throw on the northern side is sufficient to bring the Chalk (of Upper Cretaceous age) down to the level of the Kimmeridge Clay (of Upper Jurassic age). The area as a whole seems to have been affected by an episode of planation at about 700–800 ft O.D. (p. 101), which has led to the removal of the Chalk and underlying rocks (Upper Greensand, Gault Clay, and Lower Cretaceous sands and clays) and the exposure of the Kimmeridge Clay to the south of the fault. A more recent fall of base-level has initiated a phase of differential erosion of the Chalk and Kimmeridge Clay. The former now gives rise to an upland rising generally above 700 ft O.D. between Mere and Kingston Deverill, whereas the latter forms a near-level lowland at 300–350 ft, stretching southwards from Mere towards Gillingham and beyond. We have therefore an excellent example of an obsequent fault-line scarp, since the high ground today is on the downthrow side of the fault.

In detail, however, the scarp shows some unusual and interesting features. The precise line of the Mere fault runs, from west to east, through Bourton and the southern edge of Mere village to West Knoyle church, but the Chalk scarp is closely aligned to the fault for a distance of only some two miles westwards from West Knoyle. At Mere itself the escarpment, which is sharply dissected by a group of dry valleys, lies between half a mile and one mile to the north of the fault, and towards White Sheet Hill (784 ft) the divergence between scarp and fault is still greater (Fig. 43). It has been argued by some authors (notably Davis (1909) and Johnson (1919)) that fault- and fault-line scarps undergo a form of cyclic evolution. During the stage of youth there is a general coincidence between scarp and fault, though some evidence that recession of the scarp has begun may be found. By maturity the scarp may have been driven back by weathering and erosion several miles, and a gentle pediment-like slope may extend from the foot of the scarp to and across the fault-line. By the stage of old age, general lowering of the landscape by denudation will have removed the scarp altogether. It is tempting at first sight to assume that in the east the Mere scarp is at an early stage of development, whereas to the west it has attained a more mature state. The retarded nature of the scarp in the former area would not, however, be easy to explain, and in any case the remarkable outlying hills of Chalk (Long Hill and Zeals Knoll) immediately to the west of Mere suggest an alternative and more plausible explanation. As Fig. 43

shows, there are reasons to believe that at one time a succession of streams ran discordantly across the Mere fault from north-west to south-east. Because of an easterly component in the dip of the Chalk, the streams in the west were quickly able to expose Middle and Lower Chalk, and rapid valley widening (for the reasons given on p. 97) ensued. The interfluve between the Mere stream and its western neighbour was thereby reduced to a few isolated knolls, and in place of an escarpment dissected by valleys there arose a gentle plain of Lower Chalk, at approximately the same level as the Kimmeridge Clay lowland. To the east of Mere, however, the streams were forced to cut narrower valleys in the more resistant Upper Chalk, and the fault-line scarp there remains virtually intact.

The Craven district of Yorkshire

This is an area where the relationship between faults and landforms is on the face of it simple, but on closer examination emerges as highly complex. To the north, north-west and north-east of Settle rises the major upland known as the Askrigg Block (Fig. 44). This comprises in general a plateau of Upper Carboniferous Limestone (the Great Scar Limestone), in many localities exhibiting clear evidence of planation at about 1300 ft O.D., above which rise several high monadnocks (such as Ingleborough, Pen-y-ghent and Fountains Fell) formed in the Yoredale Beds—the topmost division of the Carboniferous Limestone—and capped by Millstone Grit. This upland terminates on its south side approximately along the line of the Middle Craven Fault, beyond which there is a generally more low-lying area, at the heads of Ribblesdale and Airedale. Thus the basic relief units here are structurally determined, and since the Middle Craven Fault has a downthrow on the southern side the edge of the Askrigg Block would seem to be a straightforward example of a consequent or resequent fault-line scarp.

However, a proper understanding of the landforms of this area requires some knowledge of the geological and structural history. To the north of the Middle Craven Fault the Great Scar Limestone is surmounted, as we have seen, by the limestones, shales and grits of the Yoredale Beds and the very resistant grits of the Millstone Grit series. These formations rest unconformably on a strongly folded and peneplained basement of Lower Palaeozoic rocks (mainly Upper Silurian grits and flagstones) which are exposed in the deeper valleys and on the plateau itself at Malham Tarn and to the south-west of Fountains Fell. During the Carboniferous period the Askrigg Block seems to have been a stable area of deposition, whereas to the south—beyond the already developing Middle Craven Fault—lay a subsiding trough in which accumulated contemporaneously a great thickness of lithologically distinct strata (including the Clitheroe and Pendleside shales and limestones and the Bowland Shales). Immediately adjacent to the fault itself, there developed a number of irregular masses of hard limestone (referred to as 'reef-knolls') surrounded by weak shales. It was not until Millstone Grit times that similar depositional conditions existed both to the north and south of the Craven Fault.

This very important geological contrast between the Askrigg Block and the Ribblesdale-Airedale lowlands has been accentuated by later faulting, notably along the North and South Craven Faults, each of which is again associated with a downthrow on the southern side (Fig. 44). Probably the most recent movements along the Craven fault-lines took place during the Tertiary era, in association with the Alpine orogeny, but the evidence is not clear. At the present time the morphological effects of the faults vary from one locality to another. Since in most instances easily eroded shales lie to the south of the faults and resistant Great Scar Limestone outcrops to the north, south-facing fault-line scarps are the general rule. The most striking of these is Giggleswick Scar, a 300 ft high rocky edge trending north-westwards from Settle and following exactly the line of the South Craven Fault

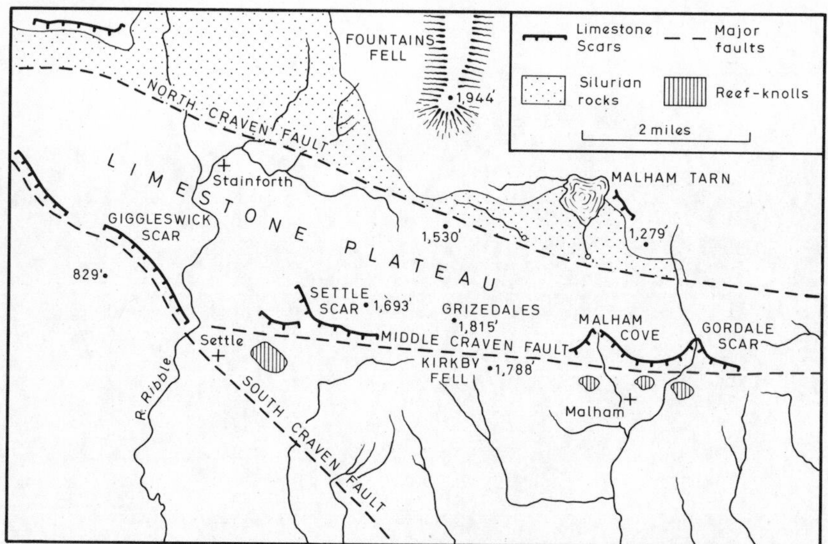

44 **The structure and relief features of the Craven district, Yorkshire**

(Fig. 45). To the east of Settle the Middle Craven Fault has been exhumed to give a similar limestone scarp at Settle Scar and immediately to the north of Malham village, where the escarpment is more complex in plan owing to the presence of deep indenting valleys at Malham Cove and Goredale Scar. In this last area the impact of the scar is somewhat diminished by prominent scarp-foot hills, which represent reef-knolls of limestone which have been isolated by erosion of the surrounding shales. At other points erosion has so far been unable to take full advantage of the differences of rock resistance on either side of the faults. A good example of this may be seen immediately to the south of Malham Tarn, where comparatively weak Silurian strata are adjacent to the more durable Great Scar Limestone lying southwards of the North Craven Fault. The Silurian rocks have been lowered by stream erosion to a level of approximately 1250 ft O.D., whereas the limestone—characterised by an absence of surface drainage—has been left standing somewhat higher, at between 1300 and 1530 ft (at Black Hill). As yet there is no well-defined steep slope marking the northern edge of the limestone, but

45 Giggleswick Scar; a fault-line scarp of Carboniferous Limestone (see p. 109). [Eric Kay]

it does seem perfectly clear that we have here an incipient obsequent fault-line scarp which, with continued downwearing of the Silurian rocks, will steadily grow in height. Finally, in some parts of the Craven district the contrasts of rock resistance between the areas north and south of the main faults are evidently insufficient to promote differential erosion and the formation of fault-line scarps. This may be admirably seen mid-way between Settle and Malham, where immediately to the north of the Middle Craven Fault the Great Scar Limestone reaches 1815 ft O.D. at Grisedales, whilst to the south the grits of Kirkby Fell rise to 1788 ft. Between the two summits lies a gentle depression, at a little under 1700 ft, which is the only surface expression of the fault.

The Vale of Clwyd, north Wales

The Vale of Clwyd, trending south-south-eastwards from Rhyl on the coast to Ruthen and beyond, is structurally a miniature synclinal rift-valley, initiated by faulting during the Hercynian orogeny (Fig. 46). The main western boundary fault, traceable from St Asaph due southwards to Denbigh, has a comparatively small downthrow to the east, and as a result there are no important lithological contrasts on either side of the fault. Hence there is no well-defined fault-line scarp, but rather a general decline of the relief eastwards across the faulted zone towards the river Clwyd. The eastern boundary fault, by contrast, is of far greater importance, and the downthrow to the west has been sufficient to bring Coal Measures and Millstone Grit of Upper Carboniferous age against rocks of the Lower Ludlow (Upper Silurian) series. A magnificent escarpment, whose summits form the Clwydian Range and rise to approximately 800–900 ft O.D. north of Bodfari and to 1400–1600 ft farther to the south, now marks the line of this fault. For the most part the scarp is developed in coarse grits and flagstones of the Ludlow Series, but in the north, between Prestatyn and Dyserth, it is formed by Carboniferous Limestone.

A very significant feature is the occurrence of Bunter sandstones and conglomerates within the confines of the Vale of Clwyd. These rest unconformably on Carboniferous rocks flooring the rift, proving conclusively that, after its formation in Carboniferous-Permian times, the vale and its scarps were affected by desert erosion in the Permo-Triassic period. The resultant landscape was in fact fossilised by the Bunter deposits, laid down in an enclosed sea or lake existing on the northern and eastern flanks of north Wales. It seems certain that the area was also covered by younger rocks, including Keuper and Liassic

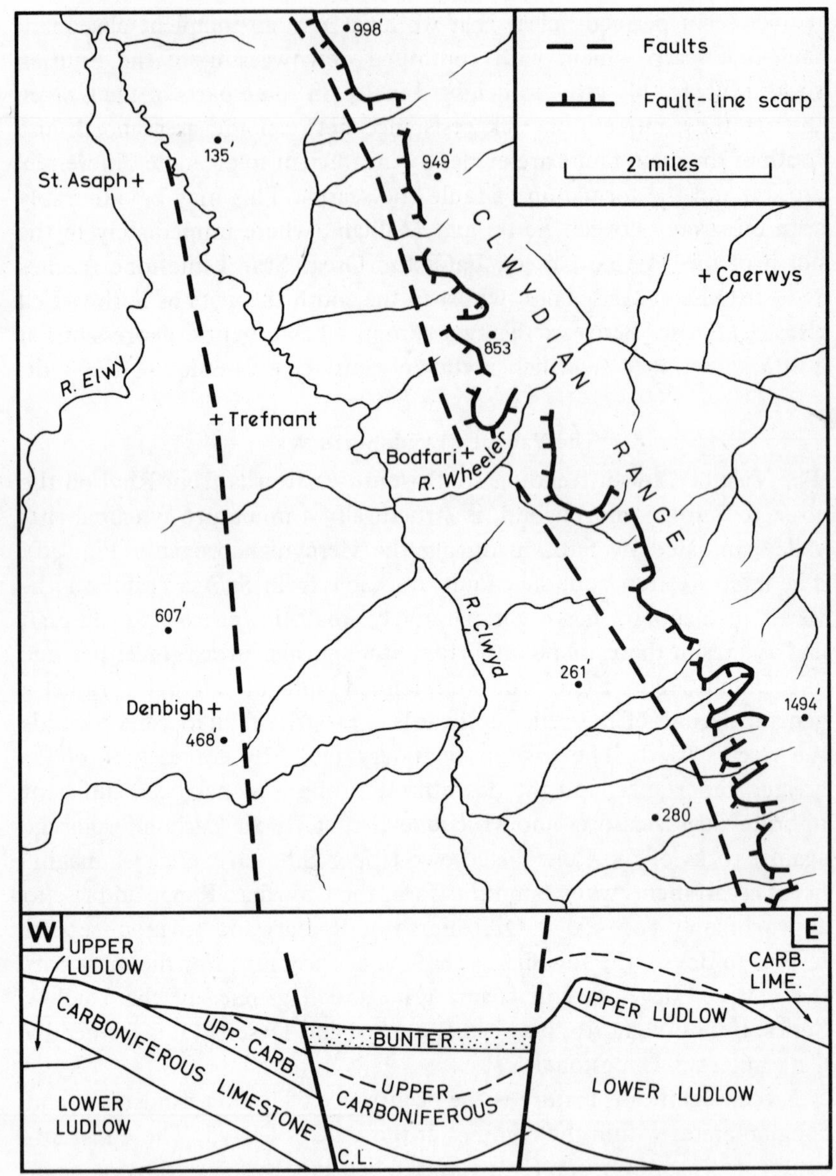

46 The structure and relief features of the Vale of Clwyd, north Wales

formations and perhaps even the Chalk, and that, as in other parts of Britain (pp. 269–70), the sub-Triassic landscape was revealed again in part by late-Tertiary erosion, carried out here by the Clwyd and its tributaries. It is also apparent that the Clwydian rift was further emphasised by faulting of Alpine date, for the remaining Bunter rocks show evidence of folding and fracturing related to an important post-Triassic period of disturbance. The extent to which these posthumous movements have modified the sub-Triassic morphology cannot be easily assessed. The great scarp east of the Vale may therefore be (i) an exhumed or resurrected fault-line scarp of Triassic age, (ii) a consequent fault-line scarp, produced in late-Tertiary times by the erosion of weak Bunter sediments faulted against resistant Silurian rocks during the Alpine period, or (iii) a composite feature, owing something to both of these modes of origin. The latter seems the most likely explanation, not only of this scarp but also of other great fault-line scarps in Britain related to faulting initiated by the Hercynian orogeny (for example, the tremendous escarpment of Carboniferous rocks bounding the Vale of Eden and marking the western termination of the Alston Block).

4

ROCK-TYPE AND LANDFORMS

INTRODUCTION

The influence of geological structure *sensu stricto* on the configuration of the physical landscape was shown in the preceding chapter to be usually though by no means invariably of considerable importance. It would seem reasonable to expect a similar causal relationship to exist between the nature of the underlying rock and the form of the land surface. As is shown in chapter 2 the various types of rock vary greatly in their resistance to weathering processes and—it might be added—to erosion by running water, waves and moving ice. As a result there is a tendency for certain rocks (granite, gabbro, limestone, quartzites, well-cemented sandstones) to form upstanding mountainous and hilly areas, whereas others (notably shales, clay, weakly cemented sandstones and unconsolidated sands) mostly form vales, lowlands and near-level plains. Furthermore, some rocks give rise to steeper slopes than others. Thus Hack (1960), considering the differences between adjacent areas of hard quartzite and soft shale, writes: 'to comminute and transport quartzite at the same rate as shale, greater energy is required; and since the rates of removal of the two must be the same in order to preserve the balance of energy, greater relief and steeper slopes are required in the quartzite area'.

Nevertheless, generalisations such as these cannot always be sustained, for erosional history and climatic influences often modify the expectable relationship between rock-type and landforms. For instance, in north-western Scotland and the Hebrides the Lewisian gneiss, an ancient pre-Cambrian rock of great durability, forms mainly low-lying and gently hummocked ground, with numerous lakes and hollows (Fig. 154). The gneissic surface in fact approximates to an Archaean erosional plain, preserved by overlying Torridonian sandstones and younger formations, and resurrected and modified by more recent erosional activity, including intense glaciation in the Quaternary (pp. 361–2). Again, the relief of limestone seems to be profoundly influenced by the prevailing climatic régime. In warm and humid areas rates of chemical weathering are abnormally high and the limestone becomes rapidly karstified, whereas in deserts, owing to the minute quantities of moisture present, rock decay by carbonation is far less active and the limestone has a correspondingly high resistance to karstic processes.

The geomorphologist is not, of course, only concerned with the general effects of rock-type on relief and slope steepness, but also with the more detailed forms associated with particular rocks. In other words, his aim is to study the precise morphology of the mountains, hills, plateaus, escarpments, valleys, lowlands or plains that a rock, because of its resistance or lack of resistance to weathering and erosion, gives rise to. In the opening chapter of this book the concept of morphogenetic regions is introduced and briefly discussed. In this it is postulated that different climatic régimes, by favouring certain processes at the expense of others, produce their own distinctive landforms and landform assemblages. Such basic elements as valley-side slopes are assumed to take on characteristic forms related to the rates of stream incision, the nature and speed of rock weathering, and the efficacy of transportational agencies such as creep and wash—all of which are, in turn, determined to a greater or lesser extent by the prevailing temperature and precipitation conditions. It would appear equally valid, however, to base such a morphogenetic classification on a lithological foundation, for denudational processes are related not only to climatic factors but may be largely determined by geological factors. An acceptance of this is implicit in much geomorphological literature, where reference is frequently made to 'granite landscapes', 'limestone (karst) landscapes', 'chalk landscapes', 'clay landscapes' and so on. An interesting example is Trueman's description of 'typical' chalk scenery given on p. 3.

However, just as there are dangers inherent in the 'climatic geomorphology' approach—for it is by no means accepted that different denudational processes, or combinations of such processes, necessarily and invariably produce different landforms—so it must not be inferred too readily that a certain type of rock will give rise to a landscape showing a majority of characteristics not associated with other rocks. It is commonly believed that there is a fundamental contrast between granite and limestone scenery—and indeed the subterranean drainage phenomena of the latter, together with the surface manifestations such as sink-holes, solution hollows and collapse forms, are not matched by the former. However, in other respects (plateau-like character, the form of valley-side slopes, the occurrence of well-jointed rock masses) the landforms of granite and limestone often show similarities rather than marked contrasts (pp. 126–56; cf. Figs. 54 and 61). In some instances geomorphologists may have been misled by the 'appearance' of the landscape formed by a particular type of rock. The Chalk is popularly supposed to give rise to 'the most easily recognisable' (A. E. Trueman) scenery in England, yet in practice it is not easy to define precisely what are the morphological characteristics that give to Chalk country its so-called 'uniqueness'. The famous dry valleys have innumerable counterparts elsewhere in England, not only in limestone areas but on many sandstone uplands, even Exmoor; the smoothly curving slopes, with no free faces or angles in excess of 35°, are again typical of other limestones (for example, the Lincolnshire hills, much of the Cotswolds, and the Purbeck hills), sandstones (the Lower Greensand country of the Weald) and clays; and the slope angles of the Chalk differ little from those of many other rock-types under a wide variety of climatic conditions. It is tempting to suppose that the uniqueness of Chalk scenery may be more a function of soil colouring, nature of vegetation, historical evolution and present-day agricultural utilisation than of pure morphology.

It is therefore necessary to approach morphogenetic classifications based on lithology with caution. In addition to the points already made, there is the fact that many parts of the earth are lithologically complex, with a variety of rocks outcropping in a very small area. Within a comparatively small mountain valley carved from a strongly folded structure one may find, closely juxtaposed, granite, gneiss, serpentine, schist, shale and limestone. To apply preconceptions of 'granite scenery' and so on in such a case would be ridiculous. Not only may lithological factors have been firmly overridden by the more general factor of intense Quaternary and/or present-day glaciation, but there is also the insufficiently stressed fact that, in a lithologically complicated area, the morphological effects of an individual rock will be determined not only by its own character but also by that of its neighbours. Thus a thin band of weak shale will not be deeply eroded if it is sandwiched between layers of very hard sandstone; it may, indeed, outcrop high on the mountain-side despite its inherent lack of resistance because of the degree of protection afforded by its neighbours. Again, the processes of weathering and transportation at work on the shale outcrop will not be conditioned solely by the physical and chemical properties of the shale, but also by the presence of overlying debris (possibly of considerable thickness and of a quite different nature from that released by weathering of the shale) which has been derived from other rock outcrops higher up the slope.

A realistic discussion of the influence of different rocks on surface landforms must be based on a brief survey of those rock characteristics which are likely to have morphological effects (through the agencies of weathering, transportation and erosion). These characteristics (for example, massiveness, jointing and permeability) are themselves often closely interrelated. It is also appropriate to consider, with reference to selected regional examples, the landforms developed on two of the most common rock-types, granite and limestone. These case-studies will be presented in such a way that the role of lithology, in comparison with other factors such as structure, climate and erosional history, can be objectively assessed.

ROCK CHARACTERISTICS INFLUENCING LANDFORM DEVELOPMENT

Rock hardness

The mechanical strength of a rock is a relatively minor factor affecting resistance to weathering, since it can be more than offset by other characteristics, notably chemical composition and jointing, which permit decay and physical disintegration (pp. 19–25). None the less, hard rocks are as a general rule more upstanding than soft rocks, for the reason that they are less susceptible to stream erosion (and therefore valley development) and to other processes such as glacial abrasion, where the ability of the rock to withstand the pressure of a great weight of ice, armed with rock fragments, is of great importance. However, the relationship between rock strength and erosive processes is by no means straightforward. Mechanical strength directly impedes corrasion by running water, but the latter may also be retarded by the nature of the stream load, which in turn is influenced by rock composition and texture and by past and/or present weathering processes. For instance, a plutonic igneous rock (such as granite, syenite or diorite) may be broken down by granular disintegration into its constituent crystals, or small agglomerations of crystals, thus forming a coarse 'sand' which is, by comparison with small stones and boulders, a very ineffective corrading tool. Waters (1957) has suggested that the streams of Dartmoor, armed mainly with fine granitic debris derived from the 'growan' which mantles much of the moor (p. 133), are quite unable to attack unweathered rock outcrops. In areas of weakly cemented sands valley development is far more easy and rapid, because although true corrasion is again hindered by ineffective stream loads the soft rock is readily washed away by fast-running water. The juxtaposition of rocks of differing hardness may also have important repercussions. For example, in the headwater area of the Shenandoah river in the Appalachians, outcrops of hard sandstone and quartzite, associated with steep slopes and high relief, are drained by streams which pass on to near-by outcrops of soft limestone, forming low hilly plains. These streams are heavily laden with coarse and hard rock fragments, which are either used in rapid corrasion of the limestone or are spread as 'pediment fan-gravels' over low-lying parts of the limestone. These gravels may protect the underlying limestone from further attack, and are therefore eventually left upstanding as bench-like features, at the junction between the quartzite and limestone, as unprotected areas are wasted by chemical weathering and allied processes.

Rock hardness not only exerts a considerable influence on corrasive processes, but may help to determine the form and steepness of such basic landscape forms as valley-side and mountain slopes. The great vertical cliffs of the Italian Dolomites (for example, those of the famous Tre Cime de Lavaredo) can only be maintained because the rock has great strength and resistance to shearing. At the other end of the scale, soft rocks such as chalk, clay and sands cannot generally sustain free faces or cliffs. In inland areas chalk slopes are rarely in excess of 30°, and are usually much lower in angle. On the coast chalk does give rise to vertical sea-cliffs, but only where there is severe basal undercutting by the waves; where less exposed, perhaps as a result of the existence of a protective shingle bank, such cliffs tend to decline in angle rapidly, not only as a result of the usual weathering processes but also because of the collapse of large masses of the rock. In extremely soft rocks, such as the boulder-clays of East Anglia and Holderness or the Tertiary sands of the London and Hampshire Basins, 'failure' of the substratum is even more likely, so that slope angles in these areas are correspondingly more gentle. It is, too, only under rather unusual conditions of climate and vegetation that the more coherent types of clay can be fashioned by intense gully erosion into badlands, in which the slope angles may be as high as 30–45°.

Chemical composition

Because of their particular chemical composition certain types of rock are inherently prone to weathering decay and corrosion by running water. The most obvious of these is limestone, whose principal constituent, calcium carbonate, is rapidly acted upon by acidulated water (p. 20), but many sandstones (particularly those bound by calcareous or ferrous cements) are also readily attacked, the more so if prevailing climatic conditions are warm and humid. It does not follow, of course, that such rocks can be classified as unresistant—indeed the very reverse is often the case. In lowland Britain a large proportion of the highest ground, in the Cotswolds, Chilterns, Yorkshire Wolds, North and South Downs, and the Dorset Downs, is formed by Jurassic limestone and Chalk, despite the fact that in a humid-temperate climate chemical weathering dominates mechanical disintegration. Evidently the factor of chemical composition, at least in helping to determine major relief outlines, is subservient to other factors, the chief of which in this case is rock permeability.

It is, in fact, difficult to generalise about the effects of rock composition on landform development. The end-products of prolonged attack on many common minerals and rocks by chemical processes are clays of various kinds (for example, kaolinite on granite and gneiss, terra rosa on limestone, clay-with-flints on chalk). The accumulation of a thick surface residue of such material will favour processes such as soil creep and related forms of mass transport, so that the landscape should become gently rounded, with many convex hill-tops and interfluves. However, this is a vastly oversimplified picture, for in certain circumstances active chemical weathering can also release large quantities of coarse detritus, including subangular or only partially rounded boulders, in which case 'grading' of the landscape will be less readily achieved. Much will depend on the 'massiveness' of the rock concerned. A soft homogeneous limestone, such as chalk, with a close and evenly spaced system of minor joints and bedding-planes (in other words, a 'non-massive' rock) will be attacked uniformly by chemical weathering. As a result relatively smooth landform profiles, with residual decomposition products being steadily evacuated by non-selective processes such as creep and hill wash, will be normal. A much harder and more massive limestone, with a wide-spaced and more strongly developed joint system and fewer bedding-planes, will react in a totally different way. Where there is rapid stream downcutting, free faces will form at the outcrops of the hardest strata on the valley sides. Weathering and chemical erosion will be selective in their attack. Large joints will be sought out and widened by acidulated rainwater, and will often guide stream erosion to give a rectangular valley pattern. The fissures on free faces will be emphasised, and the rock given a tor-like appearance—

though in time these faces will be converted into grotesque irregular and fluted forms by continued solution and perhaps even obliterated altogether, so that the 'angular' landscape will after all take on a more 'graded' appearance. The landscape of any limestone area may in fact exhibit striking contrasts. For example, in the Carboniferous Limestone upland of the Mendips there is at one extreme the deep rocky gorges of Cheddar and Burrington Combe, and at the other the almost down-like character of much of the limestone plateau, with its shallow dry valleys and smooth gentle slopes.

Rock jointing

It is arguable that joints are the most important of all rock characteristics in influencing landforms on both a macro- and micro-scale. They have an immense influence on both chemical and mechanical weathering, for they allow the penetration of corroding agents into the heart of sound rock, thus accelerating its decay, and provide lines of weakness which can be utilised by frost wedging and riving. Indeed, if rocks of great mechanical strength and totally unaffected by jointing existed—admittedly a somewhat theoretical concept—they would be virtually immune from weathering.

Joints have an equally important effect, both indirectly and directly, on erosion. A rock with a well-developed joint system will be permeable to some degree, and the amount of surface drainage—and hence erosion and transportation by streams—will be correspondingly reduced. It is not always appreciated that non-porous igneous rocks such as granite are by no means totally impermeable but support a water-table—as the Cornish tin-miners sometimes discovered to their cost. It is true that on granite outcrops, especially in upland areas of this country, there is invariably abundant surface water, but this is a reflection of (i) the sheer volume of the rainfall, (ii) the fact that the joints, except in the immediate subsurface zone, have not been effectively widened by weathering to form effective conduits, and (iii) the slow rate of run-off, itself related to the gentle relief, the layers of peat, and the boggy vegetation occupying many valley floors. On the other hand, limestone is usually far more permeable than granite, not so much because the joint system is initially better developed, but rather because the individual fissures are rendered more efficient by rapid widening through carbonation-solution. Poorly jointed limestone (and certain of the dolomites of the Jugoslavian Karst fall into this category) will, however, be impermeable or 'semi-permeable', with the result that many of the classic features of the karst cycle, both at the surface and underground, will not be formed.

It has already been suggested that surface streams may be guided by joint-lines. King (1948) has postulated that in many parts of Africa stream incision along joints has given rise to a subrectangular drainage pattern (pp. 172–3). It seems likely, however, that on the scale envisaged here it is not individual fissures which are utilised, but intersecting zones of close jointing (almost akin to shatter-belts). On an altogether smaller scale, some valleys in the English Chalk country exhibit straight sections separated by sharp changes of course, and it is possible that in such cases individual 'master' joints have been etched out by spring or stream erosion (Fig. 47). The direct effects of joints are perhaps best seen on coasts, where cliffs in well-jointed rocks normally display a multitude of narrow steep-sided inlets of the 'geo' type (p. 438; Fig. 187).

Finally, the detailed shape of some landforms may owe more to joints than to any other single factor. The obvious example is the dome-like inselberg (bornhardt), where exfoliation has acted upon pressure-release joints to give the unmistakeable curvilinear profiles. In the granite of northern Arran (pp. 353–6) the landforms on a more local

47 Rake Bottom, Butser Hill, Hampshire. This is a deeply cut escarpment valley, with a zigzag course perhaps related to major joint-lines in the chalk which have been utilised by headward eroding springs. [Eric Kay]

scale owe much to joint control. The spectacular cliffs on the east side of the A'Chir ridge coincide with near-vertical joint-planes, from the 'outside' of which the granite has been cleanly removed by ice wedging and glacial quarrying. On the ridge north of Goat Fell, isolated or semi-isolated rock masses, showing clearly the cuboidal character of the granite jointing, seem directly comparable in form and origin with the tors of Dartmoor (pp. 133–4). Finally, on the high valley slopes above Glen Rosa great slab-like outcrops of granite, tilted at between 30° and 60°, are developed where glacial erosion and frost weathering have exposed dilatation joints, possibly formed parallel to the sides of the pre-glacial or even the glacial valley.

Permeability and porosity

There is frequently confusion about the precise meaning of these terms, and some mistakenly regard them as synonymous. Permeability refers to the capacity of a rock for allowing water to pass through it. A prime factor determining the degree of permeability is the presence of bedding-planes and joints, but in some instances porosity can promote or enhance permeability. Porosity refers to the presence of small 'gaps' between the constituent mineral particles of a rock. Sometimes these pore-spaces are quite large and are in some degree interconnecting (as in coarse-grained sands and sandstone), with the result that water can pass easily through the rock; in other words it is both porous and permeable. In other instances pore-spaces are virtually non-existent (for example, in granite and slate), and in the absence of joints such rocks are totally impermeable. The case of chalk is an interesting one. It has been shown by experiment that the amount of pore-space in a piece of Upper Chalk is a little under half the total volume; or, to put it in another way, there is space in the pores of a cubic foot of chalk to absorb nearly three gallons of water. However, the individual pores are minute, and although water experiences little difficulty in getting into the chalk pores it will not readily pass through the chalk except under the influence of evaporation, when it is drawn upwards to the surface by capillary rise, or when very great pressure is applied. Thus chalk, like granite, would be largely impermeable but for the existence of joints and fissures; so indeed would other varieties of more massive and less pure limestone.

Permeability most obviously assists in the mechanical breakdown of rock when allied to well-developed jointing (as in the process of block disintegration by frost action). However, porosity is also an important aid, for water contained within pore-spaces exerts pressure as it is frozen, and either granular disintegration (for example, of sandstone) or a general weakening of the rock's coherence (as in the case of clay) will ensue. Both permeability and porosity are factors in chemical decay, particularly under hydrological conditions which favour alternate wetting and aeration of the rock interior.

The influence of permeability on erosion is a complex matter. Clearly in a highly fissured or porous rock the filtration of rainwater into the rock is facilitated and surface run-off reduced. Thus, other things being equal, a permeable rock will be resistant and form upstanding country simply because stream erosion is inhibited or prevented altogether. Conversely, impermeable rocks such as clay and shale, which promote a high degree of slope wash and channel flow, and at the same time are not mechanically strong, are eroded into gently undulating vales and lowlands. Within any single rock formation relatively small variations of permeability have apparently disproportionate effects. Thus the Carboniferous Limestone, a hard, well-bedded and strongly jointed rock, is highly permeable and forms high plateaus, escarpments and ridges. However, towards the base of the formation shaly layers are common (in the Lower Limestone Shales), permeability is reduced and increased surface erosion results, under favourable structural conditions, in the formation of a pronounced strike vale. This can be well seen in northern Mendip, between the Old Red Sandstone upland of Blackdown and the limestone ridge of Dolebury Warren (Fig. 48). An even more interesting example of permeability variation is to be found in the South Downs and the Hampshire Chalk country, where the Upper Chalk gives rise to two distinct escarpments (the 'main' escarpment rising to a maximum height of 975 ft O.D., and the 'secondary' escarpment to about 600 ft). There are no obvious differences within the Upper Chalk which readily account for these landforms, and it has been suggested by Sparks (1949) that a very slight increase in marliness (in the order of 2% or 3%), reducing permeability and causing greater run-off under past conditions, is responsible for the development of the vale, the two scarps (of rather purer chalk) being left upstanding by default.

The permeability of any rock, as expressed by its 'infiltration capacity' (its ability to take in water at a constant rate), has limits, and even in highly permeable rocks some degree of surface run-off is experienced today or, to judge from the evidence of stream-cut valleys that are now dry, has occurred in the recent past. In virtually all limestone, chalk and sandstone uplands there are systems of dry valleys, pointing to a period of active stream incision prior to the establishment of the present 'state of arrested dissection'. In the formation of such valleys, and their subsequent desiccation, a number of factors may be involved, either singly or in combination. The first is the possible variation, over a long term, in the total amount and nature of the precipitation. A greatly increased rainfall, especially if it were of a torrential type, could lead to the infiltration capacity of the rock being exceeded, with the consequence that run-off and erosion would occur widely. Secondly, a climatic refrigeration, allied to the onset of periglacial conditions and the formation of permafrost or annual taele (pp. 319–20), is capable of rendering permeable rocks impermeable, and of fostering erosion by ephemeral streams during periods of spring snow-melt or summer rainfall. Thirdly, the permeability of certain rocks (notably limestone and chalk) may well be increased with the passage of time, as subsurface weathering opens up joints and forms subterranean passages and

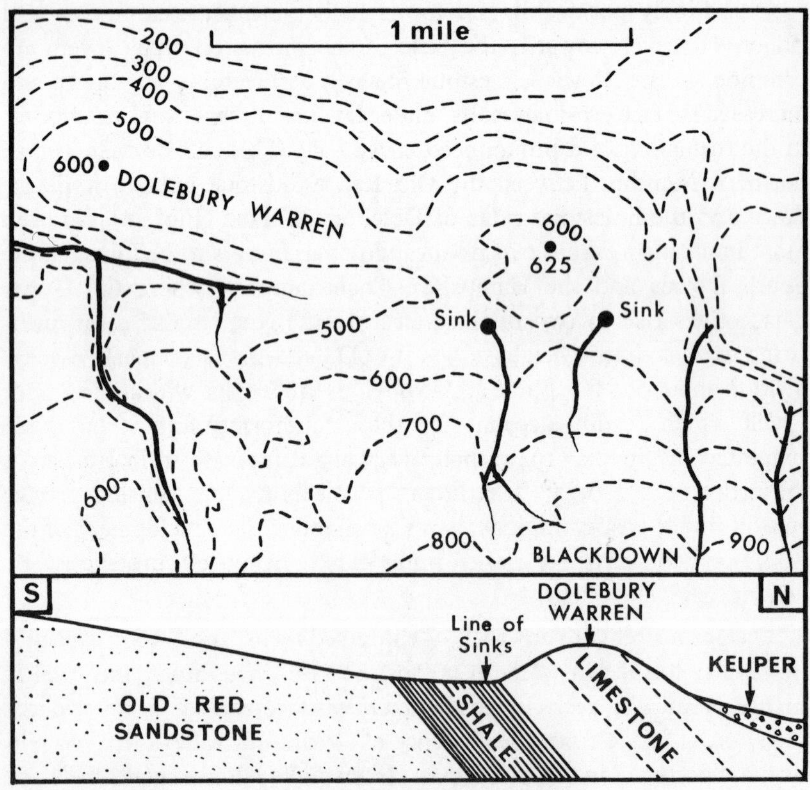

48 The relief features of part of the northern Mendips

caverns. Thus at an early stage valley development by surface streams may proceed without undue hindrance, but may be later terminated as the infiltration capacity of the rock is improved. It is, in fact, clear that for reasons of climatic change and/or progressive weathering the permeability of a rock cannot be regarded as a constant factor in landform evolution.

In rocks where surface fluvial activity is today either nil or minimal over wide areas, so that little direct modification of the landscape appears to be taking place, erosion does not cease entirely. In limestone uplands (such as those of the Great Scar Limestone of the Ingleborough area) chemical erosion proceeds underground, where it is evidently most effective at or a little above the zone of permanent saturation. Here caverns and interconnecting passages are particularly well developed. Indeed, it has been shown by Sweeting (1950) that in the Ingleborough area a number of 'horizons' of subterranean caves and conduits exists and that these can be related to former positions of the water-table (Fig. 49). The latter seems to have fallen spasmodically, either as a result of climatic change or because of successive falls in the base-level of erosion which have in turn affected the level of saturation in the limestone. In time, such underground erosion may, through the collapse of cavern roofs, modify the surface morphology of limestone areas.

Another important phenomenon of permeable rocks is the occurrence of springs, either at or near the margins of the outcrop or in the bottom of deeply cut valleys. In effect, the surface erosion which operates widely

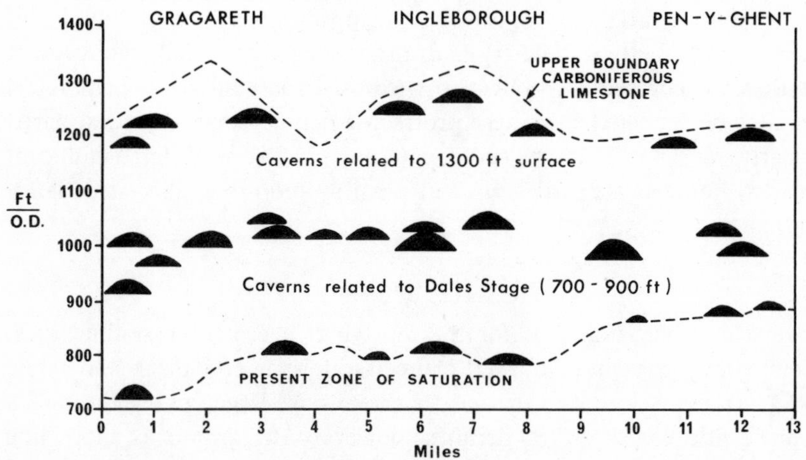

49 The relationship between caverns, water-table and base-level in the Great Scar Limestone near Ingleborough (after M. M. Sweeting)

over impermeable rocks tends to be concentrated at the sites of such springs, which as a result often occupy deep trenches. The latter are the product both of fluvial erosion below the spring and subsurface corrosion in the zone of underground water concentration above the point of issue. Valleys resulting from spring sapping are common on the scarp-faces of the English Chalk, and are recognisable (i) by their relationship with springs which are actively engaged in cutting valley-floor gullies, (ii) by their possession of abnormally steep heads, the result of under-mining of the scarp, and (iii) their occasional angularity of course, caused by springs heading back into major joints along which under-ground water flow is concentrated.

GRANITE LANDFORMS: DARTMOOR

The high moorlands of Dartmoor are formed by the largest of the five granite bosses of south-west England. These granitic masses, which were formed during the Hercynian orogeny, were emplaced in a dome-like structure in the pre-existing Palaeozoic rocks. To the north and west of Dartmoor are extensive outcrops of Carboniferous Culm Measures (a complex series of shales, sandstones, limestones and other rocks) and to the south Devonian rocks (again lithologically varied). These older rocks, which have been metamorphosed in a narrow aureole around the margins of the granite, are generally less resistant and have been lowered as a result of differential denudation, so that Dartmoor stands up to 1000 ft above the surrounding countryside. The history of south-west England in post-Hercynian times has been mainly one of erosion, by both sub-aerial and marine agencies, and the present landscape represents, in effect, the exposed roots of the ancient folds. It is to be expected that these protracted denudational episodes, particularly those of Tertiary and Quaternary times, have left significant marks, both on a broad scale and possibly too on smaller-scale landforms.

The nature of the granite

The Dartmoor granite is for the most part a coarse crystalline rock, comprising crystals of quartz, orthoclase felspar and black and white mica. In many localities the felspar crystals are large, giving the rock a porphyritic character. Sometimes, however, the granite is more fine grained (micro-granite or 'blue granite'); this seems to have been intruded into the coarse granite at a somewhat later date. In addition there are still younger dykes of aplite and pegmatite (respectively fine-grained and coarse-grained acidic rocks). At many points the granite of the South-West Peninsula has been profoundly affected by the metamorphic process known as 'pneumatolysis'. During the final cooling of the granite, volatile gases entered fissures in the consolidated rock, causing a breakdown of felspar crystals into kaolin and an allied weakening of the rock structure. Pneumatolysis was most intense in the St Austell boss, where the kaolinised granite is mined as china-clay, but was also of considerable importance in Dartmoor. Borings and excavations have revealed in some places up to one hundred feet or more of rotted granite, and in others alternate layers of sound and unsound granite (giving a 'cedar-tree' effect in profile) have been discovered.

The presence of 'live' and 'decayed' rock within Dartmoor would seem to provide ideal conditions for differential erosion, and it is in fact noticeable that whereas most of the ridges and interfluves are developed in sound granite, many valleys are underlain by decomposed rock *in situ*. The inference must be that the stream pattern of Dartmoor, though initially superimposed from overlying formations, is now highly adjusted to structure, as much so as in a typical scarp-and-vale area (pp. 233–4). A very important point is that the occurrence of rotted granite beneath valley floors militates against any theory that decomposition has resulted from chemical weathering rather than pneumatolysis. As is shown on p. 26, rocks such as granite are highly susceptible to deep chemical decay in warm humid climates, and in view of the climatic conditions known to have prevailed in southern England during the Tertiary era such weathering, to a depth of even a hundred feet in extreme cases, might be expected to have affected Dartmoor. However, the evidence from present-day tropical areas indicates that subsurface decay should normally be greatest beneath interfluves and least beneath valleys, since periodic aeration is possible under the former, whereas the latter would be associated with permanent saturation of the rock and the processes of hydrolysis and carbonation would be impossible. It must be admitted, however, that there is still controversy over this point, and that some authorities take the view that some at least of the kaolinised granite of south-west England is the product of a Tertiary weathering phase.

Another factor which has undoubtedly resulted in diversification of relief on Dartmoor, particularly on a micro-scale, is the jointing of the granite. That this is exceptionally well developed may be seen from any exposure of the rock, either on tors or in quarries. The normal rectangular pattern, the result of cooling and contraction, is everywhere evident, though over Dartmoor as a whole the two main sets of joints appear to follow no fixed regional directions. In addition there are pressure-release or dilatation joints (giving the so-called 'pseudo-bedding' of the granite), which are developed not in relation to the former margins of the granite boss but rather to individual hills and slopes; thus many of these joints are evidently post-denudational, having appeared since the 'roughing-out' of the main elements in the landscape by Tertiary erosion. One of the most significant features of the granite jointing is its variability of spacing. In general, the coarse-grained variety is the more massive, with relatively wide-spaced joints, whereas in the blue granite the joints are much more closely spaced. Since joints are a major source of weakness, it follows that the blue granite is less resistant, though not to the extent that it cannot in some areas form high ridges and interfluves. On a more local scale, too, there are important variations in joint spacing. For example, at Haytor rocks the main joints are up to 6–12 ft apart, whereas in a depression to the north-east they are from 1 to 6 ft apart. Much more information of this sort is still needed, for many theories of landform development on Dartmoor are based on the assumption that such variations are normal.

The landforms of Dartmoor

At first sight the landscape of Dartmoor (Fig. 50) has a deceptive simplicity. Over much of the moor the land rises to between 1000 and 1500 ft O.D., and many individual summits approach or exceed 2000 ft. Yet there is often little impression of real relief, except towards the margins of the granite where the high ground drops steeply away (for example, near Okehampton) or where streams such as the Dart have cut deep, steep-sided and heavily wooded valleys (Fig. 51). The main reason is that, despite variations in resistance due to kaolinisation and differential jointing, the granite as a whole is sufficiently durable to preserve large remnants of ancient planation surfaces. The interpretation of these is still a controversial matter, but a recent view (Waters, 1964) is that the following three main surfaces can be identified and approximately dated.

(i) Most of the higher summits of Dartmoor form parts of an 'Upland Surface' (at 1500–1900 ft O.D.), which has a local relief of 100–200 ft and has been attributed to erosion by west–east streams (including the forerunners of the Dart and Teign) mainly during the Eocene period.

Some residual hills (for example, High Willhays, 2039 ft, and Yes Tor, 2028 ft) rise a little above the general level of the Upland Surface. In southern Dartmoor, the surface is preserved at a rather lower elevation, and from this a slight downtilting in that direction has been inferred.

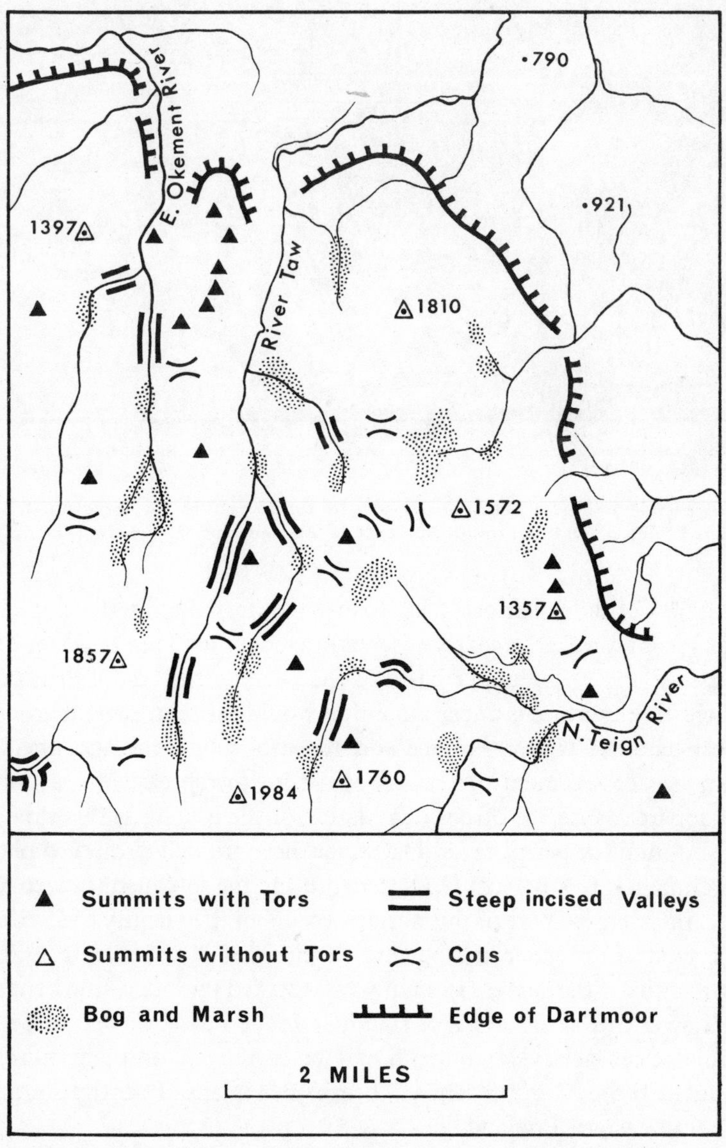

50 The landforms of Dartmoor near Okehampton (after R. S. Waters)

51 A general view over Dartmoor. Note the tor margin in the foreground, with associated 'clitter', and the smoothed aspect of the landscape as a whole. [Eric Kay]

(ii) The 'Middle Surface', at 1050–1350 ft O.D., is developed at various points on Dartmoor, but is mainly evident to the east of Moretonhampstead and in the upper basin of the Dart, to the east of Princeton. The age of the surface is uncertain, but it would seem on general grounds to post-date the Upper Surface and to be of Oligocene age; the vast thicknesses of sediment in the Bovey Basin, to the east of Dartmoor, were apparently derived from denudation of the granite at this time.

(iii) Around the periphery of Dartmoor there are well-preserved planation remnants at 750–950 ft O.D., constituting the so-called 'Lower Surface'. An extensive part of this surface occurs in the vicinity of Lydford, to the west of the moor and beyond the margins of the granite. Within Dartmoor itself the surface is mainly represented by valley-side benches. The Lower Surface, like those standing above it, is believed to be the product of sub-aerial erosion, and may be correlated, on the grounds of height, with the Mio-Pliocene (Pliocene) peneplain of central southern and south-eastern England (pp. 276–8).

In addition to these large-scale elements in the morphology of

Dartmoor there are the individual stream valleys, incised below the level of the Tertiary surfaces into zones of weakness in the granite (Fig. 52). Each moorland valley comprises a series of basins, containing usually bog and marsh, and 'gorges', and each stream profile displays several apparent knickpoints or oversteepened sections coincident with the gorges. There is little doubt that Dartmoor has been affected by numerous falls of base-level in Tertiary and Quaternary times, but these valley features cannot be simply dismissed as the result of poly-cyclic erosion. Indeed, it may be that the Dartmoor valleys are not mainly the result of stream corrasion but of the removal of decomposed granite by fluvial transport. It is certainly noticeable that where pneumatolysis has occurred or where closely spaced joints have accelerated weathering the hollowing out of valley basins by streams has been facilitated, whereas in areas of sound or widely jointed rock valley development is retarded and small 'gorges' are formed.

The interfluves of Dartmoor, besides preserving remains of planation

52 The valley of the Cherry Brook, Dartmoor. This is a good example of an infilled valley basin. [Eric Kay]

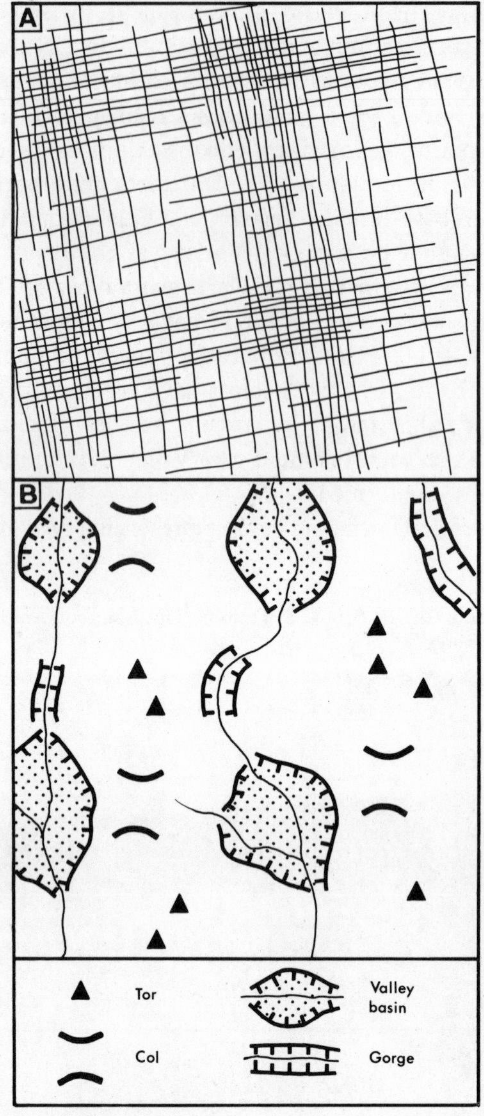

53 The relationship between jointing, weathering and granite landforms

surfaces, show in detail features of interest. Most obvious are the summits, which are often though not invariably crowned by rocky tors and associated fields and trains of boulders ('clitters'), and the long and gentle 'graded' slopes leading down to the valley bottoms. These slopes are covered, to a depth of several feet, by a granitic sand referred to as 'growan'. The interfluves themselves are not continuous, but descend at many points to low cols and boggy depressions; these do not mark the former crossing-places of streams, but result from locally accelerated wasting of the interfluve. Waters (1957) has referred to the tors as 'positive' and the cols as 'negative' elements in the landscape, and has suggested that the equivalents of these within the valleys are the gorges and basins. It is noticeable that in parts of Dartmoor the tors, cols, gorges and basins form a crudely rectangular pattern, and this leads to the inference that their distribution is related intimately to structural controls, most probably the granite joints and their spacing. The precise nature of this relationship has not been finally established, but it may be as depicted in Fig. 53.

The origin of tors

The rocky tors are the best-known and most controversial landforms of Dartmoor, and merit therefore more detailed discussion. These exposed masses of granite, which vary in scale but are often comparable in size with a house, are made distinctive by the rock jointing (Fig. 54). In the

54 A granite tor at Great Staple Tor, Dartmoor. Note the cuboidal jointing of the granite, and the collapsed masses to the right, perhaps resulting from freeze-thaw disintegration. [R. J. Small]

words of Linton (1955), 'the rock is usually traversed by bold and widely spaced near-vertical fissures, and these as well as the bounding walls are clearly the expression of joint planes in the granite. Powerful divisions also traverse the rock more or less horizontally—the so-called pseudo-bedding—and give the whole mass the rudely architectural aspect that is well described by the term "cyclopean masonry".' Tors are not, of course, confined to Dartmoor or even to granite areas, but are found widely in Britain and elsewhere and on a variety of rock-types. For example, they occur on dolerite outcrops in Pembrokeshire, on quartzite in Shropshire (the Stiper Stones), on the Millstone Grit of the Pennines (pp. 335–6), and even on Corallian limestone in Yorkshire. The one feature common to these rocks is their mechanical strength and associated well-developed joints. Linton has further suggested that the rocks on which tors are most numerous and characteristic contain felspar, and are therefore prone to slow and often incomplete chemical attack. Finally, it must be noted that, in detail, tors show important variations of site and form. A basic distinction is between summit or 'skyline' tors and valley-side or 'sub-skyline' tors, and some authorities have postulated that the two have arisen at different times and in different ways. In terms of form the main contrast is between the tors which comprise numerous cuboidal or partly rounded blocks, and the rounded or even dome-shaped features. The former are best developed where the vertical jointing is strong, and the latter where the pseudo-bedding is close-spaced and curvilinear in section and the vertical jointing comparatively weak.

A *The pediplanation theory* (pp. 170–4). It will be clear from what has been written that the Dartmoor tors rise in many instances above Tertiary planation surfaces, and some writers have attempted to demonstrate an interrelationship between the two. King (1958) has argued that the Dartmoor surfaces are not the product of Davisian cycles of sub-aerial erosion, involving stream incision and the wasting of divides, but are pediplains comparable, in terms of origin, with the great erosional plains found in parts of Africa. Above such pediplains, especially in granitic and similar rocks, rise rounded inselbergs or piles of boulders (koppies) resulting from the long-continued weathering and breakdown of inselbergs. The superficial resemblance between the Dartmoor tors and koppies is obvious, and seems to lend weight to King's interpretation. However, it is insufficiently realised that a rock with certain characteristics (of great hardness, massiveness, and strong jointing) will tend to give certain forms whatever the prevailing climatic conditions and dominant weathering processes. In other words, for once the nature of the rock becomes the dominant factor in landform development, with climate and other factors playing a subsidiary role. There is, too, the additional point that the planation surfaces of Dartmoor do not display the typical concave profiles of the pediplain—a 'multi-concave surface' (p. 174)—but rather the upper slopes, above which the skyline tors rise, often show a broad convexity. One might argue that the old pediplain surfaces have been much modified by late-Tertiary and Quaternary denudation, but in that case destruction of most of the associated inselbergs and koppies should also have occurred. Finally, as is shown on pp. 175–80, some authorities no longer regard the plains and residual hills of Africa as the outcome of pediplanation, but of a totally different process of weathering and transportation that may, in fact, have a ready application to the landforms of Dartmoor.

B *The deep weathering theory*. This theory, proposed by Linton (1955), is based on several lines of evidence, of which the following are the most significant.

(i) In Linton's view many of the Dartmoor tors are composed of 'rounded and pillowy' forms, and some even 'simulate a great heap of piled woolsacks'. It is well known that similar rounded boulders, or core-stones, result from imperfect chemical weathering of jointed granite under warm and humid conditions (p. 26).

(ii) Many tors rise above a solid rock platform, comprising sound granite which over a wide area is flat or gently undulating. It is inferred by Linton that this 'basal platform' may be equated with the basal weathering surface found in tropical lands (p. 175), and that on Dartmoor it has been revealed by subsequent removal of overlying decomposed rock.

(iii) Most tors rise some 20–30 ft above their surroundings, and from this it is possible to derive an estimate of the former depth of weathering (which is limited at its lower margins by the water-table in the granite).

(iv) The growan of Dartmoor is regarded by Linton as the remaining part of the former weathered layer. In some places it is evidently *in situ*, since it preserves the structure of the granite and indeed contains some unweathered rock, but in others has been transported downslope by wash, creep and, above all, solifluxion in the Quaternary.

From this evidence Linton has constructed the theory that the tors are the result of a 'two-stage process, the earlier stage being a period of extensive sub-surface rock rotting whose pattern is controlled by structural considerations, and the later being a period of exhumation by removal of the fine-grained products of rock decay' (Fig. 55). The decomposition of the granite, which probably took place under 'tropical humid' conditions in the late Tertiary, was closely guided by the pattern of jointing. Where this was close-spaced, complete decay of the granite down to the level of the water-table was effected, but where the joints were wide-spaced subsurface core-stones or cuboidal blocks remained unconsumed. The removal of the finely comminuted granite, to leave the tors upstanding above the exposed basal weathering surface, is assumed to be the result of periglacial processes, operating mainly during the last glacial period of the Quaternary.

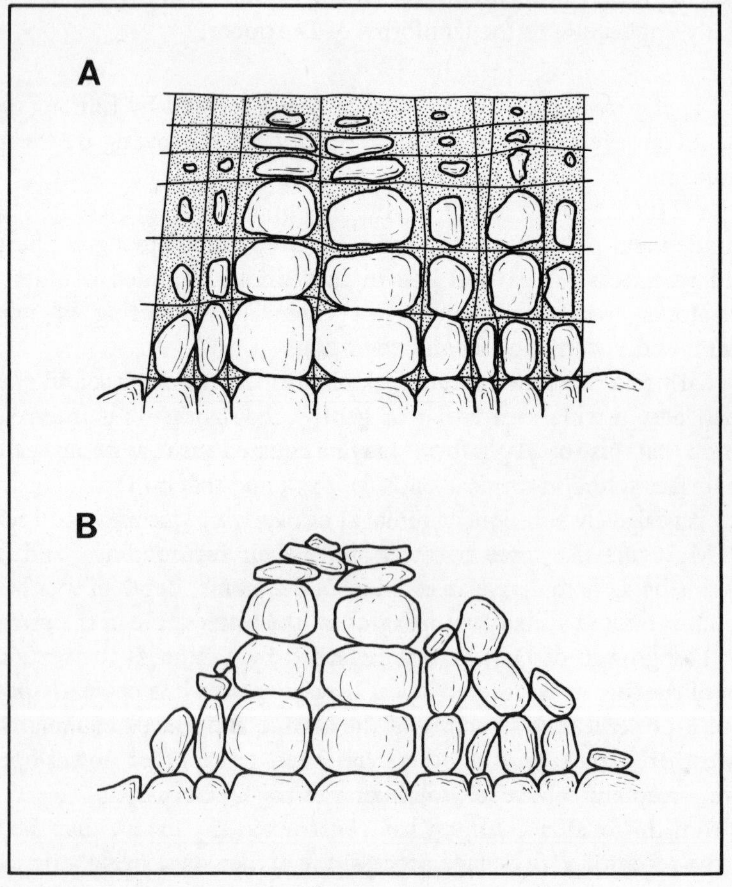

55 The development of tors (after D. L. Linton)

C *The periglacial theory.* The role of periglaciation as a subsidiary factor in the evolution of the Dartmoor landscape is clearly recognised by Linton, but other writers (notably Palmer and Neilson, 1962) regard it as the main process in tor formation. In the words of Palmer and Neilson, 'the Dartmoor tors are relict "cold" forms rather than relict "warm" forms, and may for convenience be called Palaeo-arctic tors'. These authors' views, which are in certain respects diametrically opposed to Linton's, are based on the following observations.

(i) The deeply decomposed granite at present found on Dartmoor seems to be the result of pneumatolysis rather than weathering. It has been noted above that such granite is usually associated with valley development, and that tors are most common on interfluves where the granite remains sound. The more superficial growan is composed of quartz grains, broken felspar crystals and fragments of granite, but is not clayey. It cannot therefore be the residue of a chemically decomposed layer, since chemical attack would reduce at least some of the felspar to clay, but must be the result of mechanical disintegration of granite. The only process likely to produce such granular breakdown is frost weathering acting over a prolonged period.

(ii) Palmer and Neilson admit that in some areas core-stones may be observed 'emerging' from rotten granite as the latter is slowly removed, but state that these spheroidal blocks do not give tors that remotely resemble those that are most common on Dartmoor. The typical Dartmoor tor is not, in fact, a pile of loose rounded woolsacks, but comprises dominantly angular masses of rock divided by fairly equally spaced vertical and horizontal joints (Fig. 54). 'These have been partly opened by weathering, and the greater weathering at the corners gives an incipient rounding to the constituent joint bounded blocks.' However, the weathering of the joints has usually penetrated to a small depth only, and although the tor at first sight *appears* to be a pile of core-stones, in fact it is a solid, jointed and *in situ* mass of granite.

(iii) Surrounding many tors are extensive block-fields and clitters, which are not composed of rounded boulders but of angular masses of granite broken from the margins of the tors. Some clitter blocks are as big as the joint-bounded blocks in the tors, and appear to have been isolated from the latter by water penetrating the joints and freezing. That such block disintegration by frost action does not merely represent minor modification of pre-existing tors (such as might be envisaged if the deep weathering theory were accepted) is indicated by the vast volume of many clitters. Indeed, it has been estimated that below Belstone Tor the clitter exceeds 14 million cu. ft in volume. Lastly, many of the clitters appear to have been transported downslope under periglacial conditions in the Quaternary. Some may have slipped over the surfaces of snow-patches, but the existence of 'block lines' (comparable with the stone stripes discussed on pp. 326–7) suggests the action of solifluxion on a large scale.

Palmer and Neilson have proposed the following theory of tor formation (Fig. 56). At an early stage of periglaciation existing soil was removed by solifluxion and the granite exposed over wide areas to frost weathering. This would have had its greatest effect on hill-tops and interfluves and its least in valley bottoms, for in the latter the abundance of water would have been associated with permafrost and, except in a thin surface layer, little freeze-thaw action. On the interfluves frost attack would have been differential. Areas of close jointing would be most readily affected, and widely jointed granite would be more resistant. With removal of the loosened granite by solifluxion the slopes would be gradually lowered, except where massive granite outcropped.

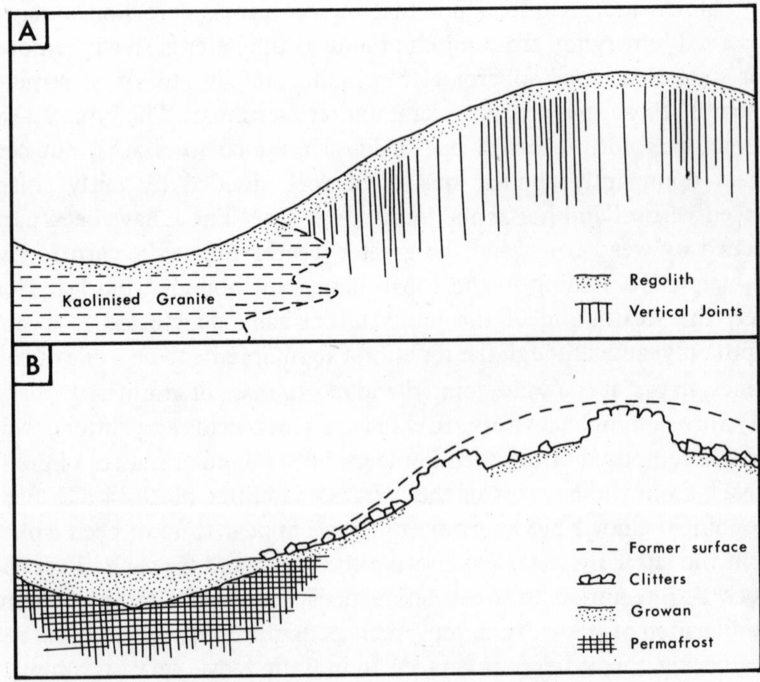

56 **The development of tors (after J. Palmer and R. A. Neilson)**

This would be left upstanding to form an incipient tor, and would be shaped in detail by 'scarp retreat' (the free faces of the granite being driven back by frost action, initiating and extending the clitters). In some localities tors may never have been formed, simply because the granite was uniformly weak, and in others they may have been consumed by scarp retreat. The more general result of periglacial denudation on Dartmoor was to produce the gentle 'equilibrium' slopes with their cover of growan.

Conclusion

The landscape of Dartmoor is often quoted as a 'typical' example of granite scenery. There is obviously a measure of truth in this, for such distinctive landforms as tors are common in granite areas elsewhere (for example, Arran, the Cairngorms, Brittany and the Massif Central). These landforms, whichever theory of their origin is accepted, are closely related to rock hardness, jointing and susceptibility to block and granular disintegration, which are characteristics of granite everywhere. Yet tors and related forms are not confined to granite areas; they are likely to be developed in any type of rock which has the characteristics listed, even some types of limestone. It is also obvious that the landforms of Dartmoor cannot be explained as the outcome of a simple interaction between rock and denudational processes. Once again there is illustration of the principle that in geomorphology full account must be taken of *all* causal factors, including in this instance denudational and climatic history.

LIMESTONE LANDFORMS: THE GRANDS CAUSSES

The Grands Causses, situated in the central southern part of the Massif Central of France, constitute one of the great limestone areas of Europe. The sediments accumulated here in a geosynclinal gulf mainly during the Jurassic period, and attained a maximum thickness approaching 5000 ft. Uplift and dissection was initiated in Cretaceous times, and continued throughout the Tertiary era, with the result that, in marginal areas in particular, the limestones have been attenuated or removed altogether to reveal ancient rocks of the underlying basal complex ('socle'). At present the limestones give rise to a series of high plateaus, rising generally to between 3000 and 4000 ft above sea-level, which are separated from each other by magnificent river gorges (those of the rivers Tarn, Jonte, Dourbie and Vis) and from adjacent areas of granitic

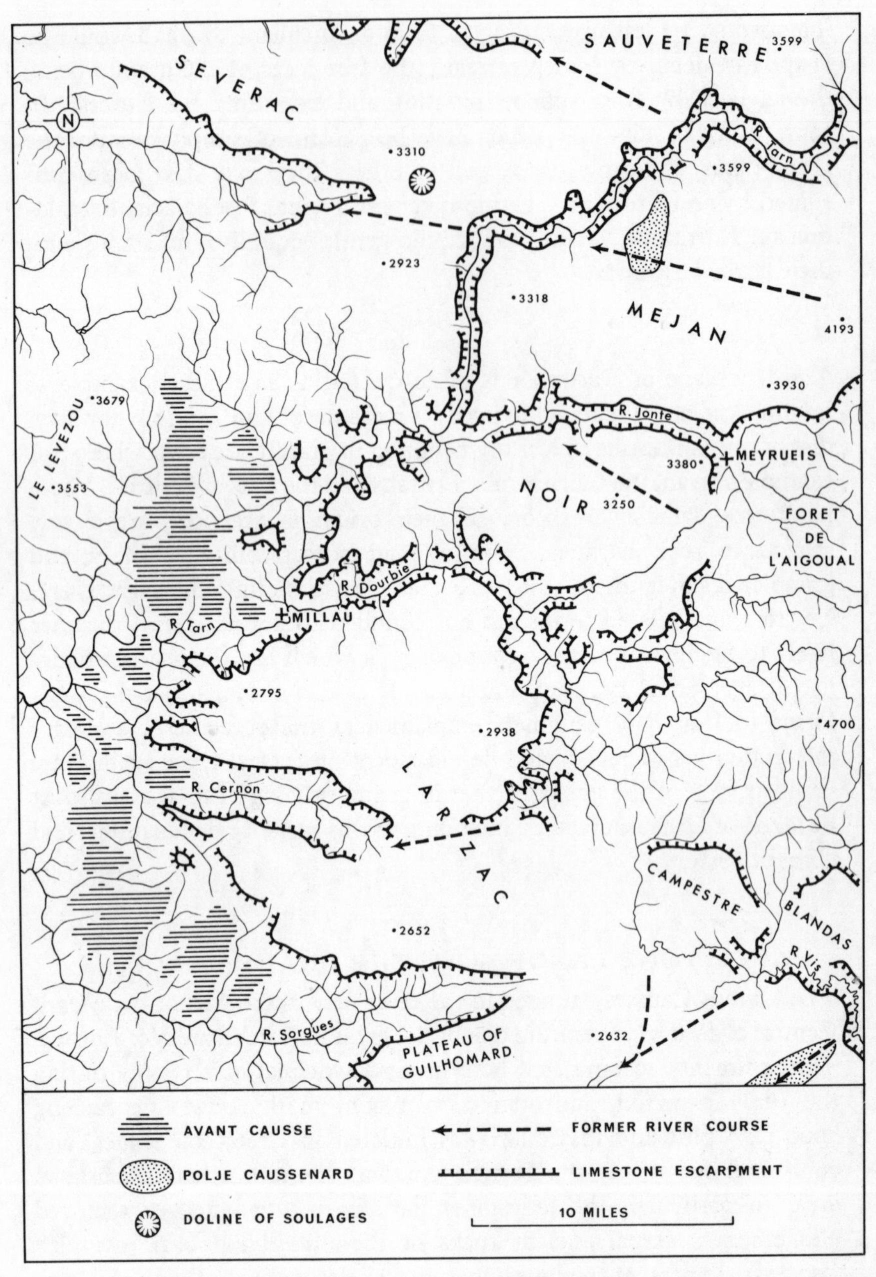

57 **The geomorphological features of the Grands Causses**

rock to west and east (Le Levezou, Mont Lozere and Mont Aigoual) by broad strike vales (Fig. 57).

The geological structure of the Causses (Fig. 58) is dominated by a north-south synclinal axis, to the west of which the dip of the limestone is gentle (5° or less), whereas to the east it is markedly steeper (15° or more). There are, in addition, smaller fold-lines, orientated generally west-east (for example, the syncline passing through the Causses de Severac and Sauveterre) and a series of major faults, aligned either west-east (the so-called 'Pyrenean' trend) or north–south (the 'Alpine' trend). The impact of these faults on the land surface varies from one locality to another. Along the eastern margins of the Causses there are magnificent fault-line scarps, related to displacements of the order of 1000 ft, and within the limestone plateaus lesser scarps are formed where rocks of differing resistance are faulted against each other. However, the effects of these structures are—considering their scale and complexity—less on the whole than might have been expected. The reason lies in a major episode of peneplanation (possibly involving several sub-cycles related to an intermittently falling base-level) in mid-Tertiary times. The fold- and fault-lines of the Causses, though closely related to structures in the underlying Hercynican basement, were developed largely in Oligocene and early-Miocene times and are therefore to be regarded as 'posthumous' (p. 103). A 'pre-Alpine' planation surface (the so-called 'Eogene peneplain'), with a covering of residual material known as 'le terrain siderolithique', was dislocated and deformed, and refashioned by erosion during the 'post-Alpine' period. Baulig (1928) has argued that the main episode of erosion occurred in Pontian (Upper Miocene) times, for on the southern borders of the Massif Central there is evidence to show that the ensuing Pliocene period was one of falling base-level and vertical stream incision—indeed, the great river gorges of the Causses were almost certainly developed at this time. The Causses peneplain, which declines in height both

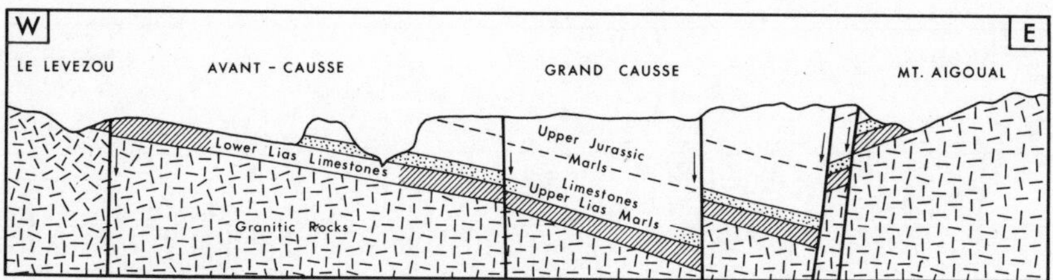

58 **Geological cross-section of the Grands Causses**

westwards and southwards, varies in the state of its perfection. In the eastern part of the Causses de Sauveterre and Mejan the relief of the limestone plateaus is comparatively mature, whereas to the west some areas exhibit advanced planation. As Fig. 43 shows, the planation is not confined to the limestone, but transects also adjacent granitic areas (though in the east the summits of Lozere and Aigoual, at 5583 and 5141 ft respectively, stand well above the general level of the peneplain).

The limestones of the Causses

Although in general classifiable as limestones, the sediments of the Causses are in detail somewhat varied. A fundamental threefold division can be recognised.

(i) At the base of the sequence, and separated from the older rocks of the Hercynian basement by relatively unimportant Triassic sandstones and clays, are some 500 ft of Lower Lias limestones. These are well bedded and jointed, but relatively impure (with sand and marl bands), and the usual karstic features, excepting dry valleys, some solution hollows, and the gorge at Bramabiau (p. 148), are not well developed. The outcrop of these rocks is greatest in extent to the west, owing to the low angle of dip, and to the north-west and south-west of Millau they form a broad and remarkably preserved dip-slope, declining gently eastwards towards the foot of the great escarpments bounding the Causses proper (Fig. 59). These areas are sometimes referred to as 'Avant-Causses'.

(ii) Above the Lower Lias limestones occurs a stratum of blue-black marls (limy clays), at its maximum approaching 500 ft in thickness, but varying in importance from place to place. These Upper Lias marls, because of their impermeability, are easily attacked by fluvial erosion, and indeed constitute the least resistant of all the rocks in the Causses area. They have been excavated into a near-continuous strike vale by subsequent streams, and are sometimes followed by the major rivers

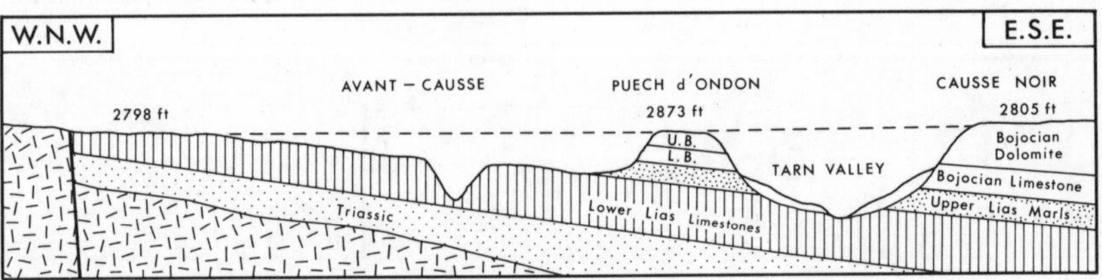

59 **Geological cross-section of the Avant-Causses**

(for example, the Tarn above Millau). These streams have been affected by down-dip shift, and have stripped the marls cleanly from the surface of the underlying Lower Lias limestones, except at a few points where striking residual hills, known as 'buttes temoins' have been accidently preserved. Even today erosion of the marls is proceeding rapidly, and in some places they are so deeply gullied as to form, in effect, badlands.

(iii) Surmounting the marls is a great mass of limestones, ranging in age from Inferior Oolite to Portlandian, which form the high plateaus of the Causses de Severac, Sauveterre, Mejan, Noir and Larzac. Taken as a whole these strata, reaching a total thickness of some 4000 ft in central Mejan, are more permeable than those of the Avant-Causses, and karstic features are far more prominent—though, as will be shown below, the Causses are in certain respects not comparable with the classic Jugoslavian Karst. However, in detail there are important lithological variations within the Upper Jurassic limestones. The main contrast is between the 'dolomites' and the well-bedded, sometimes marly, 'sub-lithographic' limestones (Fig. 60). The former are for the most part very massively bedded (the individual strata are normally at least 10–15 ft and sometimes 50 ft in thickness), and there are few joints and fractures to assist permeability. The latter, on the other hand, are

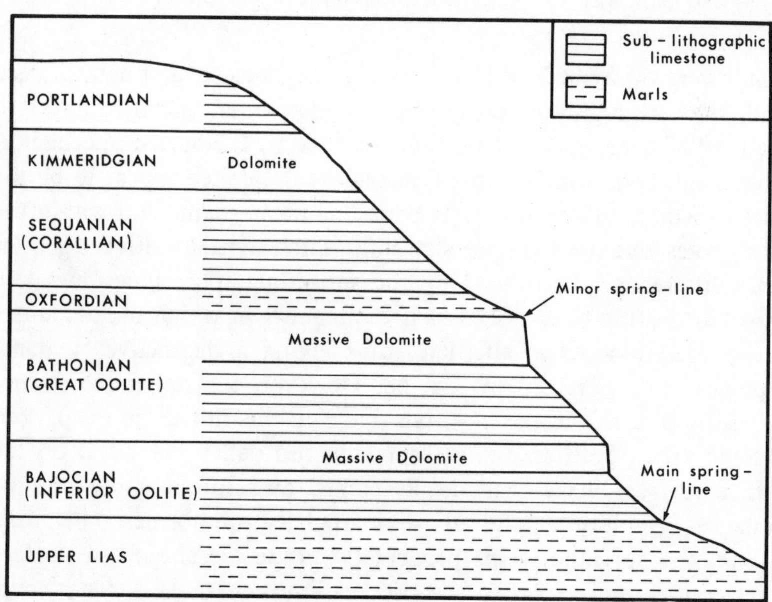

60 **Lithology and slope form in the Grands Causses**

61 'Rochers ruiniformes' at Montpellier-le-Vieux, north of the Dourbie gorge, the Grands Causses. The rock pinnacles have resulted from intense chemical dissolution of Bathonian dolomite. Note the gently undulating surface of the Causses plateau, at 2500–3000 ft, in the background. [French Government Tourist Office]

usually very thinly bedded and are affected by a close joint pattern; as a result they are highly permeable, and tend to break down readily into small slaty fragments when the joints are used by chemical or mechanical weathering. Over much of the Causses the dolomites appear to be the more resistant, and on the great bounding escarpments and within the river gorges give rise to impressive cliffs which contrast with the gentler debris-littered slopes formed by the sub-lithographic limestones. An important feature of the Causses is the manner in which the dolomites and bedded limestones alternate, thus giving a distinctive 'regional slope form' as depicted in Fig. 60. However, the dolomites do not invariably give rise to the most positive elements in the landscape, but in some areas, both on the escarpments and valley sides and on the plateau surfaces, have been deeply etched by solution into grotesque pinnacles, graphically referred to as 'rochers ruiniformes' (Fig. 61). In fact, the dolomite seems particularly prone to solution by surface water, rather more so than the bedded limestones. As a result, over much of the Causses de Sauveterre, Noir and Larzac the widely outcropping Bathonian dolomite has been carved into a multitude of minor karstic forms which are overlooked by gently rounded hills of Sequanian limestone.

The landforms of the Causses

A *The river gorges.* These are the most spectacular physical features of the Causses, and are also of importance geomorphologically because of their undoubted effects on landform development in the intervening limestone blocks. In cross-section the gorges display considerable variations, according to the particular limestone formations exposed on their walls (Fig. 62). The greatest of the gorges is undoubtedly that of the Tarn, which is incised up to 1500 ft below the level of the Causses de Sauveterre and Mejan, and is characterised in its central section by narrow almost overhanging walls of Bathonian dolomite, giving rise to the so-called 'etroits' (Fig. 63). The most interesting morphologically is that of the strongly meandering Vis, which occupies a deep trench in the arid Sequanian limestones of eastern Larzac (Fig. 64).

The river gorges pose an obvious geomorphological problem. How were streams such as the Tarn, Jonte, Dourbie and Vis, rising on the impermeable crystalline rocks of Lozere and Aigoual, not only able to maintain their courses over the permeable limestones, but also to achieve such a phenomenal amount of vertical erosion in late-Tertiary and Quaternary times? At present the water-table lies many hundreds of feet below the surface of the limestone plateaus. Furthermore, it is periodically lowered beneath the floor of the Jonte west of Meyrueis, and is permanently below the bed of the Vis in the vicinity of the aptly named Vissec. A very important factor has undoubtedly been the large amounts of precipitation, particularly in winter and spring, on the crystalline catchments to the east. Indeed, the discharge of the Tarn is even today sometimes so great that, far from maintaining its course through the limestone with difficulty, the river can cause severe flooding as far west as Millau. In addition, the main rivers, except during times of drought, tend themselves to determine the height of the water-table in the limestone. The water percolating into the upland blocks re-emerges, by way of springs, in three types of situation: (i) where thin marly layers, such as those of the Oxfordian, outcrop on the gorge walls; (ii) at the base of the escarpments of the Causses, where the permeable limestones of the Bajocian give way to the underlying impermeable Upper Lias marls; and (iii) in the bottoms of the main river gorges (for example, a number of resurgences has been located in the Tarn gorge). Since the height of

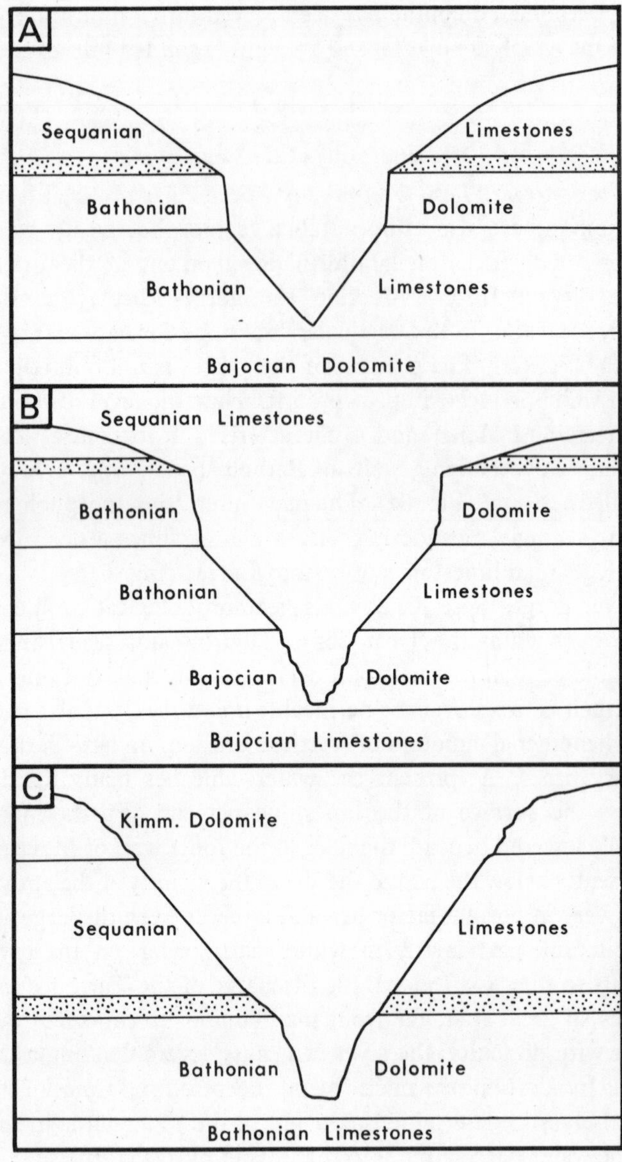

62 Cross-sections of river gorges in the Grands Causses: Jonte below Truel (A); Jonte above Truel (B); Tarn near La Malene (C)

the springs in (ii) is conditioned by the degree of incision of the strike streams, themselves tributary to the Tarn, Jonte and Dourbie, it will be apparent that the control of these rivers over the height of the water-table is both direct (as in (i) and (iii)) and indirect.

The question also arises as to the extent to which the Causses gorges are the result not of surface fluvial erosion, but of subsurface solution, cavern formation and large-scale collapse. It is impossible to give a final answer, but some evidence indicates that 'normal' river erosion has played a major role. In the first place, it is evident that early in the present cycle of erosion a larger number of streams crossed the limestone from east to west (Fig. 57). Not only are there exceptionally large dry valleys and associated closed depressions, but trains of ancient gravel, comprising large quartz pebbles and rotted granite fragments from Lozere and Aigoual, can be traced across the Causses. Once these streams disappeared underground (presumably because the more rapid incision of the Tarn, Jonte, Dourbie and Vis led to a major fall of the water-table) their valleys became fossilised and only limited karstic

63 The Tarn gorge near La Malene. Note the vertical Bathonian dolomite cliffs in the left foreground and the rectilinear slopes at 30–34° formed by Sequanian limestones in the distance. [H. Roger Viollet]

64 The Cirque de Navacelles, Vis gorge. Note the cut-off incised meander and meander core, and the waterfall on the site of the former meander neck. It is clear that a knickpoint advanced upstream to this point, causing 'abstraction' of water underground through the permeable limestone of the meander neck and accelerating its breaching. [F. Gay]

modification subsequently occurred. Secondly, the Vis gorge at least seems to be entirely the result of stream erosion. Its magnificent meanders, which become progressively more incised from Vissec south-eastwards until those at Navacelle are some 1200 ft below the level of the Causse du Larzac, its valley-side benches preserved above the incised meanders, and its sequence of knickpoints cannot possibly be explained as the outcome of subsurface corrosion. On the other hand, subterranean rivers, associated with the development of cave-systems which may ultimately be exposed by collapse, are not absent from the Causses. The finest example is that of the river Bonheur, which rises on the western flanks of Aigoual, sinks underground on encountering the Lower Lias limestones east of Causse Noir, and after half a mile re-emerges at the foot of a high cliff in the deep gorge of Bramabiau. The headward extension of this chasm must undoubtedly be facilitated by the erosion already achieved by the underground river.

B *The escarpments.* Almost as impressive as the river gorges are the high escarpments which bound the limestone plateaus of the Causses.

In cross-section these resemble the walls of the gorges, with dolomite cliffs or rochers ruiniformes alternating with rectilinear debris-covered slopes at an angle of 30–35°. East of the Tarn valley above Millau the scarp-profiles have been considerably modified in their lower parts by great rotational slips, believed to be of Quaternary age, affecting the Bajocian limestones and underlying impermeable and unstable Upper Lias marls. In this same area, large upland masses, capped by resistant Bathonian dolomite, have been isolated from Causses Noir and Mejan by the erosion of the Tarn and its tributaries; a fine example of such a butte is the Puech d'Ondon (Fig. 59). In plan the escarpments of the Causses are deeply indented by large valleys. Those of the Cernon and Sorgues (Fig. 57) are related to east–west anticlinal structures, which have brought the weak Upper Lias marls high above base-level and promoted the process of breaching (pp. 94–7). Other smaller escarpment valleys, terminating in high limestone walls at the base of which springs or seepages occur, are probably the result of spring sapping.

C *Dry valleys.* Although from afar the planation of the Causses is readily apparent, when viewed more closely the limestone is seen to be closely dissected in many parts by dry valleys or 'combes'. However, it is rare to find these valleys forming integrated and accordant systems comparable with, say, those of the English Chalk country (Fig. 65). Whilst it is clear that the valleys were originated by surface streams, during and since the process of desiccation they have been slowly modified by the development of enclosed depressions of various sizes and shapes—indeed some 'valleys' are now really no more than lines of surface hollows. In view of this it is hardly possible, as has been done in the case of the Chalk, to explain the dry valleys of the Causses in terms of (i) the development of permafrost, sealing the fissures and rendering the limestone impermeable, and (ii) erosion by snow-melt streams in the Quaternary. Periglaciation has undoubtedly left its mark on the limestone plateaus. The gentle slopes, covered by angular frost shattered fragments, developed in the Sequanian limestones of Causse Mejan north of Meyrueis, were obviously fashioned by frost weathering and solifluxion. Furthermore, there is evidence here of some recent run-off, in the form of fresh-looking vertical shafts ('avens' or 'gouffres') down which the water made its escape. However, it is clear that most of the dry valleys are of considerable antiquity, dating either from the late stages of the Pontian cycle or the early stages (in the Pliocene) of the current erosion cycle. The streams responsible for the initiation of the

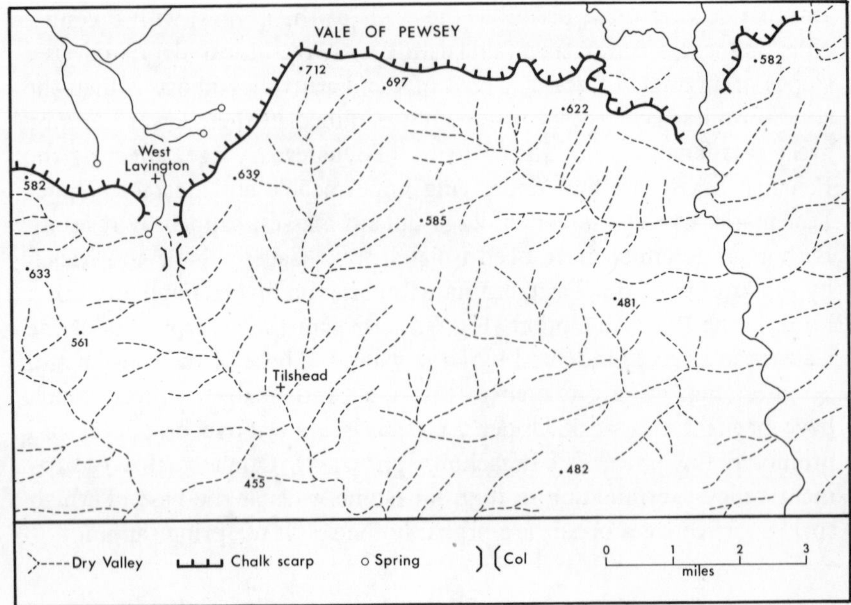

65 **Dry valley pattern of the Chalk of northern Salisbury Plain**

valleys were unable to keep pace, in downcutting, with the larger rivers and those responsible for the development of the Upper Lias clay-vales and associated recession of the Causses escarpments (Fig. 66A). As the water-table fell, stream flow became at first intermittent, then increasingly disrupted by sink-holes, and ultimately ceased altogether. Subsequently the valleys have been mainly affected by slow-acting solution processes, operating differentially, and their present 'disorganised' condition brought into being (Fig. 67).

D *The enclosed depressions*. Enclosed depressions are the most common and characteristic landforms of the Causses. The basic form is the small round or elongated hollow (referred to locally as a 'sotch') which is comparable with the 'doline' of the Jugoslavian Karst. These sotchs, which frequently occupy both valley bottoms and valley-side slopes, may possess either rocky edges or gently sloping soil-covered margins. Occasionally they are associated with avens, but more often their floors are smooth, owing to the presence of a variable thickness of 'terra rosa', a residual decomposition product washed and sludged in from surrounding areas to obscure former sink-holes. In cross-section many sotchs are asymmetrical, either because of structural control or as a result of differential frost weathering and solifluxion under periglacial conditions (see the discussion of asymmetrical valleys on pp. 339–43). Sotchs are ubiquitous in the Causses, but show local variations in form and density of occurrence which appear to reflect lithological influences. On the

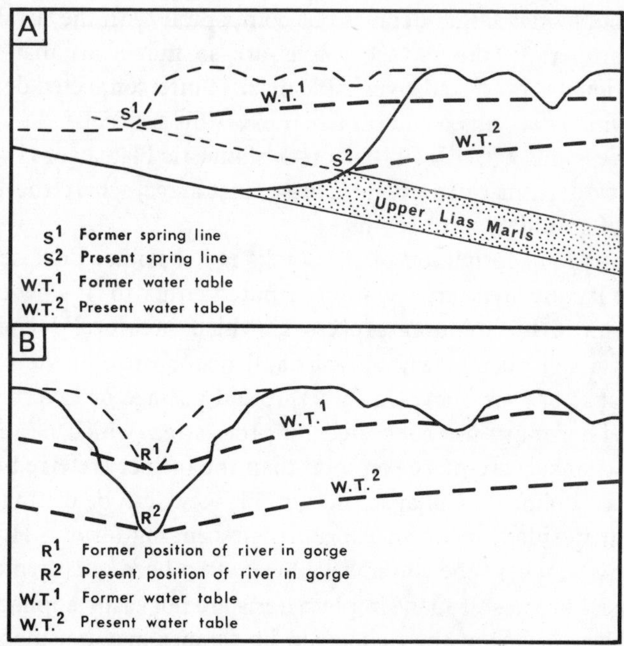

66 Causes of water-table fall in the Grands Causses: recession of the bounding scarps (A); incision of the gorges (B)

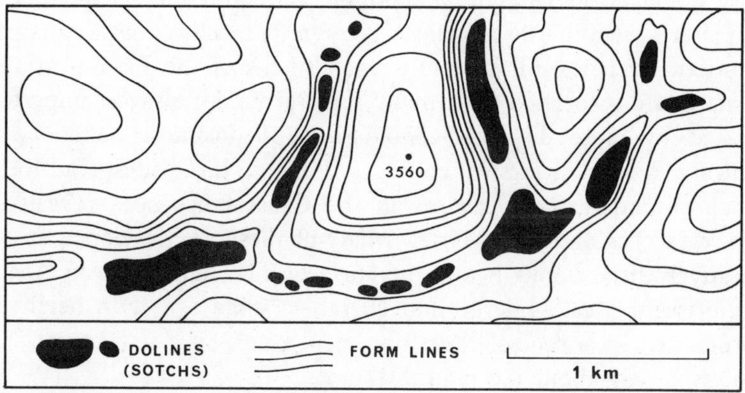

67 **Karstified valleys, Causse Mejan**

thinly bedded Sequanian limestones they are mostly circular, small in diameter, and rarely more than 15 ft in depth, whereas on dolomitic outcrops they are more irregular in outline, larger, deeper, with rocky or even undercut walls, and with floors interrupted by rocky pillars. In many areas closely adjoining sotchs have amalgamated, through lateral extension, to give larger depressions comparable with the 'uvalas' of the Karst proper. In others, the sotchs are so numerous that imperfect integration has produced a veritable maze of interconnected depressions, above which rise jagged and bizarre rocky walls and isolated rock masses. Nowhere is this better seen than on the fantastic Plateau of Guilhomard, on the western margins of the Causse du Larzac, where the Bathonian dolomite is riddled with sotchs.

The mode of formation of the sotchs is debatable. Enclosed depressions in karstic areas are usually attributed either to (i) slow downward development by solution processes, which are locally concentrated beneath a soil mantle, without physical disturbance of the rock, or to (ii) collapse of rock above an underground passage or cavern (Fig. 68). W. D. Thornbury has suggested that the latter (which he refers to as 'collapse sinks') are more common than the former (referred to by him simply as 'dolines'). Collapse sinks, in his view, can be distinguished by their 'steep-sided, rocky and abruptly descending forms'. However, in the Causses, where the sotch walls have often been greatly modified by periglacial processes, such simple criteria are not easily applied; even the gentle depressions of the Sequanian limestones may be smoothed-over hollows resulting originally from collapse. There are, in fact, some reasons for believing that collapse sinks may be common in the Causses. The very large hollows near the southern margins of Sauveterre (the great doline at Soulages has a depth of over 200 ft and is a kilometre in diameter) seem hardly explicable by reference to localised surface solution. Even the innumerable small sotchs of the Plateau of Guilhomard are associated with many underground passages, imperfectly choked by large displaced boulders and 'dolomitic sands' which indicate recent or active collapse. Finally, the widespread occurrence of underground caverns in the Causses has been revealed by the researches of E.-A. Martel. Many of these are developed at comparatively little depth below the surface. The famous Aven Armand comprises a great underground chamber, some 150 ft in height and 'protected' by a roof of limestone, penetrated by a number of shafts ('puits'), only about 100 ft in thickness.

The largest of the enclosed depressions of the Causses are the so-called 'poljes caussenards'. Only two of these are recognised by P. Marres, that at Carnac (standing high above the southern edge of the Tarn gorge on Causse Mejan) and the polje of St Maurice-Navacelles (in the south of the Causse du Larzac). The Carnac polje extends some 5 miles from south to north, where it is separated from the Tarn gorge by a col (Fig. 69). The depression, which was evidently initiated by a stream joining the Tarn at St Chely-du-Tarn, is floored by Portlandian limestones, downfaulted against more resistant Sequanian limestones forming a fault-line scarp to the east. Since the disappearance of this stream the floor of the valley has been lowered by some 200–300 ft near Carnac, and its flatness has been emphasised by the accumulation of a considerable thickness of residual clay. The latter now forms a relatively impermeable stratum, and after exceptionally heavy rain the polje may be briefly flooded to a depth of several feet. The St Maurice polje,

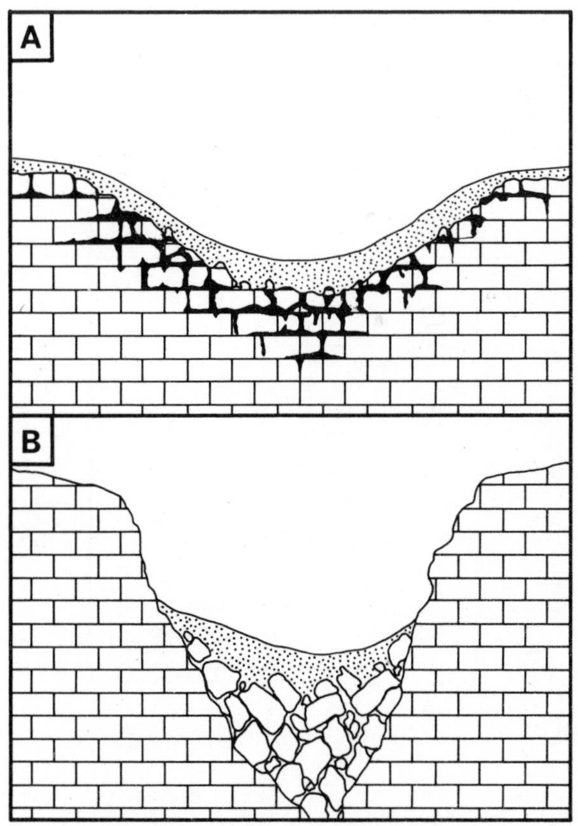

68 The formation of dolines by solution (A) and collapse (B)

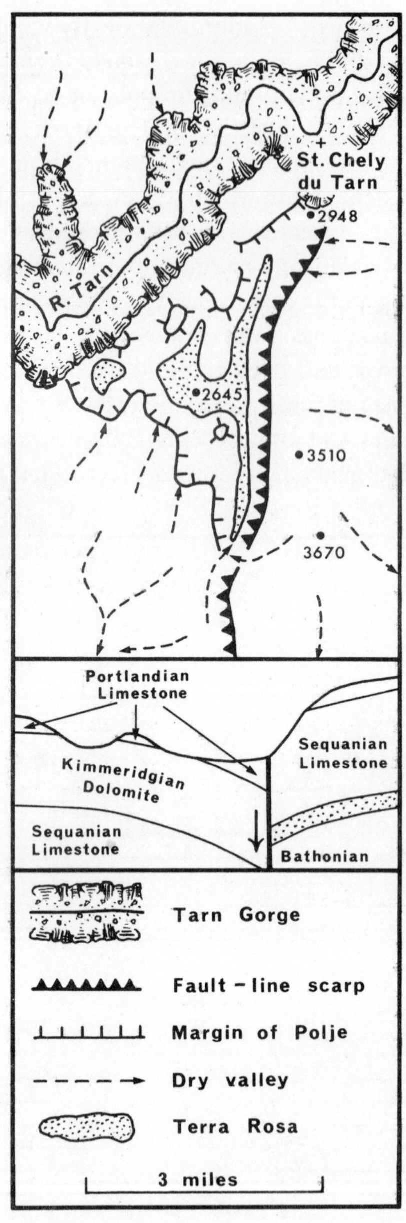

69 The polje caussenard of Carnac, Causse Mejan

stretching some 6 miles from north-east to south-west and terminated on its southern side by the fault-line scarp bounding the high ridge of the Seranne, seems to have been originated by an ancient stream crossing southern Larzac from Aigoual and the Causse du Blandas (Fig. 57). The linear form of the depression, and its occupation by rounded quartz gravels, seems to be clear proof of this. However, the fluvial development of the polje was terminated by the diversion of the upper part of the old stream, via the Vis, into the Herault. Subsequently there has been only minor deepening of the central part of the valley by the usual karstic processes.

It must be seriously questioned whether, in fact, the Carnac and St Maurice depressions are to be ranked as poljes. The typical Karst polje is an 'elongated basin with a flat floor and steep enclosing walls which owes its existence to solutional modification of downfaulted or downfolded limestone blocks' (Thornbury). The largest poljes are up to 40 miles in length and 7 miles in width. It might be argued that the Causses poljes qualify on the grounds of relationship with structural disturbance, but they certainly do not on the grounds of size. Baulig has stated quite categorically that there are no authentic poljes in the Causses. He has suggested that the Dinaric Karst of Jugoslavia has been vigorously folded, faulted and affected locally by overthrusting, so that the resistance of the rocks has been diminished and large-scale foundering favoured. In the Causses, on the other hand, the rocks have been only gently deranged, and have proved more resistant to collapse. In addition, Baulig has emphasised that poljes appear relatively late in the karstic cycle, whereas the surface forms of the Causses represent a youthful stage of karstic development.

Conclusion

It is appropriate to conclude this brief discussion of the Causses with an attempt to relate their features to those of other 'classic' limestone regions. Cvijic has suggested that these may be broadly divided into two types. Those in which all the main forms of limestone solution are found are classified as 'holokarst', whereas the regions in which such forms are limited in their development by the thinness and impurity of the limestone (for example, the Avant-Causses) are termed 'merokarst'. Cvijic argued on the following grounds that the Causses occupy a unique position between these categories. (a) The Upper Jurassic limestones of the Causses have marly intercalations (for instance, the Oxfordian marls) which have hindered full karstic development. (b) The aquifer of the Causses is not of indefinite depth, but is limited by the Upper Lias marls. (However it must be stated than an impermeable base is envisaged by Cvijic in his idealised cycle of karst erosion (pp. 182–4).) (c) Normal river valleys are far more important in the Causses than in the Dinaric Karst. (d) The Causses, as we have seen, do not possess the greatest and most characteristic karstic form, the polje.

Cvijic's views are not, however, fully accepted by Baulig. The latter has drawn attention to Cvijic's own conclusion that in a fully developed karst area there are three subsurface zones. Firstly, there is an upper zone with caves and passages abandoned by underground water flows; secondly, a middle zone with cavities used temporarily during wet periods; and thirdly, a lower zone with caverns constantly utilised by underground water. Abundant evidence exists to show that each of these three zones is present in the Causses, and in that sense the karstic development of the region has attained the fully mature stage. However, there is a paradox, for the three zones are normally found only after the surface valleys of youth have been totally obliterated and replaced by dolines and uvalas which have begun to coalesce and give extensive poljes. In the Causses, as shown above, surface valleys perhaps dating back to the Pliocene or earlier are still preserved, and true poljes are absent. Baulig attempts to reconcile these issues by arguing that special factors have influenced landform development here. Although the karstification is very deep—and in that sense there is direct comparison with the Dinaric Karst—the Causses are not affected simply by a karstic cycle. Fluvial activity has been unusually important, and has been responsible for the remarkable gorges and the rapidly falling water-table in the intervening limestone blocks, the surface forms of which were 'arrested' at an early stage. The cycle of erosion in the Causses is, in fact, a 'fluvio-karstic' one, and 'it is this double mode of erosion that constitutes the principal morphological interest of the region'.

5

THE CYCLE OF EROSION

INTRODUCTION

The cycle of erosion, postulated by Davis in a series of articles and essays published at the end of the last century, has been regarded by many geomorphologists until quite recently as the most fundamental concept in landform study. To Davis, the earth's landforms, though displaying almost infinite variety, were all closely interrelated, forming parts of an 'evolutionary sequence'. He envisaged that with the passing of time, during which they are acted upon by the processes of denudation, landforms undergo a progressive change from 'initial forms' through 'sequential forms' to 'ultimate forms'. Thus hill-tops and interfluves will not maintain for ever the same height and shape, but will be gradually lowered by weathering, rainwash and creep, with the result that the slopes leading up to them will decline in steepness.

Clearly this is a dynamic view of the physical landscape, in which the concept of change with time is central to the argument. Indeed, the detailed form of the land surface is seen as depending to a greater or lesser degree upon how long the processes of denudation have been operative. This at least seems to offer a reasonable explanation of the fact that the same rock-type, affected by similar processes of weathering, transport and erosion, can give rise in two different areas to markedly differing features. A simple example of this is depicted in Fig. 70, showing parts of the Chalk country of the South Downs and Salisbury Plain. In the former, the land reaches a maximum height of over 650 ft O.D., and the landscape is deeply cut into by branching networks of dry valleys, the slopes of which are generally steep and sometimes exceed 20° in angle. In the latter, however, the landforms are less bold, the hills and divides rarely exceed 450 ft O.D., the valleys are much shallower, and the valley-side slopes are usually less than 10° in angle. Adopting the principles enunciated by Davis, one could explain these contrasts simply by assuming that Salisbury Plain is geomorphologically older than the South Downs, and that prolonged denudation has reduced the higher hills and steeper slopes that once existed on the Plain to their present subdued form.

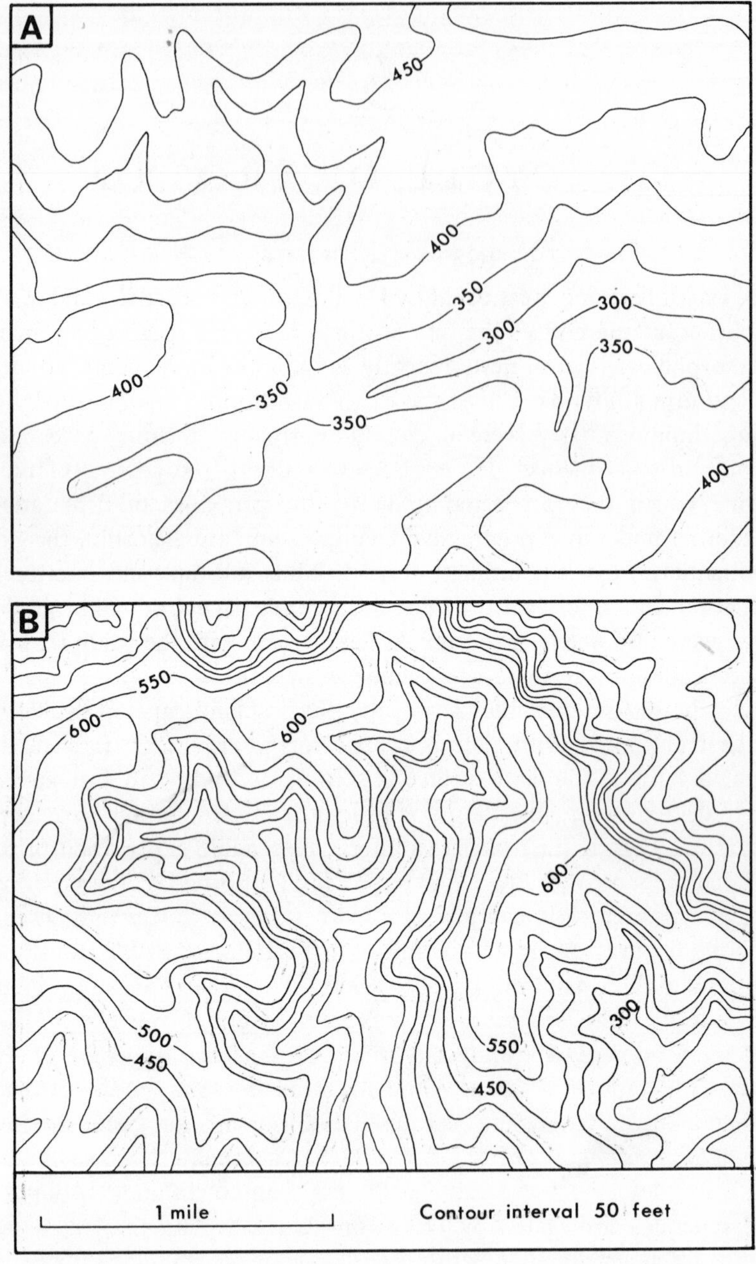

70 **The relief and valley patterns of part of the South Downs (A) and Salisbury Plain (B)**

It hardly needs emphasising, of course, that although the time factor is seen to be very important in landform development it is by no means overriding. Davis himself established the well-known principle that all physical landscapes can be analysed in terms of the three variables 'structure', 'process' and 'stage'. Structure (the underlying rocks and the manner of their disposition) obviously provides the basic materials from which landforms are fashioned, and as is shown in chapters 3 and 4 can exert a very real influence on those landforms. Process (including mechanical and chemical weathering, mass movements, rainwash, river erosion, wave attack, glacial action and so on) is responsible for the actual shaping of the landforms, and as is shown in chapter 2 in some cases its control is paramount. Stage in the Davisian usage refers merely to the length of time during which denudational processes have been operating on a particular structure.

From this it will be seen that the physical landscapes of the earth can be subdivided not only according to the underlying rocks and the denudatoinal processes they are affected by, but also according to their age. In this respect, Davis found it convenient to adopt the terminology used in describing the life-cycles of fauna, and referred to landscapes as being in the stages of 'youth', 'maturity' or 'old age'. These ages are, of course, purely relative terms, and cannot be related to an actual time-scale. For instance, it cannot be inferred that the youthful stage of the cycle will always occupy *x* million years, since so much will depend on the rate at which erosion operates. This may be greatly retarded in areas of very resistant rock-type, or may be speeded up under certain climatic conditions, such as those associated with glaciation. It may be that a cycle of erosion requires for its completion anything between 5 and 50 million years.

THE DAVISIAN CYCLE OF NORMAL EROSION

For the purposes of demonstrating his cycle concept in the most simple and persuasive way, Davis imagined as an initial form a mass of land uplifted from beneath the sea by earth-movements.

A *The stage of youth.* To avoid undue complications, of the sort later to be investigated by Penck (pp. 184–6), Davis assumed that the uplift of the land took place very rapidly, so that the processes of denudation were able to act almost from the start on what was, in effect, a stable mass. If the climate were sufficiently rainy, as would normally be the case in humid-temperate lands, a system of rivers would quickly develop on the emerged land-surface. This would comprise a number of consequent streams whose directions of flow and velocities (and thus erosional capabilities) would be determined by the gradients of the initial surface. From the stage of infancy, these streams would cut rapidly downwards, and would in due course form deep valleys bounded by slopes of 30° or more. On these slopes, weathering and slumping would operate, but at quite a slow rate compared with the speed of river incision. For a long period the valley cross-profiles would be approximately V-shaped, except in areas of complex geological structure where 'stepped' profiles would be developed. In the latter, steep and even cliff-like faces, associated with the more resistant rocks, would alternate with gentler slopes on the weaker strata (pp. 143–5).

Throughout the stage of youth, parts of the initial land-surface would be preserved on the watershed between the consequent streams (Fig. 71). In infancy the extent of this initial surface would be considerable, but would be gradually diminished later in the youthful stage as the valley-side slopes experienced retreat and as tributary streams began to extend their valleys into the interfluve areas by headward erosion. Davis envisaged that the long-profiles of youthful rivers would be markedly irregular, with waterfalls, cataracts and rapids marking the outcrops of hard rocks in the valley floors. Such profiles, which were referred to by Davis as 'ungraded' (p. 210), are in many respects analogous with the stepped slope profiles.

B *The stage of maturity*. By the onset of this stage, the deepening of the V-shaped valleys characteristic of youth would have been slowed down considerably. Through the formation of their valleys the various streams would throughout youth have lowered their channels nearer and nearer to what Davis termed 'the base-level of erosion' (which is normally the level of the sea into which the rivers eventually flow, and below which they cannot erode). In the process, the longitudinal gradients of the streams would have become ever more gentle, stream velocities would have been reduced, and the streams would possess less and less energy to use in moving their loads and attacking their beds. As we have seen in chapter 2, Davis in fact suggested that, early in the stage of maturity, streams would attain a condition of grade or equilibrium, in which the entire energy of the stream is consumed in the movement of the water and its load.

With the retardation of vertical corrasion, lateral erosion would become both actually and relatively more significant. The gentle meanders of the youthful streams, responsible for the interlocking spurs supposedly typical of youthful valleys, would become wider and more pronounced, and at many points the valley-side slopes would be undercut and driven back. However, in Davis's view, the retreat of the valley sides would be achieved more effectively by processes such as weathering, soil creep and rainwash acting over the whole surface of the slope. These would play a major part in broadening the narrow valleys of the youthful stage, and early in maturity would begin the lowering of the interfluve summits, where hitherto remnants of the initial surface had been preserved (Fig. 71). By the end of the mature stage, slope angles in general would have been considerably reduced by the process of 'divide

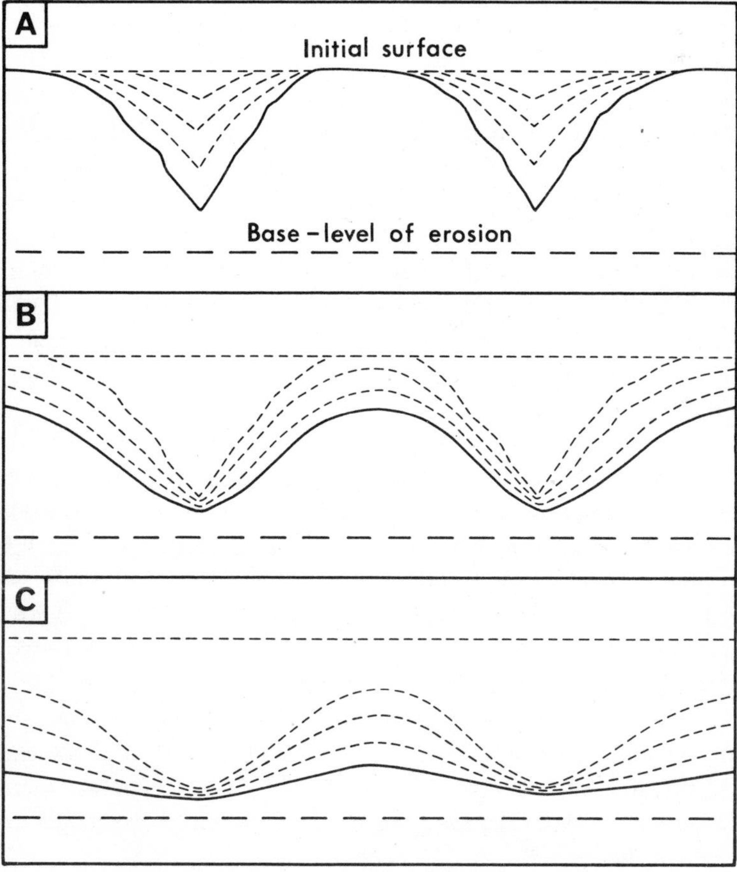

71 The Davisian cycle of erosion: youth (A); maturity (B); old age (C)

wasting', and smoothly curving slope profiles, with no major breaks, would dominate the landscape. Davis argued that such slopes had attained a condition of grade, in which the balance was between the rate of production of detritus by weathering and the rate of evacuation of this debris by slope transporting agents. Graded slopes would therefore be characterised neither by bare rock outcrops (indicating that evacuation was more efficient than weathering) nor by sections where the rock was deeply buried beneath regolith (indicating that weathering was more effective than evacuation), but would be covered by a soil layer of approximately uniform thickness. An important result of divide wasting during maturity would be the reduction of relief, or in other words a decrease in the vertical height separating interfluve summits and valley floors. This would be the reverse of the situation during youth, when the valley bottoms were being rapidly lowered by vertical stream incision and the initial surface, preserved on the divide summits, was as yet unaffected by slope retreat.

C *The stage of old age.* By this stage the processes of landscape evolution would have become extremely slow in operation. This 'running down' of the cycle would have resulted from (i) the gradual reduction of river gradients and an associated decline of stream energy, and (ii) the continued lowering in angle of valley-side slopes, so that creep and wash would become less and less active and a mantle of slope detritus, impeding mechanical weathering, would be extensive. By comparison with youth and maturity, the stage of old age would therefore be extremely protracted.

None the less, changes in the form of the land would still occur, albeit with infinite slowness. Rivers would continue to broaden their valleys by meandering, so producing near-level valley floors over which during times of flooding alluvium would be deposited to give broad flood-plains. Slopes would undergo further wasting, until angles were generally less than about 5°. By the end of the old age stage, the relief would almost totally have been destroyed, and the land-surface would assume the form of a very gently undulating plain, termed by Davis a 'peneplain', standing only a little above the base-level of erosion. Above the peneplain, a few isolated hills, as yet unconsumed by divide wasting, would remain. Such residuals were referred to by Davis as 'monadnocks', after the type-example, Mt Monadnock, which rises high above the level of the New England peneplain.

Since the time of Davis, some geomorphologists have argued that the peneplain should be regarded as a purely theoretical landform, on the grounds that the conditions of stable base-level needed for the completion of a full cycle of erosion cannot have persisted for a sufficiently long period of time. There is certainly much evidence to show that gentle earth-movements, involving both elevation and depression, are taking place today, and during the orogenic periods of the past crustal instability must have been far greater. Another argument is that, when a land-mass is undergoing erosion, it will tend to experience continuous uplift, simply because the 'unloading' will initiate compensatory isostatic movements. As a result rivers will always be incising their valleys, and attainment of the peneplain stage will be postponed indefinitely. It is true that, in an area such as the British Isles, perfectly preserved surfaces of peneplanation do not exist at or near present sea-level. However, it must be remembered that important earth-movements, associated with the Alpine orogeny and dating approximately from the middle of the Tertiary era, have affected much or all of Britain, and that during the Quaternary many eustatic changes of sea-level occurred (pp. 421–8). Indeed, the sea around the British Isles has stood at its present level for less than 10,000 years. None the less, there are some parts of the country which resemble the peneplain form, notably areas of weak Weald Clay in Kent and Sussex, Kimmeridge Clay and Oxford Clay lowlands in the south Midlands, and Keuper Marl and Lower Lias Clay in the Severn Vale. There are also reasons to believe that extensive peneplains, of considerable age, do occur in Britain, but in an uplifted and dissected state or beneath coverings of younger rock (chapter 8). Elsewhere in the world, and especially in Africa and Australia, both of which have had a far more stable geological history than the British Isles, very widespread and remarkably perfect plains of erosion have been identified.

Finally, a word must be added on the amount of time needed for the completion of the erosion cycle. Obviously it is impossible to lay down any hard-and-fast rules. So much will depend on (i) the precise amount of uplift which initiated the cycle and predetermined the volume of rock to be eroded away in the fashioning of the peneplain, (ii) the prevailing climatic conditions, which will influence the rate and type of weathering, and the frequency, volume and erosive energy of surface streams, and (iii) the resistance of the underlying rocks. It will be readily apparent that peneplanation of the highly durable Palaeozoic rocks of an area such as Wales will take far longer than that of the much weaker Mesozoic clays, sandstones, limestones and chalk of south-east England.

In fact, there are reasons to believe that the Hercynian folds of Wales, formed during late-Carboniferous and early-Permian times, had not been 'base-levelled' until the beginning of the Jurassic period, some 60–70 million years later. The English Chalk country, by contrast, may well have been effectively peneplained during the Pliocene period, which lasted only some 10–15 million years (pp. 276–8).

Rejuvenation

To judge from the available evidence it seems probable that, despite the theoretical objections which have been stated above, several cycles of erosion have run their full course, some of them during the comparatively brief Tertiary era (chapter 8). It is also clear that, for reasons of crustal and/or sea-level instability, an even larger number of cycles have been constantly interrupted by base-level changes. Often new uplifts must have occurred before the landscape had evolved beyond the stage of youth or early maturity. The landforms resulting from such rejuvenation are briefly considered on pp. 59–66. Landscapes which bear the imprint of several successive rejuvenations are variously referred to as 'multi-cyclic', 'poly-cyclic' or 'poly-phase'. In Britain there is hardly a valley of any size which does not exhibit a number of knickpoints, flights of terraces or benches, and other signs of spasmodic incision related to the changes of base-level in late-Tertiary and Quaternary times.

The process of rejuvenation can also affect a fully peneplained landscape that experiences uplift, leading to revival of the erosive powers of the streams flowing over it. In this case, the existing cycle is not merely retarded or modified, but a wholly new cycle of erosion is begun. The pre-existing peneplain will be deeply trenched by the V-shaped valleys typical of youth, but at least during the early stages of the new cycle remnants of the old surface will survive on hill- and plateau-tops and on the broader interfluves. By the stage of late youth, however, the old peneplain will be represented only by a general accordance of summit heights; and in early maturity, the renewed onset of effective divide wasting will cause the peneplain to disappear altogether as an element in the landscape. As stated already, most peneplains in the British Isles take the form of 'hill-top surfaces', and are so fragmented that they are by no means easy to identify, let alone interpret accurately. Indeed, their existence is sometimes more readily inferred from an analysis of detailed topographical maps, which contain information as to the heights of all summits over areas so wide that they cannot properly be encompassed from a viewpoint in the field.

Climate interruptions to the cycle of erosion

The interruptions to the cycle which have been considered so far are induced either by structural movements or by eustatic changes of sea-level. Davis, however, recognised that the cycle could be profoundly affected also by climatic changes, and that in certain circumstances erosion would take a wholly different course from that outlined in the 'normal' cycle.

As is shown in chapter 2, the climatic régime will exert a great influence on the process of denudation at work in any region. Davis himself considered that the modifications imposed on the normal cycle by glacial and arid climates (referred to by him as 'climatic accidents') were so great as to merit special attention. It does, however, seem a somewhat theoretical task to imagine a cycle of glacial erosion, since the Quaternary glaciations—the only ones to have left a mark on the present landscape—occupied a period of only $1\frac{1}{2}$ million years (or considerably less if due allowance is made for warmer interglacial conditions). Although glacial erosion has achieved striking effects in upland areas, the stage of youth (indicated, in Davis's view, by irregular valley profiles with basins and steps) cannot generally have been passed, and certainly old age glacial landforms cannot be recognised. At first sight it may be tempting to regard heavily glaciated lowlands such as the Laurentian and Fenno-Scandian Shields as glacial peneplains. However, it is in fact clear that these are ancient surfaces of peneplanation, produced under quite different climatic conditions in early geological times, subsequently preserved over a very long period by overlying deposits, recently exhumed by erosion, and only slightly modified by the Quaternary ice-sheets (pp. 360–1).

With desert landscapes the situation is altogether different, for it seems probable that some parts of the earth's surface have experienced arid conditions over sufficiently long periods for several desert cycles to run their full courses. Furthermore, in the past aridity was not confined to the present-day deserts, but was characteristic of many areas which today have humid, and even humid temperate, climates. In the British Isles 'desert' erosion was particularly active in Permo-Triassic times, and may also have affected any land areas that escaped submergence by the 'Chalk sea'. One could make out a good case for regarding arid and semi-arid conditions as being more normal, in the long run and over the surface of the earth as a whole, than the humid conditions envisaged by Davis. However, even if he did not appear to recognise the true importance of arid regions, Davis at least realised the need to conceive of a special cycle of desert erosion. The scheme he devised was based primarily on his knowledge of the landforms of the deserts of the south-west U.S.A. Here the geological structure consists largely of depressed and elevated fault-blocks, giving rise to so-called 'basin-and-range country'. The main stages in the evolution of such a landscape were considered by Davis to be as follows:

A *The stage of youth.* In this, the initial surface would be made up of basins, often totally enclosed, fault-scarps and plateaus, which would be attacked by mechanical weathering and surface torrents nourished by occasional rainstorms. The scarps would be gradually driven back, and dissected by gorges and canyons, whilst the ephemeral streams would flow towards the low-lying areas, forming in many instances centripetal drainage systems. Within the basins, great masses of alluvium would be deposited, and the basin floors thus progressively raised. There would be two notable contrasts with the normal cycle of erosion. Firstly, each individual basin would tend to act as a local base-level for the streams draining into it, and in the absence of a fully integrated drainage system there would be no general base-level of erosion. Secondly, since the plateau surfaces would remain at a constant elevation, except where locally dissected by valleys, and the basin floors would be built up, there would be at first a *decrease* in relief.

B *The stage of maturity.* By this stage a measure of drainage integration would have been achieved, for streams running into one basin would under favourable circumstances have eroded headwards to capture the drainage of an adjacent higher basin. Throughout maturity, with the continuation of this process, the numerous local base-levels would be of decreasing importance, and in some individual basins alluviation would be replaced by erosional regrading to a lower level. This would often involve the removal of previously accumulated sediments, and the rejuvenation of existing rock pediments on the margins of the basins must also have occurred at times (pp. 312–14). Locally, increases of relief might therefore be effected during maturity. However, erosion of the plateaus would also continue, and in this way the initial surface would be gradually consumed, residual hills would become common, and rock pediments become ever more extensive.

C *The stage of old age.* As a result of the processes described, the landscape would by now have been reduced to a series of alluvial plains, marking in the main the sites of the deepest of the original basins, and rocky plains surmounted by monadnock-type hills would occupy the sites of the former upland blocks. As we have seen, the integration of drainage would have progressed throughout maturity and perhaps have become virtually complete, so that the general base-level of erosion would at last make its influence fully felt. However, during the final stages of the desert cycle increasing aridity—resulting from a diminution of relief rainfall—would again lead to a fragmentation of the stream pattern. Indeed, it is possible that the final fashioning of the desert peneplain could be achieved by wind deflation, which would rework areas of alluvium into extensive sand-dunes, or even export the material altogether to reveal an ever-increasing area of bare rock.

MODIFICATIONS OF THE DAVISIAN CYCLE OF EROSION

It is hardly surprising that, since the beginning of this century, many prominent geomorphologists have proposed views either slightly or radically different from those of Davis, particularly regarding the detailed manner of operation of the erosion cycle. Some have gone so far as to invent wholly new cycles, whilst others have argued the case for the abandonment altogether of the cyclic approach to landform study. The views of the latter are discussed at the end of this chapter.

The cycle of panplanation

In 1933 Crickmay advocated that some revision of Davis's views, particularly of landform evolution in the later stages of the erosion cycle, was urgently needed. Indeed, Crickmay argued that the process of divide wasting, involving the weathering away of interfluves and a progressive decline of slope angles, was not of primary importance in the fashioning of late-stage landforms. Three main lines of evidence were cited by Crickmay. Firstly, he noted that the slopes of residual hills (monadnocks) are often steep and concave in profile, indicating that decline has not occurred and that protracted soil creep has not rounded the hill-summits to give dominantly convex profiles. Secondly, many surfaces which have been identified as peneplains, including the most famous example of all in New England, are not in fact gently undulating plains such as might result from divide wasting but are almost perfectly level surfaces, or in other words true plains of erosion. Thirdly, it is quite common to find in one area remains of dissected peneplains occurring at various levels, or 'in series', despite the fact that according to Davisian theory the development of a lower surface should inevitably involve the wasting and destruction of a higher surface. Crickmay himself suggested that all these features could best be explained if lateral planation by meandering streams, rather than downwasting, were assumed to be the dominant process in the later stages of the erosion cycle. In this way, true plains could be produced, and the interfluves consumed by backwearing, induced by stream undercutting, or reduced to steep-sided monadnocks. The near-perfect surface which would be formed was referred to by Crickmay as a 'panplain', and the whole process of lateral planation described as 'panplanation'.

Crickmay's theory, though based on compelling arguments, has not commanded wide acceptance, at least as an explanation of landform evolution in humid-temperate areas. One reason is that the present-day features that may be attributed to lateral erosion (such as river terraces or broad flood-plains that are veneered by alluvium) are not of any great extent. Another more telling reason is that the evidence invoked by Crickmay may be explained in another way (see the discussion of pediplanation on pp. 170–4). However, it must be added that many authorities have regarded lateral planation by running water as a major factor in the formation of desert surfaces (chapter 9), and a process akin to panplanation has been assumed by some to account for savanna plains.

The cycle of marine erosion

During the nineteenth century it was firmly held by British geologists that wide surfaces of planation could only be the product of wave erosion. Davis, on the other hand, argued that marine action could only have limited effects by comparison with sub-aerial denudation, which can operate over the whole of the land-surface instead of merely along its margins. During and after the First World War the American geographer D. W. Johnson did much to focus interest once again on the importance of coastal processes. Indeed, Johnson argued that, since the very largest ocean waves can produce some disturbance on the sea-bed down to a considerable depth, it is at least theoretically possible to imagine a marine peneplain forming over an immensely long period of time well below sea-level.

According to Johnson the youthful stage of the marine cycle would operate somewhat differently from one locality to another, depending on the detailed configuration and gradient of the initial coastline and on whether the cycle was begun as a result of submergence or emergence of the land-mass. For example, a submerged upland coast, much indented, would be characterised during youth by concentrated erosion of headlands, and the material thus released would be moved by longshore drift into bays and inlets where a wide variety of beaches, spits and bars would be formed. Gradually the shoreline would be straightened in plan, and in the mature and old age stages of the cycle cliff recession, leading to the eventual destruction of the land-mass and its replacement by an extensive submarine platform, would be the dominant process. By way of contrast, on an emerged lowland coast the youthful stage would involve the development of offshore bars, supplied with material eroded from the sea-bed by waves breaking well offshore. These bars would enclose lagoons, which in time would be silted up to give mature salt-marshes. Actual erosion of the land-mass itself, protected by these features of accretion, would not begin until the stage of maturity had been reached, when the bars would have been driven 'inland' over the marshes, the latter destroyed by wave attack, and the initial shoreline exposed to waves that no longer expended the greater part of their energy on the shallow sea-bed. Thereafter, the development of cliffs and abrasion platforms would, as in the first example, proceed uninterruptedly.

Despite the persuasiveness of some of Johnson's arguments, it is still not wholly clear whether the cycle of marine erosion has more than theoretical value. Many geomorphologists have argued that wave erosion can only be really effective when the sea-level is rising, so that the shoreline migrates steadily inland and the results of wave attack spread over a wide area. If there is no change of level, the wave-cut platforms may not grow to a width of much more than a mile (pp. 443–4), since wave energy will be almost entirely dissipated through friction with the sea-bed and erosion at the base of the cliff will grow weaker and weaker. None the less, it must be admitted that there is abundant evidence of some planation by the sea. For example, around the coasts of western Britain there are very obvious raised coastal platforms, at up to 600 ft O.D., which are cut across very resistant and often strongly folded Palaeozoic rocks. These have been attributed by most authorities to episodes of marine erosion in the Pliocene period, when the land stood at a lower level in relation to the sea. Again, there is evidence of an important transgression of the sea, affecting much of lowland England up to a height of 690 ft, in late-Pliocene and early-Pleistocene times (pp. 278–9). It has been suggested that this sea achieved the planation of quite wide areas, as in the Chalk country of central southern England. However, there must be some dispute on this point, for remaining fragments of the 'Calabrian' bench are rarely more than a mile or two across and in any case the transgression was induced by a rise of sea-level, perhaps of 300 ft or more. In some areas very extensive and perfect plains of erosion are found on which deposits of marine origin rest. In the Massif Central of France, the so-called 'Post-Hercynian' surface has in many localities been exhumed from beneath a cover of sedimentary rocks, mainly Jurassic marine limestones. However, it is apparent, from the occurrence on the surface of an ancient regolith of sub-aerially decomposed rock, that this is not a marine peneplain. It was in existence prior to Jurassic times, and was slowly submerged beneath the calm waters of the Jurassic sea, which at best locally trimmed the surface and for the most part did nothing to damage its essential character. The sub-Chalk surface of southern England, transgressed by the upper Cretaceous sea, almost certainly originated in a similar way (p. 275). Indeed, it will be obvious that extensive marine transgressions are always likely to affect land areas which have already been base-levelled, so that only a slight rise is needed for broad tracts of the land to be inundated.

The cycle of pediplanation

In 1948 the South African geomorphologist L. C. King proposed a wholly new cycle of erosion to account for the remarkable surfaces of planation, surmounted by isolated hills (inselbergs) and piles of rocky boulders (castle koppies), that are such an obvious feature of the landscape in arid, semi-arid and savanna parts of Africa. It had long since become apparent that the arid cycle of erosion put forward by Davis was inadequate to explain the erosional plains of the African deserts (and, indeed, those of other regions such as Arabia and Australia too). For the most part it could be clearly shown that the current cycle had not been initiated by block-faulting, as in the south-west U.S.A., but simply by the uplift of previously formed planation surfaces. In this context, it must be emphasised that many parts of Africa have not been affected by folding or faulting (with the notable exception of the Rift Valley zone of the east) since very early geological times. Much of the continent is in fact underlain by extremely ancient rocks, mainly of igneous or metamorphic origin, forming what is known as a 'basal-complex', which must have been affected during Mesozoic and Tertiary times by a whole succession of erosional episodes.

72 A scarp and pediplain landscape, the Karamoja district, Uganda. [G. R. Siviour]

King's examination of the African landscape led him to the conclusion that it was made up of two basic elements (Figs. 72 and 127). Firstly, in the valley bottoms and flanking either existing streams or old watercourses, gently concave slopes, varying in angle from $\frac{1}{4}°$ to 7° and cut into solid rock, were very common features. These rock pediments occur, in fact, in many parts of the world, and have excited much geomorphological controversy, particularly in the U.S.A. (pp. 307–8). The available evidence suggests that, with the passage of time, these low-angled slopes are gradually extended in area at the expense of adjacent uplands, from the steep marginal slopes of which they are normally separated by a sharp 'knick'. The processes involved in pediment formation are collectively referred to as 'pedimentation'. The second main element in the African landscape comprises the steep slopes bounding most upland blocks. These are termed by King 'scarps'—but in this context the term has no structural connotation (or in other words these steep slopes do not form component parts of cuestas and are not associated with faults). The scarps, wholly erosional in origin and usually varying in angle between 15° and 30°, appear not to undergo any decline in angle as they are wasted back by weathering and rainwash.

This process of parallel retreat is referred to by King himself as 'scarp retreat' or 'backwearing'.

The twin processes of pedimentation and scarp retreat are considered by King to combine in the cycle of pediplanation. The main stages of this cycle are as follows:

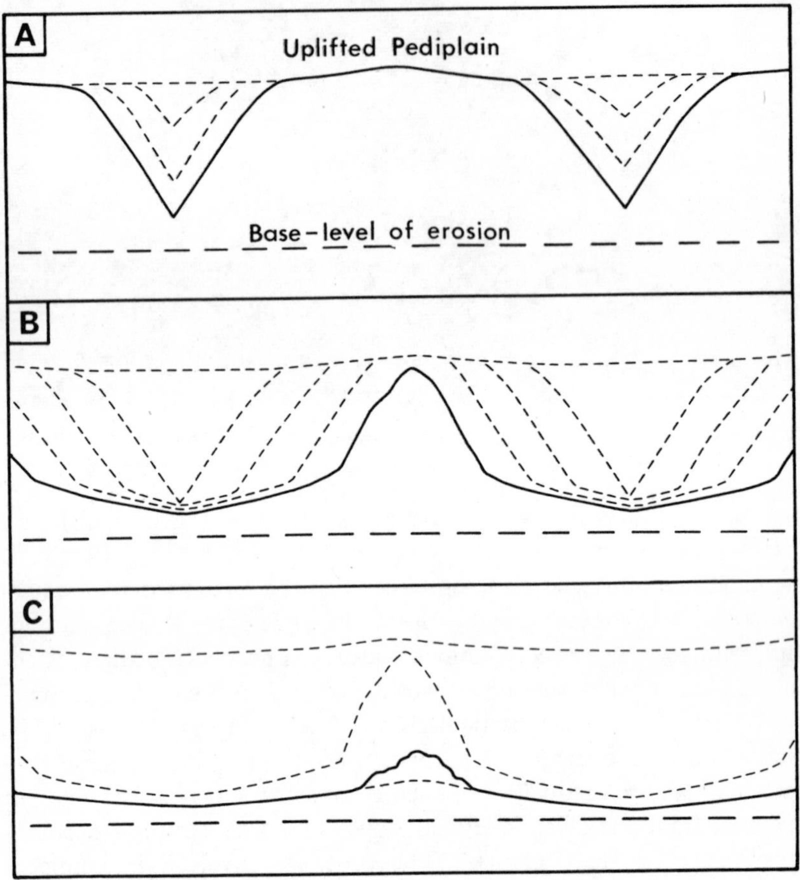

73 The cycle of pediplanation: youth (A); maturity (B); old age (C)

A *The stage of youth.* The cycle of pediplanation is initiated by the uplift of a previously formed pediplain, rather than by structural movements of a more complex kind. Existing streams, either permanent or ephemeral, cut rapidly downwards, much as in the Davisian cycle of normal erosion, towards the new base-level. In areas of granite or gneissic rocks, common in Africa, the pattern of valleys so formed may be crudely rectangular, reflecting the influence of major joint-lines. Eventually, downcutting will become less active, and small pediments will begin to appear in the valley bottoms. These will become more extended as interfluve and upland areas are consumed by scarp retreat (Fig. 73). It is important to remember, however, that remnants of the original pediplain will exist on all summits. By the stage of late youth, many interfluves will have already been converted to inselbergs, many of which would take the form of remarkably rounded domes ('bornhardts'), and castle koppies (Fig. 74).

B *The stage of maturity.* This will see a progressive reduction in the number of inselbergs, as these are weathered into koppies and finally destroyed, and the widening pediments of adjacent valleys will begin to coalesce. However, even during late maturity the summits of the few remaining inselbergs, which may tower as much as 1500 ft above the surrounding lowlands, frequently preserve small vestiges of the former

74 A granite bornhardt, Uganda. The massively jointed granite dome has been reduced by weathering, and will eventually be transformed into a koppie. [G. R. Siviour]

pediplain. Thus relief, which had been increased during early youth by valley incision, will either be decreased or locally remain constant during the mature stage.

C *The stage of old age.* By now the residual hills will have become very rare, and everywhere the relief will at last have been diminished. The whole landscape will now be dominated by low-angled pediments; the 'multi-concave' surface is the ultimate form of the cycle, the pediplain itself.

The theory of pediplanation in fact seems to explain satisfactorily all the problem features noted by Crickmay, without invoking lateral stream planation as a process of any importance. It is not surprising to find that King has gone on to apply his concept not only to the African landscape, but also to regions which today experience climatic conditions quite different from those of Africa and which exhibit 'peneplains' not readily accounted for by the Davisian theory. It does not necessarily follow that these surfaces were shaped under arid or savanna conditions—though there is strong evidence that the Tertiary plains of western Europe were formed under what may be loosely described as 'tropical' and seasonally humid climates, and that even earlier surfaces here (for example, those dating from the Triassic period) were the product of desert erosion. Rather King is implying that pediplanation is the basic process of landform evolution, operating under all but glacial conditions. This concept, with its acceptance of 'climatic uniformitarianism', is needless to say not readily acceptable to those who believe that slope form and development are closely influenced by the prevailing climatic conditions.

Another important aspect of the pediplanation concept is as follows. The cycle of pediplanation, like the Davisian cycle which King wished to supplant, is initiated by a fall of sea-level. In a large continent such as Africa, the pediplain will be first completed successfully near the coast, and only very slowly extended inland, via individual drainage basins, by the backwearing process. This may take a very long period of geological time—indeed King has suggested that certain inland surfaces of Africa, apparently developed quite recently, are actually related to a major cycle of pediplanation (the so-called 'African' cycle) which began to operate, after continental drift had roughed out the African continent, some 120 million years ago, in the Cretaceous period. This great duration of the cycle of pediplanation may be contrasted with those suggested for the Davisian cycle earlier in this chapter.

The cycle of savanna erosion

Some geomorphologists have contended that the landscapes of savanna lands, which experience high annual temperatures and contrasting rainy and dry seasons, must be differentiated from those of other climates in terms of their cyclic evolution. For instance, Cotton (1942) stated that the savanna is characterised by true plains of erosion (evidently the result of powerful lateral corrasion, rather as in Crickmay's cycle of panplanation) surmounted by steep-sided and often dome-like inselbergs. In his view, such 'inselberg-and-plain' landscapes rarely exhibit rock pediments, which seem to be landforms developing only under conditions of greater aridity. As we have seen, this is refuted by King, who claims to have found pediments in abundance in the savanna areas of Africa, and who has accordingly invoked pedimentation as one of the two dominant processes forming the savanna landscape.

King's apparently convincing application of the pediplanation concept to savanna landforms is, however, at present under some attack. A third basic process at work in these areas, of which he seems to have taken insufficient note, is intense chemical weathering, promoted by the high prevailing temperatures and the seasonal abundance of surface and ground water. This weathering, particularly in areas of low relief, may penetrate to considerable and sometimes remarkable depths. For example, it has been suggested that over much of Uganda the underlying rock may be rotted to an average depth of between 100 and 150 ft, and a depth approaching 200 ft has been recorded recently in the Jos plateau of Nigeria. This weathered layer commonly assumes one of two forms: either it is an imperfectly decomposed zone, comprising numerous rounded core-stones or woolsacks (p. 26), or it is a more completely rotted layer, consisting of clayey soil and 'soft' rock which is separated at the base from fresh 'hard' rock by a well-defined line of contact, the 'basal weathering surface'. In detail, the form of this basal surface is complex and irregular. 'Basins' occur where the rock is strongly jointed (and therefore more prone to decomposition by percolating acidulated rainwater) or where the rock minerals are chemically unstable, whereas 'domical rises' are associated with areas of greater resistance. The subsurface domes sometimes lie at no great depth, or are even exposed at the surface, in which case their summits are affected by exfoliation weathering. In form and dimensions there is often a very close resemblance between subsurface and partially revealed domes and also the rounded bornhardts which stand high above many savanna plains. It is significant that whereas these inselbergs are formed of sound unweathered rock (except at their outer margins where exfoliation is operative), the adjacent lowlands are usually underlain by chemically decomposed rock.

It is now suggested by some students of savanna landforms that 'downwearing' rather than 'backwearing' is the dominant process in the evolution of plains and inselbergs. Under 'stable' conditions, chemical weathering may act uninterruptedly, producing a deep rotted or partially rotted layer, the surface of which is removed only very slowly by wet season streams and floods. However, falls of base-level, either local or regional, or even climatic change can lead to a great increase in the effectiveness of stream 'erosion' and transportation, and the weathered layer may then be partially or perhaps even wholly removed. The result will be some exposure of the basal weathering surface, particularly where there are pronounced domical rises. The latter will, in fact, be revealed as incipient inselbergs, and their dome-like appearance will be further accentuated by exfoliation weathering. It is possible that over a long period the inselbergs may be reduced to 'koppies' or 'tors' (Fig. 74), but there is evidence to suggest that these are more usually accumulations of core-stones which have been left on the surface by the washing away of the finer weathering products (Fig. 55). Indeed, it seems more likely that once subsurface domes are revealed, far from being reduced by atmospheric weathering, they are actually accentuated with the passage of time. This is because rainwater will tend to run quickly off the bare rock surface of the dome, which will therefore dry out and not be subject to intense chemical attack. In surrounding areas, on the other hand, where the mantle of rotted rock has not been removed, the water will sink into the ground and, if anything, lead to an acceleration of subsurface weathering. One widely noted feature of the savanna landscape is the occurrence of 'linear' or 'marginal' depressions at the bases of scarps and inselbergs. These are usually presumed to mark zones of concentrated rock decomposition which have subsequently been etched out by surface streams or springs. It is now thought that, as a result of the mechanisms described, inselbergs are virtually indestructible landforms, which are left standing higher and higher as the surrounding plains are lowered by successive phases of weathering and removal (Figs. 75 and 76).

Pugh (1966) has recently argued that in savanna areas the cycle of erosion may operate in the manner depicted in Fig. 77. In the first place, an initial surface eroded across weathered and fresh rock alike is assumed (A). With a fall in base-level, the rotted rock is partially removed by

75 A bornhardt dome in northern Nigeria. Note the curvilinear profile, related to pressure-release joints, the massive exfoliation, and the sharp break of gradient at the foot of the dome. [M. F. Thomas]

stream incision and downwearing, giving an incipient plain above which former subsurface domes rise as inselbergs (B). Once in existence, the plain may be extended laterally by scarp retreat, which is most effective in the decomposed rock. It may also lead to the obliteration of some minor bornhardts, but will have comparatively little effect on the major residuals (C). A further uplift will lead to a second cycle of downwearing and exhumation, the initiation of a second lower plain, and the development of 'multi-cyclic' or 'dome-on-dome' residuals (D). The role of pedimentation in this whole process seems to be a minor one, despite King's observation that all bornhardts in Africa are surrounded by pediment slopes. In fact, what are evidently pediments in savanna areas are usually underlain by a considerable depth of weathered material, and in that sense they are not strictly comparable with the rock pediments of the desert. Recently Thomas (1966) has suggested that what have been identified as pediments in the Nigerian savanna are not truly basal slopes formed below a retreating scarp (as the pediplanation theory demands), but simply concave 'wash' slopes marking stages in

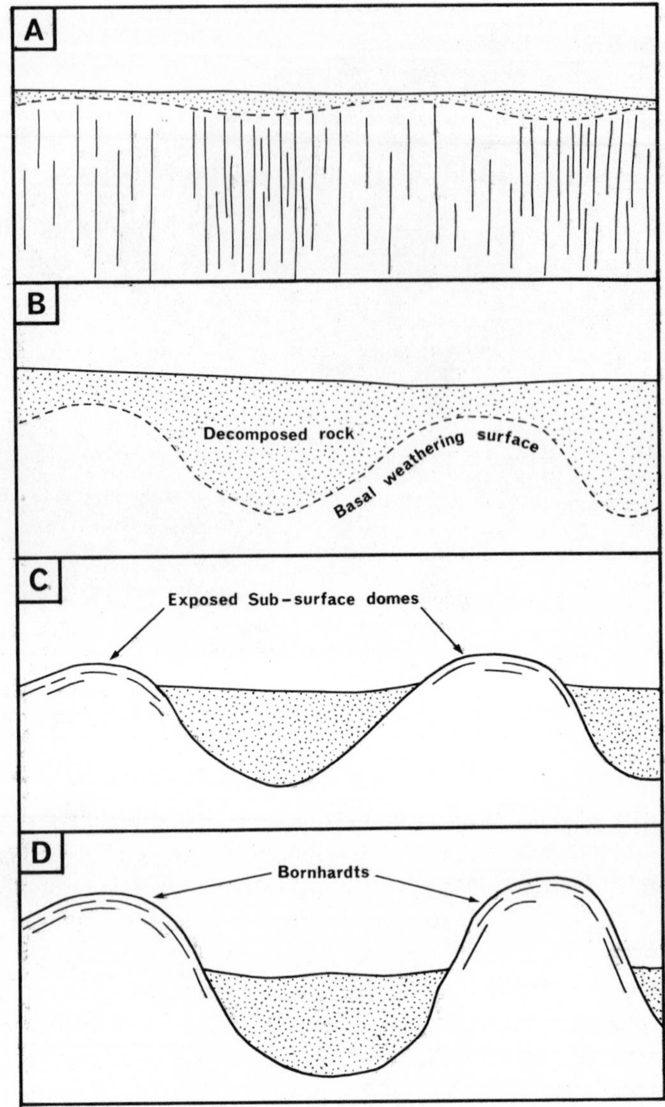

76 Deep weathering and dome formation in savanna areas

the removal of the weathered rock by downwearing processes. Thomas goes on to argue that savanna landscapes are not produced by scarp retreat, but result from a combination of etching, stripping of the basal weathering surface by streams and wash, slope retreat within the weathering profile, and weathering processes affecting rock domes and tors.

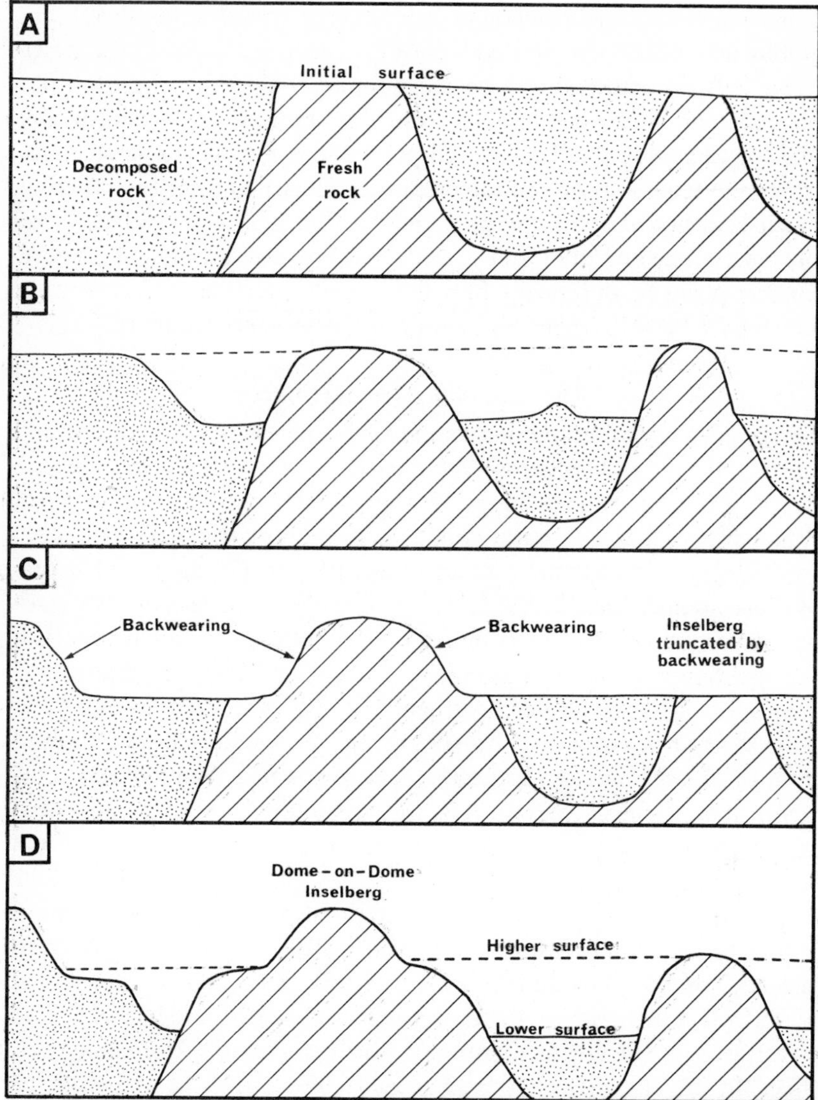

77 **The formation of dome-on-dome inselbergs (after J. C. Pugh)**

The outcome of these processes operating over a long period is not a pediplain, but an 'etchplain'.

Finally, it must be said that the processes at present observable in savanna regions may in the past have operated elsewhere. For example, around Fort Trinquet, Mauritania, where the annual rainfall is now only

2 inches, the desert landscape displays fine rounded inselbergs and granite is found rotted to a depth of 65 ft. Again, in the Australian desert chemically decomposed rock has been found at a depth of up to 75 ft, a phenomenon which has been explained as the result of a considerably moister climate in the Tertiary. In present-day humid-temperate areas, where the temperatures are too modest to favour advanced weathering, much evidence of similar deep decomposition has been described, and the German geomorphologist J. Budel has suggested that the high-level summit plains of Europe are in reality savanna surfaces of Tertiary age which have been dissected as a result of Quaternary erosion (p. 349).

The cycle of periglacial erosion

The importance of weathering and erosion in areas of high altitude and latitude today, and in areas that during the Quaternary experienced 'arctic' climates, is now properly appreciated (chapter 10). In 1950, Peltier proposed a theoretical cycle to account for periglacial landforms. In this cycle the essential process is that of 'cryoplanation' (planation by frost), which leads to the levelling of the land-surface by two component processes: frost shattering of the solid rock and its regolith ('congelifraction', derived from the Latin *congelare* = to freeze, and *fractare* = to break), and disturbance of the weathered layer by the growth of ground-ice and by solifluxion ('congeliturbation', derived from the Latin *congelare* and *turbare* = to stir up). Many of the comments that were made regarding the possibility of a cycle of glacial erosion would seem to be appropriate to the periglacial cycle. However despite the short duration of the Quaternary glacial periods some rocks that are particularly prone to frost attack seem to have been reduced locally to near-level surfaces, and even in upland areas 'altiplanation terraces', produced by differential frost attack and solifluxion, have been recognised. Furthermore, Peltier himself envisaged, quite logically, that a refrigeration of climate, initiating the periglacial cycle, could well affect an already dissected landscape, so that much of the work of levelling the surface would have already been accomplished by other processes.

A *The stage of youth.* Peltier envisaged that early in this stage intense frost action would convert the pre-existing slopes into steep bare rock faces, at an angle of 25–30° or even more and undergoing a form of parallel retreat. At the base of these 'frost-riven' slopes would develop at a distinctly lower angle (15–20°) cryoplanation surfaces, which would resemble steep pediments and over which frost debris derived from above would be moved by solifluxion to the valley bottoms (Fig. 78). This debris would tend to accumulate at the lower margins of the cryoplanation surfaces, for the reason that the existing streams, fed only by seasonal meltwater, would be insufficiently powerful to transport it away.

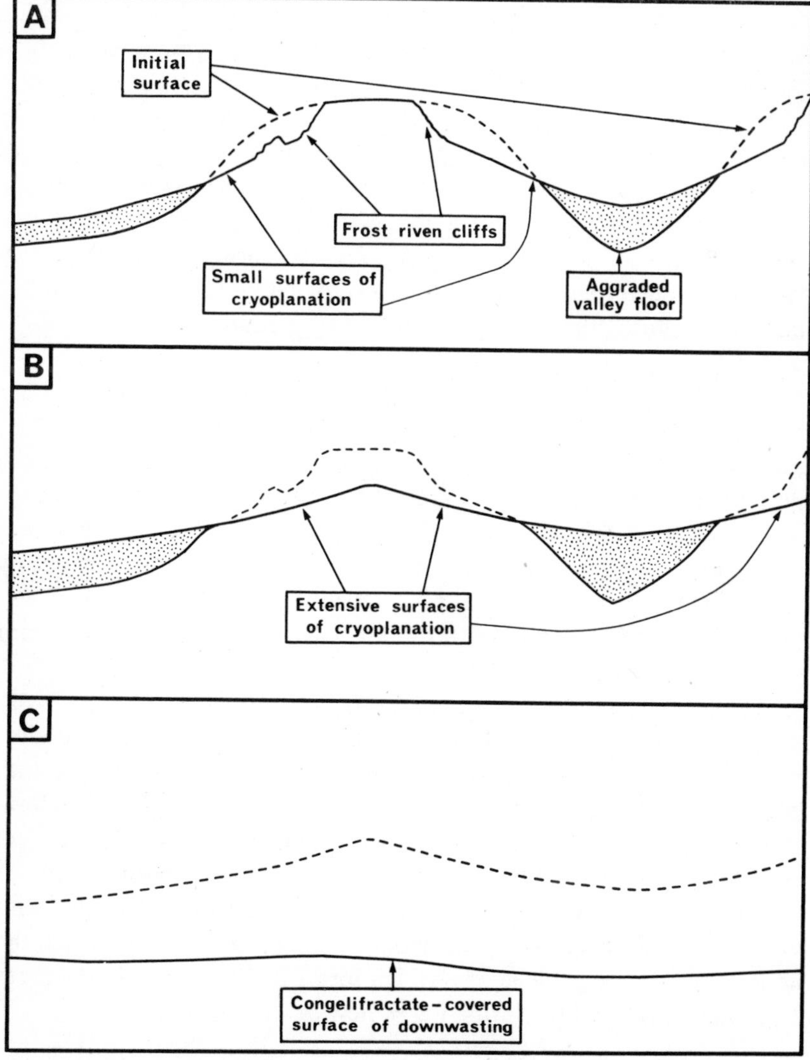

78 The cycle of periglaciation: youth (A); maturity (B); old age (C) (after L. C. Peltier)

B *The stage of maturity.* During this stage, the interfluves will be gradually consumed by the retreat of the frost-riven faces, and extensive cryoplanation surfaces will begin to emerge. Such surfaces will themselves be further lowered in angle by a combination of frost weathering and solifluxion.

C *The stage of old age.* By now, all slopes will have declined to an angle of 5° or less, and their detrital cover will have been comminuted to the state where aeolian transport can begin, 'sweeping' the cryoplanation surfaces by the process of deflation and producing 'wind-swept pebble pavements'.

The cycle of karst erosion

In the early years of this century the Jugoslavian geologist Cvijic postulated that the landforms of major limestone areas evolve in a manner that cannot be accounted for by the normal cycle of erosion. Cvijic envisaged that the karst cycle is initiated by the uplift of a thick, pure mass of limestone, underlain by an impermeable layer.

A *The stage of youth.* At first, a well-developed system of surface drainage will exist, either because prior to uplift the water-table is at or near the surface or because another impermeable stratum overlies the limestone. However, the development of sink-holes by solution (which will result from the breaching of, say, a clay cover to expose the limestone at the surface, or from a fall in the water-table brought about by stream incision) will lead to the steady disruption of this system, and dry valleys and collapse features (such as dolines and uvalas) will become common (Fig. 79).

B *The stage of maturity.* By now a fully developed system of underground drainage, utilising fissures and caverns, will have come into being, and the surface drainage will have disappeared altogether. However, the lowering of the limestone by solution processes will continue, and in the course of time underground caverns and passages will be revealed, in some instances by the collapse of their attenuated and weakened roofs. Eventually, the underground streams will approach the underlying impermeable layer—which forms in effect the base-level for karstic erosion—and the formation of new caverns will cease.

C *The stage of old age.* During this stage all remaining underground

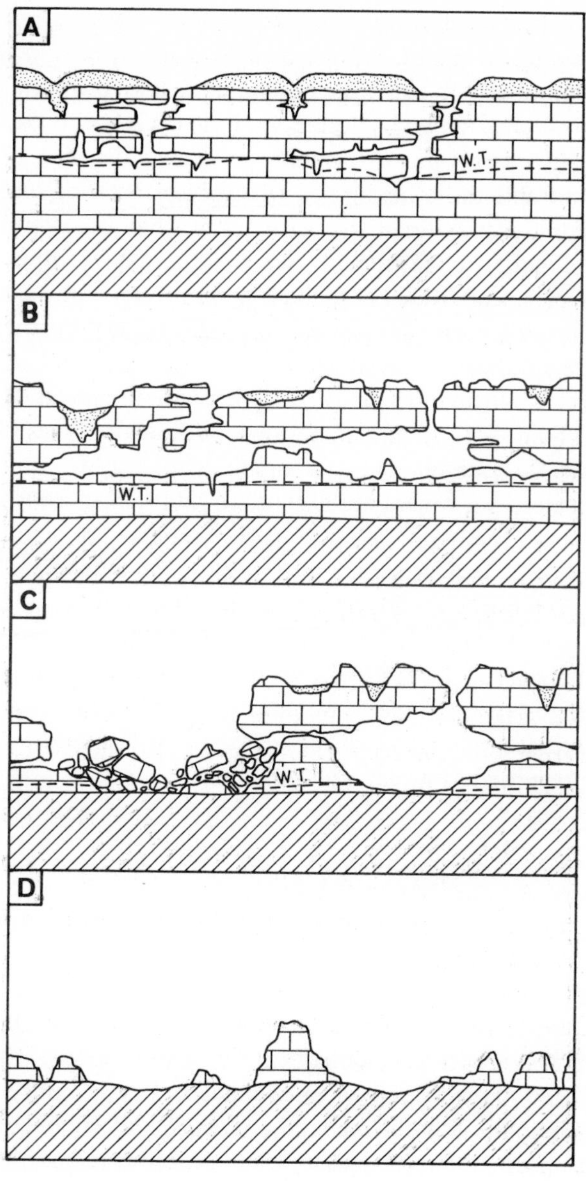

79 The cycle of karst erosion: initiation (A); youth (B); maturity (C); old age (D)

caverns will be revealed by continued surface wasting, and overland streams, flowing on the basal impermeable layer, will reappear in larger and larger numbers. In time, the remaining masses of limestone will be destroyed by corrasion and solution, and the cycle will be terminated.

A clear distinction must obviously be drawn between the cycle of karst erosion and the other cycles which have been discussed in this chapter. For the first time, the term 'cycle' has been applied to the evolution of landforms fashioned from a certain type of rock, instead of under particular conditions of climate (humid-temperate, arid, savanna or glacial) or by a particular process (marine erosion). Furthermore, the structural conditions involved are precisely defined, and in fact these are not met with in all limestone areas, where impure layers of marl often disrupt the homogeneity of the limestone. Some major limestone regions, such as the Causses of southern France (pp. 139–56), possess some karstic features, but others are absent. In addition there may be large surface rivers which do not owe their existence to underlying impermeable rocks. It is very difficult to place such landscapes within the general framework of the ideal karstic cycle. The validity of the concept must therefore be regarded as somewhat doubtful, even though it may help to explain the features of *one* great limestone area, the Dinaric Karst of Jugoslavia. A cycle is, in effect, a model, or in other words an idealised representation of reality; and unless it has wide application it cannot be regarded as a very useful model.

CRITICISMS OF THE DAVISIAN CYCLE OF EROSION

The first really serious criticisms of the Davisian approach to landform study to attract wide attention were those made in 1924, in his book *The morphological analysis of landforms*, by the German geomorphologist Penck. The theories that Penck proposed as a substitute for the cycle concept are complicated, sometimes difficult to grasp, and in some respects have been widely misunderstood. Penck himself took the unusual standpoint that the main objective of geomorphological research is to gain information that might be of assistance in the elucidation of earth-movements (or the 'diastrophic history of the earth's crust'). He considered that these movements left their mark on the landscape not only in obvious ways, by the production of folds, faults, mountain chains and so on, but also through their effects on denudational features, in particular valley-side slopes. Penck based this inference firstly on deductive reasoning and secondly on the fact that certain well-defined structural units studied in the field each possessed its own characteristic slope form (convex, rectilinear or concave) and maximum slope angle. Clearly these slopes did not merely reflect the influence of the prevailing climate (which might affect a much larger area, over which slope forms and angles might show a good deal of variation) or the local rock-type (which within the structural unit concerned might be extremely variable). In Penck's view, this phenomenon could best be explained by the assumption that the valley-side slopes were controlled primarily by the rate of river erosion—itself largely a function of the rate of uplift which the structural units were experiencing.

Penck argued that Davis's misconceptions about the evolution of landforms in general, and of slopes in particular, stemmed from his basic tenet that the cycle of erosion is initiated by a very rapid uplift, and that thereafter occurs a period of structural stability during which erosion can proceed without hindrance. In reality, the uplift of land-masses may be very protracted, and more often than not is contemporaneous with erosion. Uplift must inevitably affect the course of that erosion and the landforms produced. It is true, of course, that Davis was well aware that his assumption of initial uplift was oversimple—though a very convenient one for the purpose of demonstrating the main principles of his cycle concept; but it is equally true (i) that although Davis frequently admitted that some uplift could continue after the start of the cycle he did not consider in any detail the full implications of this fact, and (ii) that many geomorphologists who followed Davis applied the concept with undue rigidity. An example of the latter is the common assumption that a surface cutting across folded structures must clearly post-date the formation of those structures, when in fact erosion and diastrophism could well have been contemporaneous.

Penck's basic thesis was that slopes do not necessarily change form, in an evolutionary manner, with the passage of time (as they were assumed to in the 'slope decline' theory of Davis), but rather that slope forms—and the manner in which those slopes are altered—are determined by one overriding control, the rate of downward erosion of the rivers flowing at their bases. As is shown in more detail on pp. 210–15, Penck argued that (i) concavity of profile occurs above rivers which are incising their courses at a decelerating rate, (ii) rectilinear slopes are developed above rivers that are eroding at a constant rate, and (iii) convex slopes are formed where the rivers are cutting downwards at an accelerating rate (Fig. 80). Moreover, Penck considered that the precise rate of river erosion was of fundamental importance. Thus a river cutting down at a rapid constant rate will give rise to a steeper rectilinear slope than a river incising its course at a slow constant rate (Fig. 81).

It will be apparent that Penck's theory is not wholly incompatible with that of Davis, in the sense that Penck himself admitted that, in the case of decelerating erosion or in the total absence of river erosion (a condition that might be met with in the very ultimate stages of the cycle), all slopes must undergo decline and relief be diminished with the passage of time. However, it is equally clear that Penck envisaged circumstances that are divorced entirely from the cycle concept, with its emphasis on progressive change of slope form and angle. Thus, in the case referred to above of a river incising its valley at a constant rate, the valley-side slopes will change neither their rectilinear form nor their angle as time passes, nor will there be any diminution of relief (as is implicit in 'divide wasting'). The concept of the parallel retreat of slopes is commonly associated with Penck's name, though from what has been written above it will be evident that Penck by no means regarded such retreat as a universal process, operating under all conditions. Rather Penck argued that, except in the case of constant erosion, only component parts of the slope would experience parallel retreat; and that since, in the case of a slope comprising individual units of differing angle, some parts of the slope might be consumed owing to the parallel retreat of other parts, the slope form as a whole could be modified. For a fuller discussion of 'slope replacement' see pp. 212–15.

At the time of their publication Penck's views did not command wide acceptance, and for a long time afterwards were attacked strongly by leading British and American geomorphologists. Although certain of Penck's theories (notably that slope form is controlled only by the rate of river erosion, with climate, vegetation, weathering, rock-type and transportational process apparently playing subservient roles) are still regarded as untenable, others (such as the concept of parallel slope retreat, adapted by King in his cycle of pediplanation, and the theory of constant erosion) have been revived, modified and applied successfully by modern geomorphologists. In fact, since 1945 it is Davis rather than Penck who has come under fire, and in some quarters the view is held that not only should the cycle concept be abandoned, but that in the past its dominance has seriously hindered the development of geomorphology. The main criticisms that are levelled against the Davisian cycle are as follows.

Firstly, adherents of the cycle concept have been too concerned with

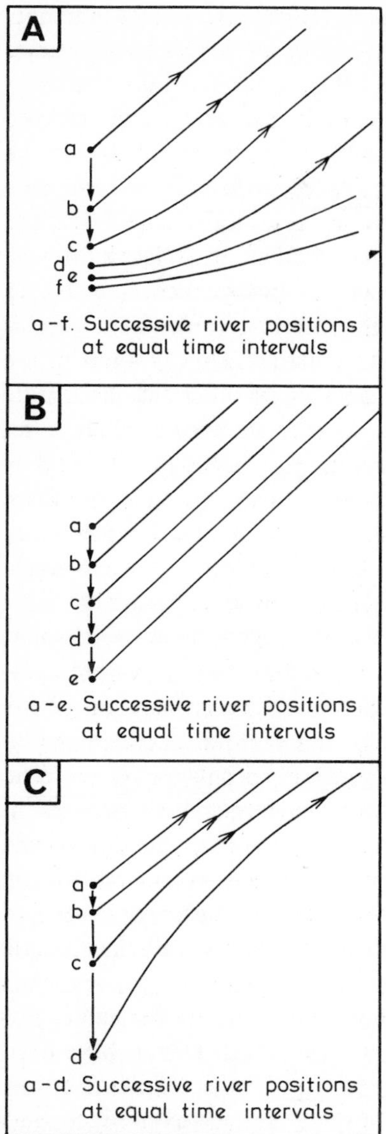

80 The development of concave slopes above rivers cutting down at a decelerating rate (A); rectilinear slopes above rivers eroding at a constant rate (B); convex slopes above rivers cutting down at an accelerating rate (C) (after W. Penck)

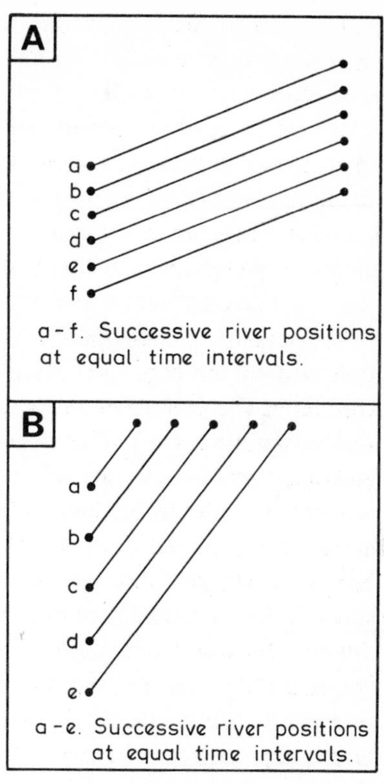

81 The relationship between slope angle and river erosion

making generalisations about landform development, without attempting to measure those landforms objectively or to make a proper study of the processes moulding them. The latter have often been 'taken as read', yet as soon as careful analysis of these processes is undertaken their great complexity is realised, and what has often been regarded as axiomatic is revealed as untrue. Secondly, acceptance of the cycle concept has often led to an over-emphasis on 'historical studies' of landforms, in which reconstruction of the development of the latter through time to their present condition is seen as the primary objective of geomorphology. The concentration of many British workers on denudation chronology (pp. 9–10) is one aspect of this. Thirdly—and in the context of the present chapter most serious of all—there has been little real attempt to show that landforms actually do evolve, along a set course, towards an inevitable end-form, the peneplain. One may certainly observe in the field slopes of a wide variety of angle, but there are usually no grounds at all for the assumption that the steeper slopes are 'younger' and the gentle slopes 'older'—indeed, within a single small valley, apparently of one 'age', slope forms and angles may differ greatly. Returning to the example given at the beginning of this chapter, we may state that it is quite possible, or even probable, that the landforms of Salisbury Plain (evidently 'older') were never like those of the South Downs (apparently 'younger'). Furthermore, it would be dangerous—and indeed indulging in needless speculation—to assume that the landscape of the South Downs will necessarily be changed in due course in such a way that it will come to resemble that of the present-day Salisbury Plain. Yet this sort of attitude is inevitable if the Davisian approach to landform study is followed. Although these two areas are similar in the sense that both have virtually the same climate and both are underlain by Chalk, the factors that have determined landform development in the past have not been precisely the same. For example, throughout the late Quaternary the South Downs have been nearer to the sea than Salisbury Plain, and the various episodes of downcutting brought about by recent falls of sea-level have naturally made a greater mark in the Downs. Again, there are even significant geological differences between the two areas. The precise horizon of the Upper Chalk which outcrops over the southern half of the South Downs (the fossil-zone of *Gonioteuthis quadrata*) is not the same as that which underlies most of the Plain (the zone of *Micraster cor-anguinum*). The Downs too are structurally more complex, and are affected by several sharply folded anticlines and synclines, whereas Salisbury Plain is made up of almost horizontal strata. It is these and other factors, rather than stage of evolution, which must be invoked in any meaningful comparison between the two areas.

Some modern geomorphologists (including Hack (1960), Strahler (1952) and Chorley (1962)) have developed the thesis that, so long as the factors which control denudational processes remain the same, the form of the land need undergo no change with time (or in other words landform evolution will not occur). Such landforms will be 'time-independent' (whereas the cycle concept, with its emphasis on stage, implies that landforms are always 'time-dependent'). In a sense this non-evolutionary interpretation was foreshadowed in the writings of Penck. As we have seen, his theory of rectilinear slope formation at a constant angle, in response to constant stream erosion, envisages no progressive change of form. It cannot be too strongly emphasised that such landforms are not to be regarded as static; they will continue to be affected by weathering and slope retreat, and the land-surface as a whole will be lowered, but throughout the whole process no change in the dimensions and angles of the landforms will be effected.

These arguments stem largely from an entirely new theory of landform development, known as 'dynamic equilibrium'. In this it is assumed that all aspects of the landform geometry of an area (such as relief, slope length, mean and maximum slope angles, channel gradient, and so on) are intimately related to each other. Indeed, it has recently been suggested that in favourable circumstances *all* the main components of a landscape (valley bottoms, valley-side slopes and divide summits) will experience lowering at a uniform rate. In order that this should happen, a delicate condition of 'energy balance' (hence the term dynamic equilibrium) must be established. On a slope, for instance, the 'energy input' (related to the intensity of the active denudational processes) must be such as to produce the continuous and efficient evacuation of all detrital material; such a slope, like a stream in a condition of quasi-equilibrium (p. 59), is referred to in modern terminology as a 'system in a steady state'. In order that the energy input should be just sufficient, the slope itself will undergo adjustment of form until, under the prevailing conditions, its height and angle are 'correct'. A relevant example of the application of the dynamic equilibrium theory may be seen in the quotation, referring to landform development in quartzite and soft shale, on p. 114.

It will be apparent that the energy balance of a landform or landscape will be affected by a number of 'controlling factors'. Among these are rock-type, jointing, angle of dip, permeability, climate, vegetation, rate of uplift, and so on. Since these factors are not necessarily constant, so the conditions of balance may from time to time be altered, together with the form of the land itself. However, there may well be times when no such changes occur, in which case the form of the land-surface will remain constant or time-independent. Furthermore, when changes do occur they need not result in landforms evolving always in the same direction towards an inevitable form. Instead, the landforms will merely undergo adjustment to meet the requirements of the new conditions.

The theory of dynamic equilibrium may be further illustrated by a very simple example. Fig. 82 depicts an area underlain by three horizontal strata, the upper and lower comprising resistant sandstone and the middle composed of weak clay. It will be assumed for convenience

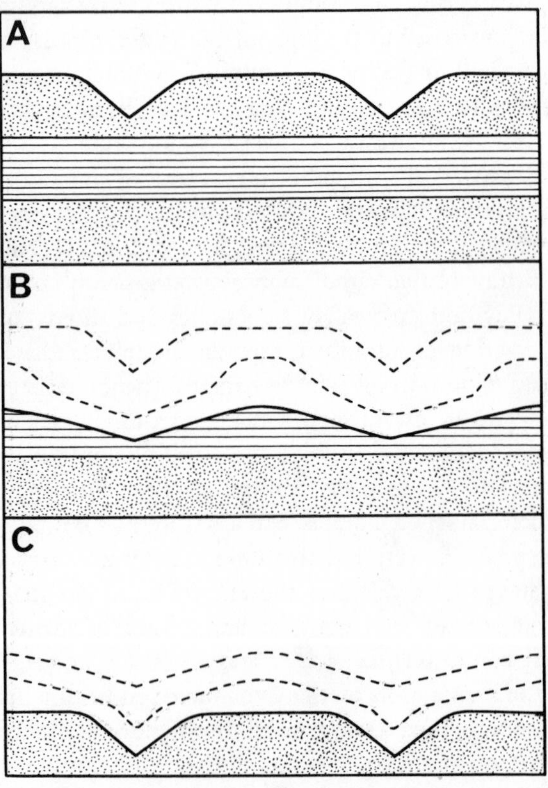

82 Landform development in horizontal rocks according to dynamic equilibrium theory

that all controlling factors other than rock-type remain constant. At first, the land surface will be developed only in the upper sandstone (A). Because of the resistance of the rock, the valleys will be narrow, the relief considerable, and the valley-side slopes steep. These features will remain unchanged so long as only the upper sandstone is exposed. In time, however, this will be removed by river erosion and slope retreat. In the now uncovered clay, because of the lack of resistance the valleys will become broader, relief will be reduced, and the valley-side slopes will become gentle (B). Finally, with the removal of the clay and the exposure of the lower sandstone, the landscape will revert to its original form (C). This sequence of changes is non-evolutionary. What has happened is that the landscape has merely adjusted its form to maintain a state of equilibrium in relation to the controlling factors. In this particular instance, rock-type—the only one to change—was in effect the most influential. Similar effects might well have been produced in, say, an area of homogeneous rocks by a change in climate or changes in the rate of stream erosion. In reality, of course, the changes that do occur are more complicated, and several factors may be undergoing alteration concurrently. Theoretically at least, it is possible for these changes to counterbalance each other, and for the form of the land to remain the same.

Two final points must be discussed briefly. Firstly, it was assumed in the example described that no impediment to downward river corrasion existed, so that the rivers were able to occupy successively positions within the three rock layers. However, if the structure concerned had lain close to sea-level, and were not affected by progressive uplift, this would not have been possible. The rivers might, for instance, have penetrated only to the clay before proximity to base-level halted downward erosion. However, the clay interfluves would still be subject to weathering, creep and wash, and a reduction of relief involving an eventual decline of slope angle would be inevitable, as in the Davisian cycle. On the one hand it could be argued that this would be a special case of the dynamic equilibrium theory, in which the stationary base-level became for once the dominant controlling factor, replacing the rock-type influence which would have remained paramount under conditions of unrestricted downward erosion. On the other hand, one could contend that, once base-level control is admitted, the cycle concept becomes valid and its total abandonment could not be justified. The crux of the matter seems to lie in whether or not stability of base-level can exist over a long period and thus impose its influence on landform development. The supporters of the dynamic equilibrium theory would evidently regard a 'fixed' base-level as a rare occurrence. It is true that, in the recent geological past, changes of base-level have been almost continuous, but it is surprising to discover how many features of the British landscape (graded river profiles and low-lying clay plains) show a close relationship to present sea-level.

Secondly, it might be considered that the widespread occurrence of erosional plains, in many parts of the world and under various climatic conditions, must vindicate the cycle concept, proving as they do that the factors which control landscape form *have* altered in such a way that landform evolution has taken place. However, even this is disputed, and protagonists of the dynamic equilibrium theory, such as the American Hack (1960), have suggested that many so-called 'peneplains' (particularly those indicated by accordance of hilltops and interfluve summits) have in fact arisen in a different way. They argue that in any area of rocks which are reasonably uniform in terms of resistance, where the stream spacing (drainage density) is uniform, and where the slopes are at the same maximum angle, it is to be expected that the summits and the divide crests will all reach to about the same height and so give the impression of a former level surface which has, subsequent to its formation, been dissected by valleys. Hack has even gone so far as to propose that such a landscape, which he refers to as 'ridge-and-ravine topography', is the *normal* expression of a condition of dynamic equilibrium (Fig. 83). There would still seem to remain the problem of explaining,

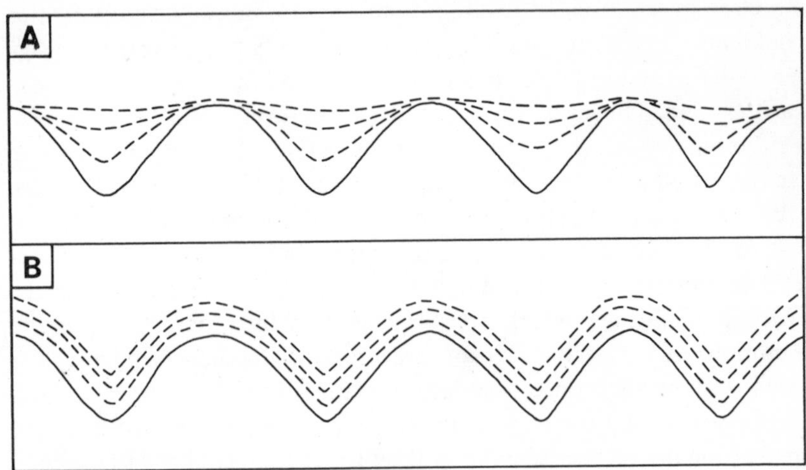

83 A dissected peneplain (A) and ridge-and-ravine topography (B)

say, the pediplains of Africa, many of which are in a remarkable state of preservation and suggest that base-levelling of the landscape on a large scale can occur. However, if these are indeed for the most part stripped basal weathering surfaces, as students of savanna areas have postulated, then to explain them one does not need to invoke a cycle of erosion in the accepted sense of that term.

Perhaps it would be most appropriate to conclude that the theory of dynamic equilibrium opens up interesting possibilities in the analysis of landforms, and that it has already served a particularly useful purpose in focusing attention on the relationship between process and form rather than the historical development of landforms. The time may or may not be ripe for the abandonment of the cycle of erosion, but there can be little disagreement that it has held sway without really serious challenge for too long.

6

SLOPE DEVELOPMENT

Most geomorphologists would probably agree that the form and origin of valley- and hill-side slopes pose the most fundamental problems in landform study. The physical landscape is obviously no more than an assemblage of such slopes, and the dimensions and appearance of the slopes give to any area its essential morphological character. At the same time it must be admitted that the study of slopes, though vital to geomorphology, has until recent times been somewhat neglected. The main reason for this probably lies in the great difficulties which are encountered in attempts to determine the nature, rate of operation, and precise effects of the processes at work on slopes, and to trace the changes of form and angle that slopes undergo with the passage of time. As a result, in no other field of geomorphology are the facts apparently quite so confusing and the conclusions reached so contradictory and widely disputed. H. Baulig's oft-quoted dictum 'truth in geomorphology, indeed, is seldom more than increasing probability' is nowhere more appropriate than in the context of slope study.

METHODS AND PROBLEMS OF SLOPE STUDY

It is impossible in the space available to do more than summarise the more basic difficulties which confront any student of slope development. From the start it needs to be emphasised that much will depend on which particular approach to slope study is being pursued, for there is no one slope problem which is readily definable and allowing of one method of investigation only. It is necessary to seek the answers to many questions, among them the following. What are the various forms assumed by slopes? How (if at all) do slopes change their form and gradient with the passing of time? What is the precise relationship between slope form and the various denudational processes operating on the slopes? Are slopes, like rivers, able to attain or even approach a state of equilibrium or grade? What is the influence of geological structure and rock-type on slope form and steepness? Is there a relationship between the prevailing climatic conditions and slope form? This list is by no means exhaustive, and the questions it contains are not of equal importance. Rightly or wrongly, geomorphologists have tended in the past to concentrate on the second and third questions, and rather surprisingly they have neglected the first, which is obviously the most basic of all and, in theory at least, relatively easy to find the answer to. In pursuing these two approaches, geomorphologists have found themselves faced by certain difficulties, now to be discussed in more detail.

The 'slope evolution approach'

This approach is concerned with tracing the historical development of the slope from its initiation to its present-day form. It also involves the forecasting of future slope form when the objective is the formulation of comprehensive theories of slope evolution. Certain problems stem immediately from this approach. Firstly, it is necessary to reconstruct the initial form from which, over a very long period, the present slope has been fashioned. But how can this be done with any certainty? Many writers have simply evaded the issue by assuming that the slope of today has been developed from a near-vertical cliff face, which has been weathered back and in the process has been modified both as regards form and steepness. Occasionally conditions like this are actually met with in nature, as when a precipitous marine cliff is cut off from further wave undercutting by a fall of sea-level or an episode of marine deposition. The majority of slopes, however, border river valleys, the sides of which were almost certainly never vertical cliffs. It is an oversimplification to assume that, in the beginning, the valley was created by downward river corrasion alone, thus creating a gorge-like form. In certain circumstances gorges are developed (where river incision is exceptionally rapid, and the rocks are so resistant that weathering back of the side walls is very slow), but it is manifestly wrong to regard the river valleys of, say, the Weald as once having been gorges. More normally, river erosion will tend to produce cliffed edges, but weathering and transportational processes will modify those edges as it proceeds, converting them almost immediately into perhaps more subdued and certainly more complex slope forms. The concept of 'initial slopes' is perhaps a totally unrealistic one. Even so, it may be of some use in the development of hypotheses (as in that of W. Penck discussed below), in which such slopes are 'assumed', though it is recognised that they never actually existed.

A second commonly encountered difficulty is that of slope dating. Ideally, the only completely satisfactory method of determining whether and/or how slope form and angle change with time is to examine numerous slope profiles in the field, discover the actual or relative age of each and place them in a 'time-sequence' (or in other words a 'developmental sequence'). In this way it might be found that the oldest slopes are the gentlest and the younger slopes appreciably more steep. A theory of slope decline could then be applied to the area of study. Unfortunately, the determination of slope age is far from easy, though in some instances it can be done rather crudely. Thus in a rejuvenated river valley the slopes of the incised section are younger than those standing above them; furthermore, within the incised section the slopes near the knickpoint, which progresses by erosion towards the valley head, are younger than those found farther downstream. Precise dating, however, can rarely be satisfactorily achieved. In an area such as southern England today, many denudational slopes are covered by a veneer of soil, geologically very young and formed since the present slope profiles were formed during the Quaternary. In such cases dating by reference to deposits on the slope is impossible. Even those slopes which bear older periglacial deposits, or which lead down to valley-bottom deposits which are clearly related to the main episode of slope formation, cannot always be dated accurately, since the deposits may consist of unstratified or crudely stratified frost rubbles with no included fossils or carbon.

Despite these problems, some geomorphologists have formulated theories of slope evolution. For instance, Davis has argued the view that slopes tend to decline in steepness as the cycle of erosion proceeds towards the stage of old age (pp. 159–62). Thus steep slopes are designated as 'youthful' and gentle slopes as 'old'—though the adjectives as used here have a general and relative meaning, for the rate of slope development will vary from rock to rock and from area to area, and, theoretically at least, a gentle slope in an unresistant rock may actually be younger, in the strict sense of the term, than a steep slope in a very resistant rock. Unfortunately there are pitfalls in the Davisian approach. As some geomorphologists have since conclusively proved, all slopes do not necessarily decline with time, but may either maintain their angle (undergo parallel retreat) or steepen. Also it must be remembered that Davis's theory was based on a subjective examination of slope forms in the field, the making of certain assumptions regarding the relative ages of these slopes, and a series of purely deductive arguments. Yet perhaps a majority of geomorphologists, strongly influenced by Davis's 'classic' views, have approached the physical landscape with the belief that slope decline is inevitable, and not realising sufficiently that Davis's theory was unproven.

It is perhaps fair to say that the reconstruction of previous slope forms, and the formulation of theories of slope evolution, must remain largely speculative tasks—and for that reason, and others (pp. 189–93), some modern geomorphologists would add useless tasks. Nowadays there is, in fact, a growing preoccupation with the relationship between slope processes and slope forms. As we have seen, the proponents of the dynamic equilibrium theory contend that time may not be an important factor in slope modification, but that slope form is merely 'adjusted' in response to changes in the controlling factors. The implication is that an examination of the 'process-form' relationship is a far more valid approach than the older 'evolutionary' one.

The 'process-form approach'

This approach is based on the reasonable assumption that there is a direct causal relationship between the processes of weathering, transportation, erosion and deposition and the form and gradient of slopes. The immense variety of slope form and steepness observable in the field is regarded as due to the fact that the processes of denudation operate in varying combinations and with differing relative effectiveness in areas of different rock-type, structure, climate, vegetation, relief and so on. A typical, if very simple, example of this approach is the attempt to show that limestone slopes in humid areas are dominantly convex for the reasons (i) that limestone is a permeable rock, so that rainwash, widely taken to be the cause of slope concavity, is greatly diminished, and (ii) that in the absence of effective wash another process, soil creep (believed to have a rounding effect and to produce summital convexity), becomes relatively more important.

Unfortunately, attempts to investigate process-form relationships in a more realistic way, using evidence rather than mere assumptions, come up against serious difficulties.

First of all the processes at work on slopes (weathering, soil creep, rainwash and so on) generally operate very slowly indeed, certainly by comparison with those associated with rivers, and in many cases intermittently. Both patience and highly refined methods of recording are therefore required if any accurate assessment of the nature and speed of operation of these processes is to be made.

Secondly, it is not easy to demonstrate that all slope processes, and in particular wash and creep, have any direct effect on the form of denudational slopes. The latter are eroded from the solid underlying rock, and are covered by a thin weathered layer of approximately uniform thickness in many cases. The surface of this weathered layer is slowly removed by rainwash which, except where there has been rapid soil erosion or deep gullying, does not come into contact with the bedrock. Further, the slope detritus will undergo slow downhill creep, but the movement at the base is virtually nil, so that there can be no abrasion, however slight, of the unweathered rock surface. Some writers, notably W. Penck, have argued that soil creep, rainwash and other allied processes perform only the role of 'weathering removal' (or in simpler terms transportation). They are in fact agents of denudation in the strictest possible sense of that term (that is, they denude the slope of its weathered mantle), and have no *direct* effect on slope form and steepness, which are determined by other factors, such as the rate of erosion of the river at the foot of the slope.

Thirdly, one of the most teasing problems associated with the process-form approach is that in many, and perhaps even the majority, of instances present-day slopes are not the product of present-day process, or have been only slightly modified by those processes. In fact, there is not the precise balance between process and form that is implicit in the dynamic equilibrium theory. To take one simple illustration, there are good reasons to believe that the slopes of the southern English Chalklands were largely shaped under periglacial conditions, when agencies such as frost shattering and solifluxion operated rapidly to produce the smoothly curved profiles regarded as typical of these areas. In post-glacial times, slowly acting solution, soil creep and hill-wash have done little to alter these older forms, which are therefore to be regarded as 'relict'. It is indeed within the bounds of possibility that the form of the periglacial slopes actually influences the way in which present-day processes act. Thus the distribution of soil creep as a dominant process over the profile may simply reflect the distribution of the steeper slope sections, and rainwash might have relatively greater effects near the crest and base of the slope, where the gentler gradients are insufficient for effective creep to occur. Thus it is dangerous to assume that a certain form necessarily results from the process associated with that form today.

Non-relict slopes, however, can and do occur. For example, in very weak rocks where present-day erosion is proceeding rapidly and slope-forming processes are in no way impeded, the slopes are usually truly modern features. This is especially true of certain areas experiencing a semi-arid climate, where the vegetation cover (which slows down the action of both creep and wash) is ill-developed and where the rainfall is infrequent but often violent in nature. The so-called 'badlands' of parts of the U.S.A. are a case in point. Here geomorphological development is so rapid that the form of slopes in the past is of no interest at all in any assessment of the present-day features.

The last problem raised by the process-form approach is concerned with the concept of climatic geomorphology. If slope forms are controlled directly by slope processes (either present or past), it is logical to suppose that in different climatic regions different slope forms and landscapes should result. For instance, it is often said that there is a fundamental contrast to be made, in terms of weathering, between humid and arid climates. In the former, chemical breakdown of the rock may reach an advanced stage (pp. 26–7), and preponderantly fine detritus, including much clay, will be formed; whereas in the latter, the physical breakdown of the rock, produced mainly by temperature changes with the aid of some restricted chemical decay, produces generally coarser material. In turn, the efficacy of the processes of wash and creep in the two climates will be affected, and since these agents are responsible for the evacuation of slope debris and so influence renewal of exposure of the underlying rock, slope forms will accordingly be modified. Observations in the field often support the view that the landforms of one climatic régime are in some respects different from those of another; and it is for this reason that geomorphologists have found it necessary to formulate separate cycles of erosion to account for the landscapes of humid-temperate, desert, savanna and periglacial areas. Again, many writers have commented on the marked divergence of form between the 'humid' landscape of lowland England, with its integrated systems of valleys and smoothly graded valley-side slopes, and the arid landscape of, say, the south-west U.S.A., where there are steep and rocky mountain fronts, exposed rock pediments, great alluvial fans and spreads, and a general impression of angularity (pp. 303–9).

However, it may be that the relationship between slope form and climate has been overstressed. Critical examination of all the evidence does not always support the contention that a certain type of slope will be formed only under one set of climatic conditions. Davis has been partly responsible for the view that the gently curving convexo-concave (graded) slope is typical of a humid-temperate (normal) region in the mature stage of the erosion cycle. However, such slopes are also found in other climates—indeed, as already suggested, the slopes of this form found in many present-day temperate areas may be a relic of a former periglacial episode. For long pediments, or concave basal slopes, have been regarded as peculiar to desert landscapes, but in the view of many modern authors occur very widely. King, for instance, has referred to them as occurring in the Chalk country of Wiltshire, where again they are probably periglacial forms. Indeed, it is appropriate to comment, at this juncture, on the many resemblances between the cycles of pediplanation and periglaciation as formulated by King and Peltier (pp. 170–82). King himself is largely responsible for the concept of climatic uniformitarianism, which states that there are only four basic slope forms (the convex or waxing slope, the free face, the debris or constant slope, and the concave slope or pediment) that are found, in varying combinations, under all climatic conditions except glacial. It is certainly necessary to stress the fact that the form of any slope is affected by a number of factors (including chemical composition of the rock, jointing, permeability, angle of dip, rate of erosion of the river at the foot of the slope, climate, nature and rate of weathering, nature and rate of transportational processes such as creep and wash, nature of the vegetation cover, contemporary earth-movements, and so on). It is not realistic to assume that one of these factors must necessarily be totally dominant, even though, if the factor is climate, it can exert some control over certain of the others. It is always possible that a factor such as rock-type (which itself can influence the weathering processes that are likely to be most effective) can to a greater or lesser extent 'mask' the admittedly important climatic factor. For these reasons it must be said that the climatically controlled slope form—the expectable outcome of a total dependence of slope on process—remains a somewhat theoretical concept.

THE FORM OF SLOPES

As a prelude to a detailed discussion of theories of slope formation it is essential to examine in some detail the actual forms assumed by valley- and hill-side slopes. Some information, particularly regarding mean and maximum slope angles, can be obtained from the study of accurate topographical maps, but the most useful method is to survey profiles and gradients in the field, and then to analyse carefully the data obtained. Curiously enough, the collection of such information in the field has been undertaken on a large scale only recently. This is a reflection of the trend in modern geomorphology towards quantification (p. 4) and away from the generalised and usually imprecise descriptions undertaken in the past. Many early students of slopes based their theories more on visual impressions of the landscape than on accurate information, and it is not surprising therefore that certain misconceptions have persisted unchallenged for many years. Among the main types of slope form recognisable are the following.

A *The cliff*. Perhaps the simplest, though not the commonest, type of slope is the cliff, developed along the coast by wave undercutting, in some deeply cut river valleys, on escarpment faces in massive rocks, in glaciated mountain areas, in faulted landscapes, and so on. Cliffs are so steep (40° or more) that the products of weathering for the most part fall immediately to the base. There is little or no accumulation of detritus on the cliff itself, and it is therefore commonly—and meaningfully—referred to by geomorphologists as a 'free face' (Fig. 84A). The weathered material may, of course, accumulate at the foot of the cliff, unless removed by a transporting agent such as waves or a stream. A depositional feature (a talus or scree slope) will then develop at an angle controlled by the size and shape of the weathered fragments. The coarser and more angular this is, the steeper will be the scree slope, and a gradient of 35° or more may be attained. In time, the growth of the scree slope will lead to a reduction in the height of the free face, the lower parts of which will be submerged beneath detritus and protected from further weathering. Eventually, the cliff may disappear entirely, to be replaced by a wholly aggradational slope at 20–35°.

B *The concave slope*. In its lower part, a slope profile will commonly exhibit a concave section (Fig. 84B and C). In some cases this concavity will result from depositional processes. Thus the lower part of a talus slope may become lowered in angle through the washing downslope of finer material which has an angle of repose less than that of the majority of the scree fragments. However, it is more usual to find denudational slopes, covered only by a thin layer of soil or exposing bare rock, with marked basal concavities. In arid and semi-arid lands such slopes are characterised by a sharp break of gradient between the concave section (the pediment) and the much steeper slope section above (Fig. 124). Under more humid conditions, the basal concavity may be equally characteristic, but usually grades more smoothly into higher slope sections.

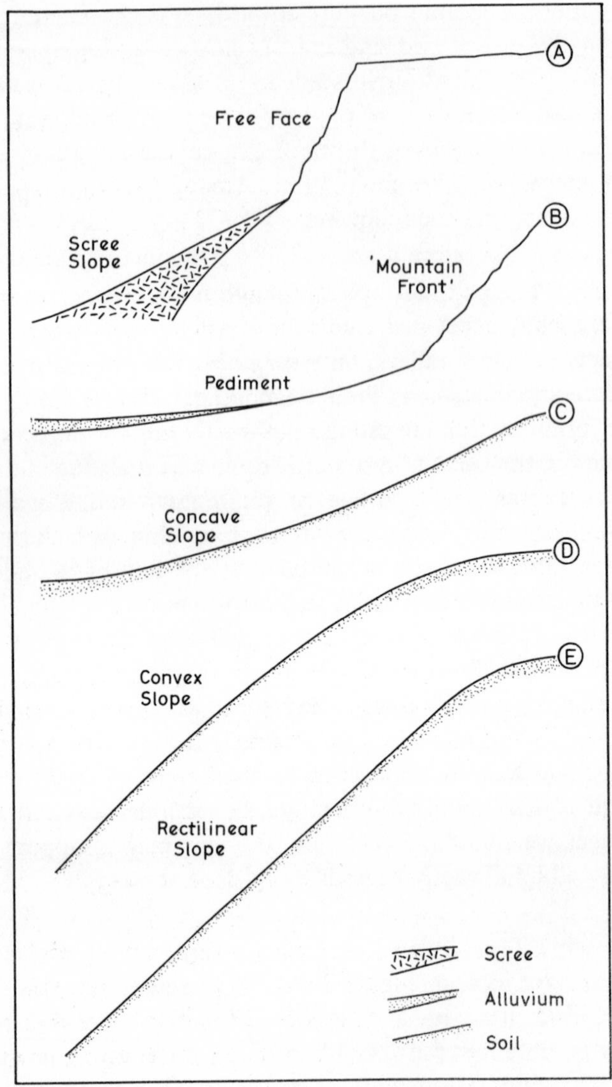

84 **Simple slope forms**

C *The rectilinear slope*. Many slope profiles display a marked rectilinear (or straight in profile) section, which may vary a good deal in its dimensions and sometimes dominates the whole slope (Fig. 84E). It is quite common to find such a major rectilinear section leading down to the very bottom of the valley and upwards to a limited summital convexity. On other slopes, the rectilinear section is restricted to the central part of the profile, where it separates a broader convexity above from a large concave section below. One of the most significant features of rectilinear sections (which normally form the steepest part of the whole profile) is that in any particular area, especially one underlain by a uniform rock-type, they tend to develop at or close to a certain angle. Careful examination shows that this angle is often determined by the angle of repose of the weathered fragments derived from the underlying rock and occupying its surface. The American geomorphologist Strahler (1950) has in fact referred to such slopes as 'repose slopes' (p. 8). The term is perhaps a little misleading, in that it implies that rectilinear slopes are the result of aggradation only. In many cases this is certainly not so. Rectilinear slopes can be essentially denudational forms, underlain by solid rock and bearing only a veneer of detritus, either at rest or moving very slowly downhill owing to disturbance by frost and other agencies. It would seem more appropriate to describe such slopes as 'debris-controlled'. Another similar term is 'boulder-controlled slope'; this was coined by Bryan to refer to slopes covered by weathered blocks in the arid south-west of the U.S.A.

D *The convex slope*. Convexities of slope profile, as mentioned already, are found on many slopes (Fig. 84D). Occasionally the whole of the slope will assume a convex form, but it is more usual for the convexity to be developed only on the upper part of the slope; hence the term 'summital convexity' is widely used by geomorphologists. Convex profiles are almost invariably the result of denudational processes, and are rarely covered by more than a thin layer of soil. They are often assumed to be most characteristic of humid-temperate climates and certain types of rock (especially limestone and chalk), but this is by no means always so. Sometimes summital convexities are referred to as 'waxing slopes', after the terminology introduced by W. Penck, just as rectilinear slopes may be described as 'constant slopes' and concavities as 'waning slopes'. However, the adoption of these Penckian terms for general use is rather unwise, as they may lead to certain assumptions (for instance, that convex slopes are necessarily increasing in length or height, or actually steepening in angle) that are not really justified.

Obviously this list of slope types is not exhaustive, and in fact what may be termed 'composite slope forms'—and these are the most widespread of all—result from the combination in one actual profile of two or more of these simpler elements (Fig. 85). An almost infinite number of permutations can be imagined theoretically, but in reality certain combinations are much more likely than others. Three common examples of composite slopes are as follows.

(i) The convexo-rectilinear-concave slope comprises an upper convexity, a central rectilinear section, and a lower concavity, the three grading imperceptibly into each other to give a smoothly curving profile (Fig. 85A). Such slope forms are typical of weak rocks (such as sands and clays) in lowland England, where the actual variety of landscape from place to place is a reflection of differences in the lengths and heights of the slopes, in the maximum angles of the rectilinear section, and in the relative importance of the three component slope elements. In

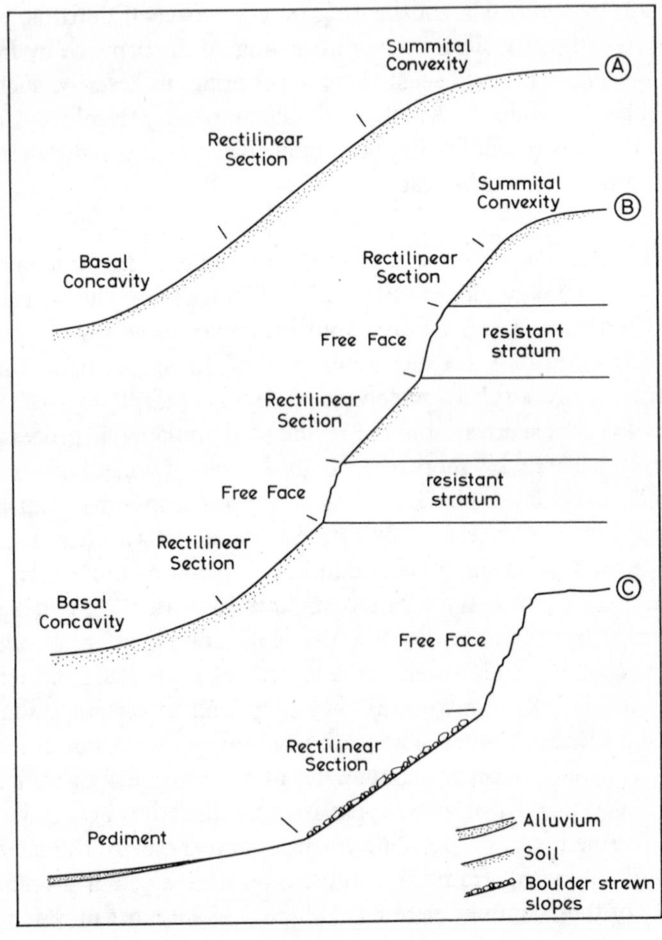

85 **Composite slope forms**

areas where the rock-type is varied, comprising alternating resistant and unresistant strata, or where base-level falls have caused successive rejuvenations, there may be a whole sequence of convexities, rectilinearities and concavities, giving a 'complex slope form'.

(ii) In an area of alternating massive and thinly bedded weak strata, where the relief is considerable, valleys are deeply incised, and weathering very active, a totally different composite slope profile may occur (Fig. 85B). This may comprise numerous free faces (associated with the massive strata) and rectilinear debris-controlled slopes (in the more easily weathered thinly bedded rocks), and summital convexities and basal concavities may be very limited in extent or absent altogether.

(iii) In arid areas of hard crystalline rock, a composite slope form may be developed with an upper free face (at an angle of 40° or more), a central boulder-controlled slope (at over 25°, and littered by masses of rock either weathered *in situ* or detached from the cliff above), and a lower concave slope, the pediment, at between 7° and ½° and often revealing bare rock (Fig. 85C). Such profiles are discussed more fully on pp. 303–8.

PROCESSES AND SLOPES

Many of the early attempts to recognise the precise influence of process on slope form were aimed merely at relating individual basic processes, such as rainwash and soil creep, to the simpler types of form, such as concavities and convexities. For example, in 1908 the American physiographer N. M. Fenneman attempted to explain the formation of convexo-concave slope profiles in terms of the action of running water alone. He argued that on the upper slope the amount of surface run-off during a period of rainfall would be small, and would take the form of a thin sheet of running water ('unconcentrated wash') which would quickly become fully loaded with small particles picked up from the slope. As there is normally a downslope increment of surface water on a slope, as a result of run-off from higher up the slope being added to that derived from rainfall at a point lower on the slope, the total pick-up of load will increase downslope. It is possible to envisage this greater 'erosion' away from the slope summit producing a convexity of profile over a long period. There are many arguments which may be set against Fenneman's hypothesis, but it is to a certain extent vindicated by the more recent work of Horton (1945). Horton, who has made a more scientific analysis of the action of running water on slopes, has demonstrated that at the summit of a slope, where the gradient is very gentle and the catchment small, surface run-off possesses insufficient energy to effect erosion. Lower down the slope, however, as the run-off increases this 'belt of no erosion' is left behind, and the erosive action of the sheet wash becomes of importance.

Fenneman considered that the lower concavity of the slope marked the section where surface flow had increased to the extent that small-scale channelling could occur. Thus the unconcentrated wash of the upper convexity becomes concentrated into numerous small rivulets which, by cutting close-spaced gullies, impose on the slope the normal concave curve of water erosion. An important point is that the larger a body of water becomes, the more efficient it is (p. 37), so that its load can be transported over a lower and lower gradient. If the increase of efficiency is sufficiently rapid, even a load which is growing in volume as a result of further erosion (presumably the normal case as one moves down a hill-side slope) can be moved successfully over a concave slope section.

It is clear that Fenneman's hypothesis is not wholly adequate in the sense that it takes no account of the possible effects of such processes as soil creep, and cannot therefore be applied to the slopes of humid climates where mass movements of the flowing type are important. In many cases, such movements play a major role in the downslope transportation of detritus, since they affect material that may be highly permeable and on which surface wash is seriously diminished in volume by infiltration. One of the earliest authorities to emphasise the results of soil creep was Gilbert (1909), who invoked it to explain the rounding of many hill-top summits and the development of summital convexities. Gilbert's argument was basically very simple. He took, as the basis for discussion, the case of a slope bearing a layer of soil material undergoing downhill creep. Towards the summit of the slope the amount of soil lying above a point (A in Fig. 86), and eventually to be moved by creep past that point, is quite small. Farther downslope (at B) this amount has substantially increased, and at C, towards the base of the slope, it is greater still. Thus, the further a point on a slope is from the top of that slope, the larger will be the amount of soil which must pass the point during the total evacuation of a single layer of soil. Gilbert postulated that, for this removal to be achieved, the slope will in fact have to assume a steeper and steeper gradient downslope, in order to ensure more effective creep and so the transportation of the increasing 'slope load'. This notion that a landform can adjust its form in response to certain controlling factors has been widely applied in geomorphology—indeed it is central to the modern theory of dynamic equilibrium—but it is not always easy to visualise just how the change is effected. This is true of Gilbert's hypothesis. It could well be argued that, in the situation described, an alternative result is the gradual build-up of soil in a downslope direction, perhaps to give a depositional rectilinearity or concavity. The fact that accumulations of weathered material on the lower parts of slopes do exist shows that the land does not always adjust its form so perfectly that there is rapid and effective evacuation of weathered waste. Another weakness in Gilbert's explanation is his assumption that the soil on a slope is of constant mobility. If, towards the base of the slope, the soil cover can move more easily than on the upper slope, either because the soil-water content is higher or because there is more abundant fine material moved down by rainwash, an increased amount of material might be transported over a constant or even a decreasing gradient.

There is no doubt that Gilbert's arguments are oversimplified, and it is not surprising that geomorphologists have sometimes invoked other mechanisms to account for slope convexity. Lawson, for example, revived in 1932 the idea that rainwash is an important formative process on the upper slope. However, he differed from Fenneman in suggesting

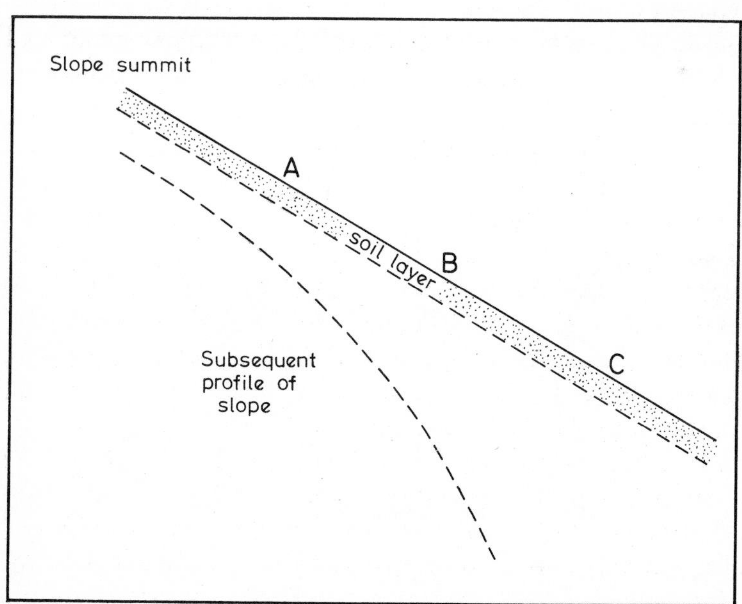

86 Soil creep and the development of convex slopes (after G. K. Gilbert)

that wash was most effective near and at the summits of the slope, and that downslope, as the wash became more heavily loaded with sediment, its erosive effects were at first diminished and eventually halted. Below the point of non-erosion, the slope convexity would be replaced by a basal concavity, resulting from deposition of material washed from the upper slopes. In Lawson's view at each stage of slope development a 'lune-shaped' mass of material will be removed from the hill or interfluve, in such a way that the radius of curvature of the latter (in profile) is continually increased (Fig. 87). A corollary of Lawson's theory is, therefore, that with time relief is diminished, divides are wasted, and slope angles steadily decline, much as envisaged by Davis. There are, however, some fundamental difficulties raised by this. In the first place, both Horton's observations and deductive reasoning make it hard to accept the idea of maximum summital erosion, in a zone where surface run-off must be at a minimum in volume and velocity. Secondly, it is not clear from Lawson's theory how the upper convexity is actually initiated by surface wash. Thirdly, since the volume and velocity, and therefore energy, of running water must normally increase in a downslope direction, a gradual diminution of erosion in that direction does not seem logical. Fourthly, as stated already, basal slope concavities are by no means always the result of deposition, but are very commonly denudational forms.

The theories so far considered have all invoked one process only, and it was inevitable that composite theories should be developed.

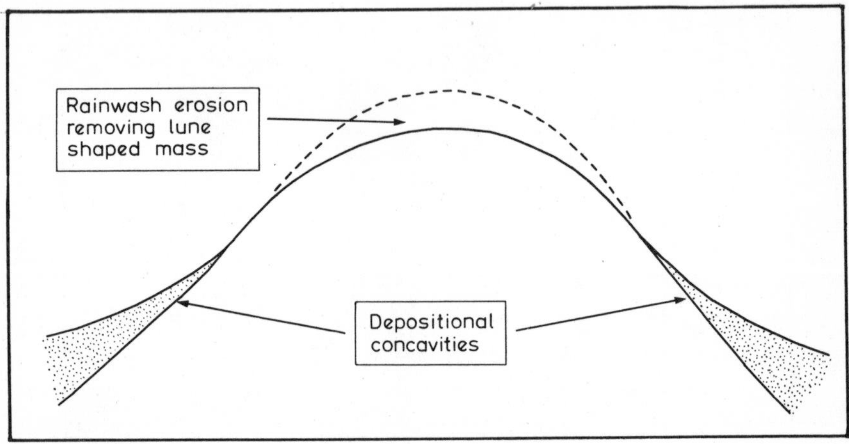

87 The lowering of divides by rainwash (after A. C. Lawson)

The French geomorphologist Baulig proposed in 1950 the apparently reasonable view that both rainwash and soil creep are active processes on most hill-side slopes, at least in humid climates, but that on different parts of the slope one or other of the processes becomes dominant. Thus, on a convexo-concave slope in a humid-temperate climate running water is least effective towards the summit, where the amount is small and the type of flow ('sheet-flow') is very inefficient in terms of erosion and transportation. Towards the slope base, however, the surface water is more abundant, owing to the downslope increment of wash and, perhaps, because of decreased soil permeability, and as it is concentrated into 'rills' becomes far more efficient. These rills are able to effect considerable erosion, and to impose on this lower slope its characteristic concavity of profile. Clearly soil creep, which is sufficiently active on the upper slope to mould the profile unaided, also occurs on this lower slope, but in Baulig's opinion is relatively less effective than wash.

Baulig's theory represents, therefore, a combination and modification of the conclusions of Fenneman (1908) and Gilbert (1909). An interesting development of the theory is that, with the passage of time, the balance between the two main slope processes changes, and that this will be reflected in the altered form of the slope profile as a whole. Late in the cycle of erosion, as relief is diminished and slope angles decline, soil creep will be reduced, the weathered mantle will become more comminuted, and wash (which can operate on very gentle slopes where normal creep is inhibited) will become the dominant process on more and more of the slope profile. The basal concavity will therefore be steadily extended, while the summital convexity decreases in extent, with the result that the peneplain will comprise mainly concave slopes.

An alternative approach to assigning different slope forms to different dominant processes is to assume that rainwash and soil creep, instead of working against each other, actually combine to produce different forms on different parts of the profile. As shown already, in the discussion of Gilbert's theory, it is theoretically possible to envisage soil creep promoting concavity of profile towards the foot of the slope simply because the soil there is finer, contains more moisture, and is therefore more mobile. At the same time this may also be the zone of maximum rill activity, which also promotes concavity. Similarly, on the upper part of the slope the increased 'erosional' effect of wash away from the crest may tend to produce convexity of profile, and this may be exaggerated in some way by soil creep.

A slightly different approach is to regard both rainwash and creep as forming one transporting agent which has the task of removing all the material produced by weathering on the slope. Since the amount of material ('slope load') to be moved increases with distance from the summit, the combined processes may require a steeper and steeper slope angle for the evacuation to be complete. Convexity of profile will therefore result. However, this is oversimplifying matters, for the two processes will themselves increase in effectiveness downslope (because of the greater volume and velocity of the wash and enhanced soil mobility). Thus a constant or increasing load of slope debris could, at least in theory, be transported over a concave slope. Perhaps in reality the rate of load increase on the upper slope tends to be greater than the rate of increase in efficiency of the slope transporting processes, so that a progressive increase of gradient is needed. On the lower slope, however, the reverse may be the case, so that slope concavity will result. Where the upper convexity and lower concavity are separated by a rectilinear section this may reflect a condition of approximate balance between the downslope increment of load and the downslope increase of efficiency of slope transportation. The various possibilities are expressed in graph form in Fig. 88. Clearly these suggestions are only theoretical, and require the support of field study of slope processes, and in particular an assessment of their rates of operation on different parts of the slope.

THE EVOLUTION OF SLOPES

Many geomorphologists since the time of Davis have followed the evolutionary method of slope study. To them, the most important question to be answered is: how have the present-day slopes developed through time to their existing forms? Some attempted answers have been contained within the concepts of erosion cycles. Thus Davis, in his 'normal' cycle of erosion, postulated that with the passage of time slope profiles become progressively smoother and, during the stages of youth and maturity, steadily decline in angle. More recently, in the context of his cycle of pediplanation, King has argued that slope steepness is maintained almost to the stage of old age, and that the most important evolutionary change in slope form entails the extension of concave basal slopes or pediments.

Some authors have proposed much more precise and detailed theories of slope evolution. Perhaps the most influential of these has been the German geomorphologist, W. Penck, some of whose views are discussed

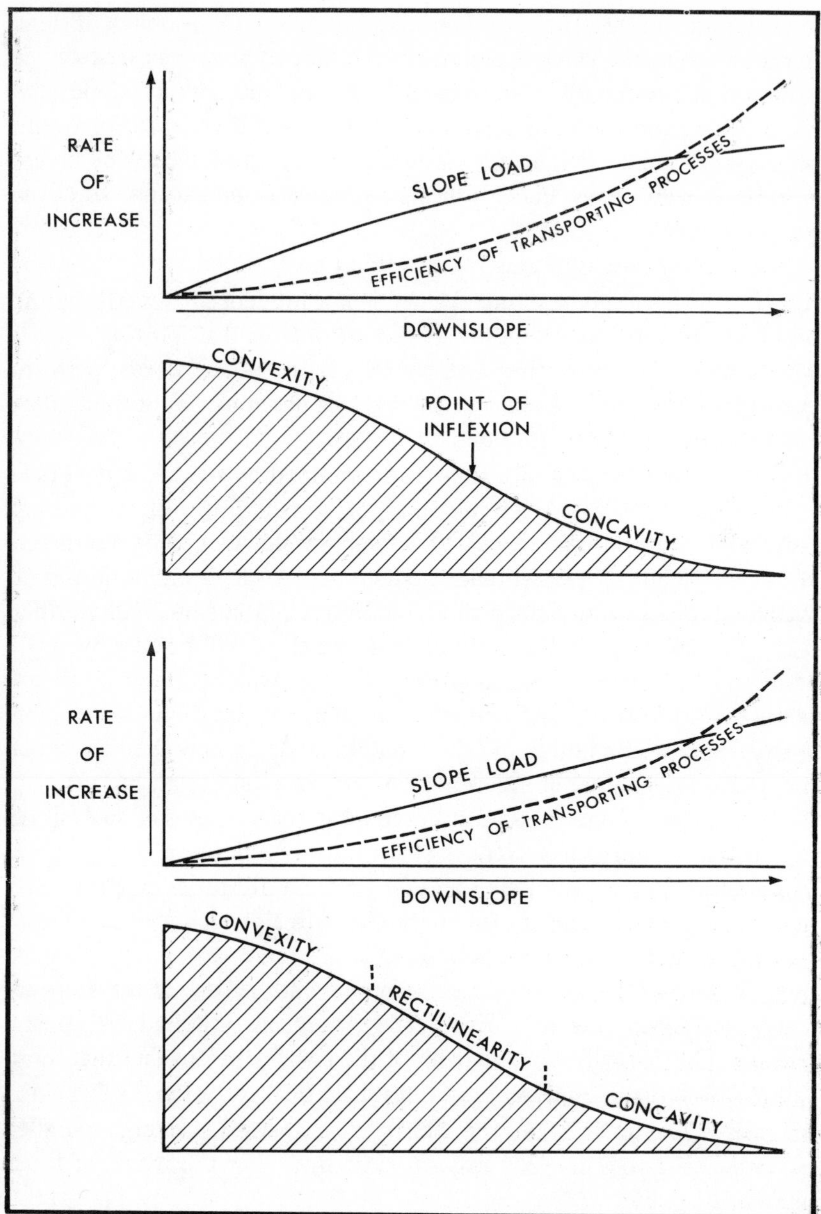

88 Possible relationships between slope process, load and form

in outline on pp. 184–6 above. Penck's approach to the problem of slope formation was, like Davis's, essentially deductive; he made certain basic assumptions, especially about initial forms and the role of slope processes, and from these inferred logically how and if slope forms would undergo change with time and in response to changing factors of stream erosion. In some ways, Penck's thesis was a mathematical analysis of the slope problem.

Penck's basic assumptions were as follows.

(i) The form of a denudational slope is not influenced directly by the processes at work on the slope. These act only as transporting agents, which move away weathered material (or in other words perform 'weathering removal') and so cause 'renewal of exposure' of the underlying rock to further weathering.

(ii) The rate of retreat of a slope is determined by its gradient. Thus, a steep slope will undergo more rapid weathering and weathering removal than a gentle slope, on which the more slowly moving weathered layer will tend to persist and cushion the rock from atmospheric agencies. As a result, a steep slope beneath a gentle slope will weather back so quickly that in due course the upper slope will be destroyed. In addition, the steeper slopes on either side of a divide will destroy themselves by rapid retreat, and will be replaced by gentler slopes developing at their bases. The notion of 'slope replacement' is now seen as one of the most fundamental of Penck's contributions to slope study.

(iii) Any slope, even if smoothly curved, can for purposes of analysis be regarded as comprising a number of rectilinear sections or 'slope units'. These units may be infinitely small (as on a curvilinear slope profile) or quite considerable in extent (as on a slope with a pronounced rectilinear section, or on a 'faceted' slope).

(iv) These slope units are normally initiated at the base of the slope as a whole. In the case of a river valley, they are initiated by stream erosion. This is really a restatement of the rather obvious fact that most valley-side slopes would not exist but for vertical corrasion.

(v) All slope units, whatever their size and angle, undergo parallel retreat. This is because of 'uniform exposure' over the whole of their surface.

As an introduction to his theory, Penck took the simple case of a steep slope unit bordering a valley in which the river (though responsible in the first place for the initiation of the unit) had reached a stage where it was neither eroding or depositing (Fig. 89). In this admittedly theoretical situation, the unit would undergo parallel retreat, owing to weathering

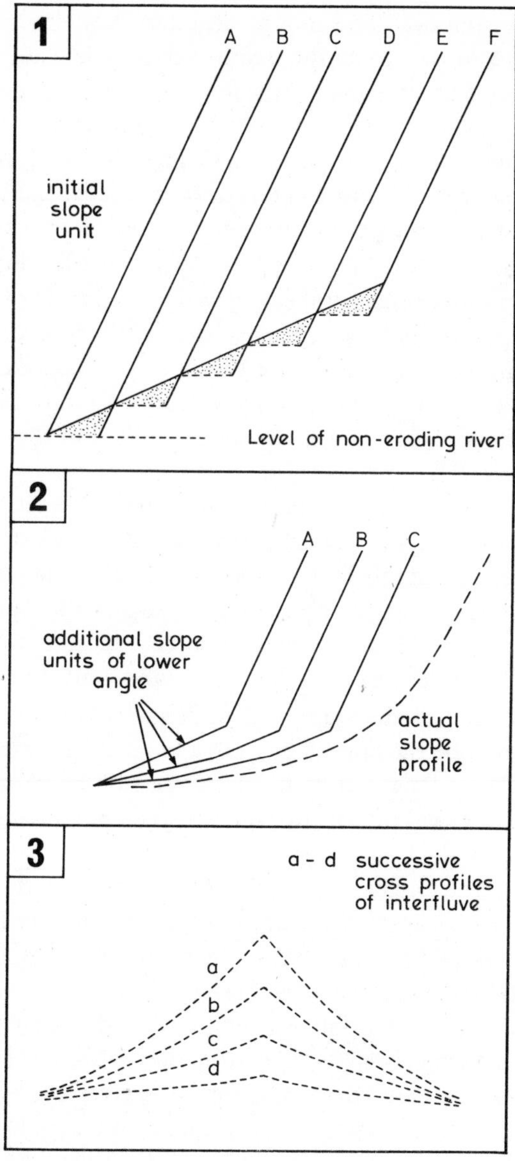

89 The development of concave slopes above non-eroding rivers (after W. Penck)

and weathering removal, and in a given interval of time would retreat from position A to B. Not all the weathered material can be evacuated, however, for some must remain at the foot of the unit to provide a 'slope of transport' down to the non-eroding river. Subsequent stages of development are indicated by positions C, D, E and F of the main slope unit. It will be seen that, as slope retreat proceeds, the length of the original unit will be progressively reduced, and a new basal unit, of ever-increasing dimensions, will be formed. It must be emphasised that the step-like nature of this basal slope, as depicted in Fig. 89A, will not actually exist. It results merely from the arbitrary division of slope development into 'given intervals of time'. In the field the slope would be a rectilinear one, and would be occupied by a uniformly thick layer of weathered detritus moving down from the base of the original slope unit to the river.

Subsequently, development of the slope will proceed as shown in Fig. 89B. Both the upper (older) and lower (younger) slope units will continue to undergo parallel retreat. At the foot of the latter, yet another basal slope of transport will be needed, and so a third slope unit, of still lower angle, will be added. In time an almost infinite number of new units, each of gentler gradient than the next unit above it, will be formed at the foot of the slope and undergo migration upslope. The overall result will be the development of a basal slope concavity. On the upper part of the slope, where it forms one side of an interfluve, the original slope unit will eventually be destroyed, and a lowering of relief and a decline in slope angle (since the remaining slope units will be younger and gentler) will ensue (Fig. 89C). Thus contrary to what is often stated, Penck considered that slope wasting and decline of angle is actually an inevitable process, *except* where the situation is complicated by the factor of river erosion at the foot of the slope.

To illustrate this last point we may take Penck's analysis of slope development above a river that is cutting downwards at a constant rate (Fig. 81). It will be seen that, providing the river is lowering its bed at precisely the correct rate, no development of new slope units can occur, and the original slope unit will, as it retreats, become steadily longer but maintain its rectilinearity and angle. In this case, parallel retreat of the whole slope in effect occurs, and Davisian decline can only intervene if river erosion slows down or ceases altogether—in which eventuality the slope will change in the manner described already. It might be argued that the situation postulated above is of only theoretical interest, in that it demands an exact, and on the face of it rather unlikely, balance between the rates of slope recession and river erosion. However, Penck was able to demonstrate quite convincingly that, even where an imbalance at first exists, the rectilinear slope must eventually assume an angle such that its rate of retreat would be precisely correct for its foot always to lead down to the constantly eroding river.

Penck also considered the development of slopes under conditions of accelerating and decelerating erosion. The first led in his view to the formation of slope convexity (the stage of 'waxing development') and the second to slope concavity (the stage of 'waning development'), much as under conditions of no river erosion. Fig. 80 illustrates the essentials of Penck's arguments in a very simple fashion. It must be added that Penck himself accepted that limited slope convexity could result from other causes. Thus the uppermost part of a rectilinear slope, as developed in response to constant river erosion (the stage of 'constant development'), will become rounded simply because weathered material is continually moved away, and never replaced by any coming from above, so that the solid rock is far more exposed here and is weathered back more rapidly than on the lower parts of the slope.

The essence of Penck's theory remains, however, that slope form and angle are primarily determined by the rates of erosion by rivers. It is with this idea in particular that many geomorphologists have quarrelled. Whilst there is undoubtedly a tendency for slopes to react to river incision in the ways described by Penck, other factors must also be taken into account (p. 200). Many authorities would argue that of these factors rock-type, climate and structure (all of which influence the processes at work in some way) could be locally or even regionally dominant. A particular difficulty concerns the concept of graded (or, to use a more modern term, equilibrium) slopes. A condition of grade or equilibrium cannot be related to downslope controls alone (that is, river erosion) but must also be affected greatly by upslope factors (in particular, the amount and calibre of the load, and the volume of the surface wash, arriving at any point on the slope from higher up). Gilbert's theory of slope convexity, for instance, implies that these upslope factors may be of first importance in determining slope form and angle; and even if the theory is not wholly acceptable, the principle on which it is based seems a sound one. Yet upslope factors are considered to be of virtually no importance in Penck's theory, in which it is assumed that the slope angle at a point is related only to the speed of weathering and the effectiveness of the transporting processes at that point. In a sense, a condition of balance, perhaps akin to grade, is implied in the acceptance of parallel retreat of slope units, for this is possible only because the rate of weathering back of the unit exactly equals the rate of removal of detritus. However, it does not seem realistic to ignore the possible influence of material moving from a higher to a lower unit. This point is suitably emphasised in the theory of slope development proposed in 1942 by Wood.

Wood took as his initial form a steep cliff (free face), which could have resulted either from erosional processes or earth-movements (for example, faulting). As weathering proceeds, the cliff is driven back, and the weathered fragments collect at its foot (Fig. 90). In time, as the scree continues to accumulate, the height of the free face is progressively reduced by the burial of its lower part. The gradient of the scree will be determined by the angle of repose of its constituent materials. So long as the calibre of the rock fragments remains unchanged, the scree slope will maintain the same angle; thus it will form what Wood termed a 'constant slope'. Beneath the surface of the scree there will form, in the manner described on pp. 312–13, a convex rock slope, the upper part of which will approximate to the angle of the constant slope.

Obviously, this situation will be complicated by the presence of an agent, such as a river, which is capable of removing some of the scree. For example, let us conveniently assume that, after an initial period of scree growth, erosion suddenly begins to remove from the foot of the scree slope exactly the same amount of weathered material as is being added at the top of the scree through retreat of the free face. The volume, or cross-sectional area, of the scree will henceforth remain unchanged, but since the recession of the free face will continue the constant slope will be extended upwards, passing on to solid rock; it will thus comprise both depositional and denudational elements. The extended section of the constant slope will be a true slope of transport, and its angle will be determined by the calibre of the debris arriving from above and passing across it. It will thus be an excellent example of a debris-controlled slope, and will typically assume a rectilinear form. Eventually, the constant slope will be developed to the point where the free face is totally consumed. According to Wood, rounding of the junction between the constant slope and the plateau surface or interfluve crest will give a summital convexity or waxing slope. In addition, the lower part of the constant slope will become concave, thus giving a waning slope, in the first instance as a result of the washing of the finer scree material to the foot of the slope profile as a whole. In this way a convexo-rectilinear-concave slope will develop. Finally, the rectilinear

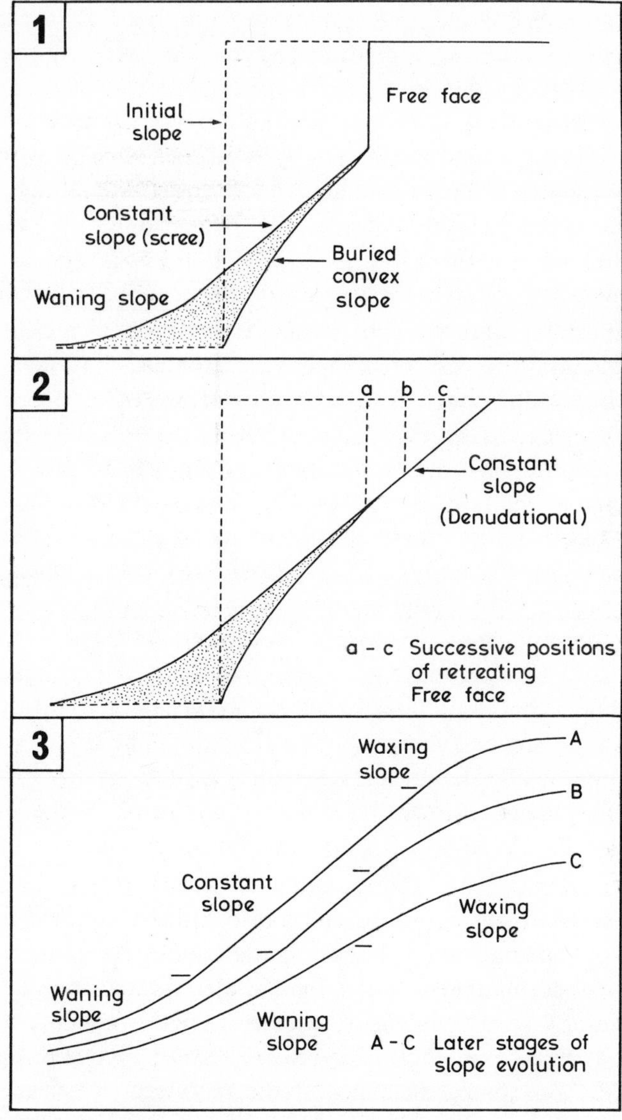

90 **The development of slopes (after A. Wood)**

element, the steepest part of the slope, will disappear as a result of the growth of the convex and concave elements downwards and upwards, and thereafter the convexo-concave slope will be affected by wasting and decline.

In criticism of Wood's theory it might be said that his approach is a highly geometrical one, and seems oversimplified in the sense that certain basic forms, such as convexity and concavity, are assumed to appear only at a rather late stage. Furthermore, it is by no means clear how, say, a concavity formed at the foot of a scree slope becomes transformed into a concavity underlain by solid rock—for, as we have seen, slope concavities are often denudational forms. Lastly, much of Wood's terminology (such as the use of the Penckian terms 'waxing' and 'waning') seems unnecessary.

At this point it is relevant to consider more recent work which throws light on slope evolution. This is usually based more on examination of field evidence and less on deductive reasoning. Savigear (1952) has drawn interesting conclusions from the analysis of certain cliff profiles at the head of Carmarthen Bay, south Wales. A noteworthy development in this area has been the recent growth, from west to east, of extensive salt-marsh at the foot of the Old Red Sandstone cliffs. Marine undercutting has therefore ceased, though at an earlier date in the west than in the east. Subsequently the cliffs (providing for once initial forms that actually existed) have undergone sub-aerial weathering and recession, and important changes of form have occurred. In the east, the younger slopes are characterised by a basal free face, the product of comparatively recent wave attack, which is surmounted by a rectilinear section at an angle of about 32° (A in Fig. 91). The latter seems to be a good example of a debris-controlled slope, for 32° is approximately the angle of repose of fragments of Old Red Sandstone. At the lower end of the rectilinear slope, debris is quickly evacuated by gravity fall over the former marine cliff, and thus no build-up of scree occurs. Farther to the west, the older slopes are of convexo-rectilinear-concave form (B in Fig. 91). The concavity is largely aggradational, consisting of material which has accumulated at the foot of the former sea-cliff since the cessation of wave attack. The rectilinear section is now appreciably less steep, having a mean angle of approximately 28°. Savigear has inferred from this that the maintenance of the rectilinear slope at a constant angle (32°) requires effective removal of weathered debris from the slope foot. Where this can accumulate, the lower part of the slope receives some protection from weathering, whilst the upper slope is subjected to unhindered recession; in such circumstances some decline of angle is inevitable. Another possible inference from this western area is that the summital convexity is a relatively late development, leading to a reduction in the extent of the 'declining' rectilinear slope. It does seem possible, to judge from the evidence in Carmarthen Bay, that both parallel retreat and slope decline are tenable theories, and that the two mechanisms can exist together in one small area in response to different conditions of slope-foot erosion or deposition.

Especially valuable work on the analysis of field evidence has been carried out by the American geomorphologist Strahler (1950), who has concentrated particularly on the measurement of maximum angles attained by slopes. An area of study, as far as possible uniform in terms of geology, climate, vegetation, soil and relief, is chosen and, at carefully selected sample points, slope readings are taken with the aid of a clinometer. These readings of maximum angle are then averaged to give a 'mean maximum slope angle' for the area as a whole. Within any one locality, it is common to find by this method that there is very little deviation of individual readings from the mean. For example, in an area where the mean is 32·5°, the vast majority of the slope angles may be between 30° and 35° (Fig. 92). Strahler has argued that such a low dispersion of values cannot be the result of coincidence. The slopes have all developed at approximately the same angle for the reason that this is the angle allowing the steady and efficient removal of slope debris by slumping, creep and wash. Such slopes are, in fact, in a delicate state of equilibrium. Debris-controlled slopes, which as stated already are developed at one particular angle in one rock-type, fall into the category of equilibrium slopes. However, one must avoid the inference that equilibrium slopes are controlled by only one factor (rock-type). Strahler

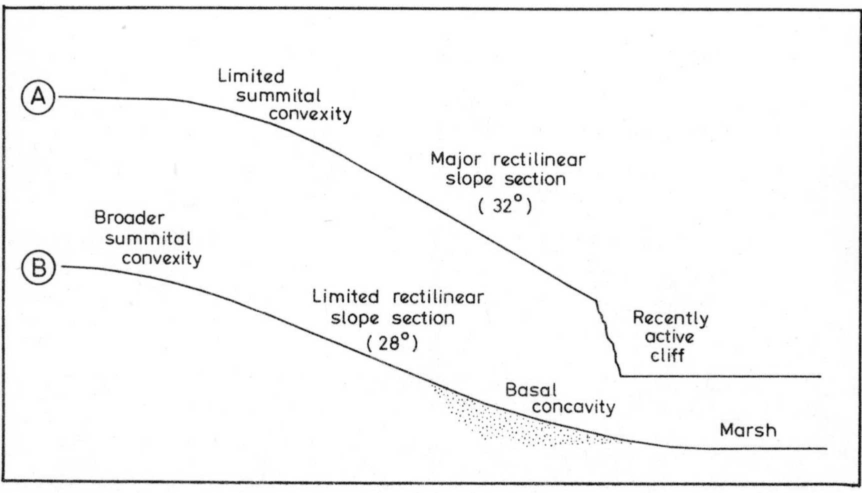

91 **Cliff profiles, Carmarthen Bay (after R. A. Savigear)**

has emphasised that the equilibrium angle represents a condition of balance between all the slope controlling factors (listed on p. 200 above). If one of these controlling factors undergoes change, then if necessary the equilibrium angle will be adjusted.

The implications of Strahler's work in terms of slope evolution are very important. The first and most obvious point is this: since (i) in any one area, on slopes which are actively retreating, maximum slope angles are uniform, and (ii) it is impossible to believe that all these slopes are of precisely the same age, parallel retreat rather than slope decline would seem to operate generally. However, this is something of an over-simplification. In addition to the studies described, Strahler also made a

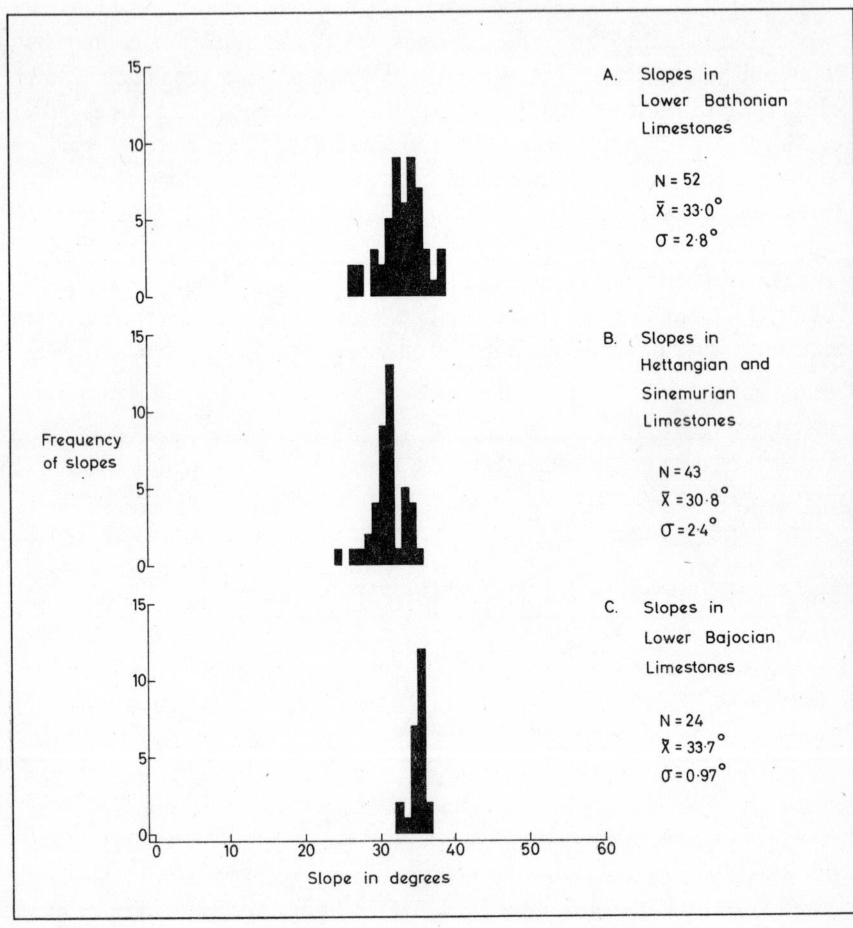

92 **Frequency histograms of maximum slope angles, Grands Causses**

field investigation of the relationship between the angle of the valley-side slopes (ground slopes) and the gradients of the rivers at their foot (channel slopes). On general grounds one would expect a close relationship to exist between the two. Where ground slopes are steep, and thus shed a large amount of detritus into the river, the gradient of the channel slope will also need to be steep, to provide the energy for the transportation of the load downvalley. On the other hand, gentle ground slopes will feed a small load into rivers, whose gradients can therefore be comparatively gentle. Strahler's measurements in fact confirmed this relationship, though he showed that the correlation was not a perfect one (or in other words a certain ground slope angle is by no means always associated with a channel slope at a fixed angle). He found, for instance, that an unvegetated slope tends to shed a greater load than a vegetated slope of the same angle, with the result that the gradient of the channel beneath the former will need to be steeper than that below the latter. This is an interesting example of how the angle of the stream channel, as well as the valley-side slope, is modified in response to a change in one controlling factor, in this case the vegetation cover.

So far as slope evolution is concerned, there are two main lessons to be learned from the relationship between ground and channel slopes. Firstly, it allows us to envisage that the progressive steepening of valley-side slopes instead of parallel retreat can occur. This would be true, for example, of a valley where the channel slope had been steepened by rejuvenation or perhaps as a result of climatic change altering the discharge-load ratio. Another case is where an unvegetated slope becomes vegetated, so that the movement of soil is impeded and its angle of rest effectively increased, and at the same time the processes at work in the river will be modified. Secondly, if one assumes that because of base-level control there is a tendency for river gradients to become progressively gentler with the passage of time, then it follows that in the long run slope decline must also occur.

However, yet another complication has been revealed by Strahler. In one area he examined (in the Verdugo Hills of California), the maximum slope angles measured grouped themselves about two means at 38·2° and 44·8°. By careful analysis Strahler discovered that the steeper slopes existed where river flow was at the immediate foot of the slope, thus removing all the debris arriving from upslope. The less steep slopes, on the other hand, were located where, in the absence of a slope-foot stream, accumulation of detritus led to the burial and protection of the very bottom of the slope. This evidence resembles that found by

Savigear, and again the implication is that attempts to explain slope behaviour in terms of one theory alone are totally unrealistic.

CONCLUSION

It will be abundantly clear by now that the various problems posed by the study of slopes have not yet been fully solved. Many theories have been proposed, and most seem to have some applicability to certain cases or in some areas. Certainly, the main factors involved in slope development are appreciated in general terms. What is needed now is an extension of the quantitative approach, typified by the work of Strahler and other modern workers, involving a greater and greater amount of recording in the field, not only of slope forms and dimensions but of slope processes too. In this way existing hypotheses, so often based on deductive reasoning or cursory field examination, can be adequately tested and, if need be, either modified or abandoned.

Some of the useful lines of research being pursued at present are as follows:

A *Slope form.* Scientific attempts are being made to relate slope forms not simply to the action of certain processes or to stages in a theoretical scheme of evolution, but to readily observable factors such as lithology and slope dimensions. On the basis of subjective observation, Baulig suggested that the nature of the rock is of fundamental importance in the development of slope convexity and concavity. He argued that concavity is promoted by rocks weathering to give a permeable soil layer (such as limestone, chalk and sandstone), and that concavity is best developed where the detrital cover is fine grained and impermeable (as on clays and schists). However, actual measurements carried out in the Chalk country of England have revealed that on convexo-rectilinear-concave profiles (which form 45% of all the profiles measured), the rectilinear section is in reality the most extensive. In fact, on average it occupies 41% of the total profile length, and the convexity only 37%. On the other hand, on the convexo-concave profiles (32% of those studied), the convexity occupies on average 63% of the total profile length. Figures such as these reveal that the form of chalk slopes is not simply an expression of the nature of the rock. Recently, Belgian geomorphologists have demonstrated the apparently curious fact that concavities occurred on all limestone slopes surveyed by them in the field, on 88% of the chalk slopes, on 87% of the slopes formed in clay, and on only

72% of slopes underlain by schist. This is the very reverse of what would be expected if Baulig's theory were correct.

If slope form cannot be related to lithology alone, other factors must be taken into account, such as the actual length and height of the slope (which in turn are influenced by the spacing and depth of individual valleys). Clark (1965) has considered this relationship in the English Chalk country, and has found that, as valleys become deeper, the summital convexity becomes relatively less marked, whereas the rectilinear section grows in importance. It is of interest to construct 'downvalley sequences' of slope profiles in order to show more clearly the effects of valley deepening and changes in the dimensions of the profile (Fig. 93).

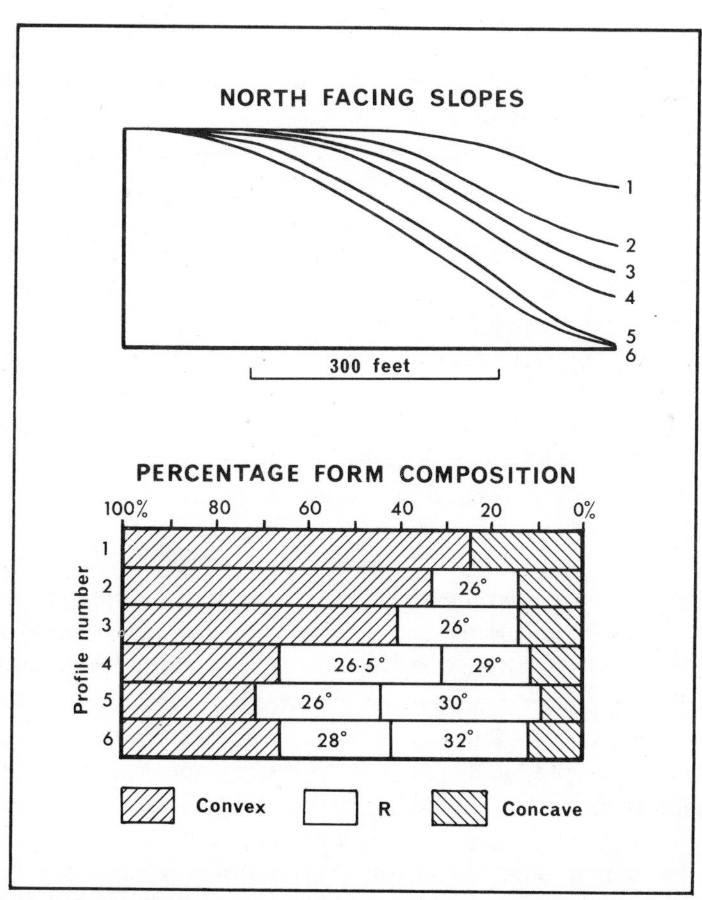

93 A downvalley sequence of slope profiles, Devil's Dyke, near Brighton (after M. J. Clark)

For example, in the Devil's Dyke, a steep dry valley north of Brighton, the slopes are at first convexo-concave (the convexity occupying 75% of the total profile length), but downvalley a small rectilinear section appears and subsequently increases until it occupies over 65% of the total profile length (at which point the convexity is restricted to less than 30%). It must be added that, since one cannot say that any individual profile in the Devil's Dyke is actually older than any of the others, this downvalley sequence cannot be readily converted into a 'time-sequence', and so it is impossible to test existing theories of slope evolution by this method. Instead, it might be argued that there is some support for the Penckian view, except that total amount of river erosion (that is, valley deepening) rather than rate of erosion is a dominant factor in the Devil's Dyke.

B *Slope Processes.* There is a growing concentration of study on the the mechanism and rates of operation of basic slope processes such as weathering, soil creep and rainwash. Field measurements of these demand patience and precision, but with care useful results may be obtained. In due course it may be possible to settle finally such issues as whether or not different processes operate more effectively on one part of a slope than another, and thus to test whether there is any real relationship between type of process and slope form.

The rate of soil creep can be accurately determined by knocking small pegs into the soil and measuring their displacement with reference to a marker embedded in solid rock. Young (1960) has used this method in the Pennines, and has discovered that downslope creep is of the order of 0·5 to 2 mm per year in the topmost organic soil layer, and of the order of 0·25 to 1 mm per year in the upper 10 cm of the soil. Oddly enough, the effects of gradient appeared to be comparatively unimportant, and creep on a 7° slope was only slightly less than on a 26° slope. The effects of rainwash can similarly be determined by the insertion of pegs so that their tops are flush with the soil surface. The extent to which the pegs subsequently project above the ground affords a measure of the surface erosion carried out by wash. In the area already referred to, in the Pennines, Young found that on well-vegetated slopes soil creep is about ten times as effective as rainwash in the downslope transportation of soil.

Such observations need to be multiplied many times, and carried out under as wide a range of climatic, vegetational and lithological conditions as possible, before the true role of process in slope development can at last be assessed.

7

DRAINAGE DEVELOPMENT

Geomorphologists have devoted much time and effort to the study of drainage systems and their evolution. The reasons for their interest are twofold. Firstly, there is the obvious point that a drainage system is a major feature of the physical landscape. Indeed, the form of that system, and especially the orientation and spacing of its component streams, does much to determine the essential character of the landscape. One thinks immediately of the striking contrast between, say, an area of badlands, with its very close network of literally thousands of streams and rivulets each carving its own valley or gully, and a broad limestone plateau, largely dry and traversed only by the occasional large stream occupying a deep valley or gorge. Secondly, evolutionary studies of drainage systems may afford valuable information about the denudational history of an area. For instance, it is often useful to attempt a reconstruction of the initial form of a river system in order to gain evidence of the nature and mode of origin of the land surface on which that system began its existence. Even major geological events such as marine transgressions of present-day land-masses may be inferred as much from a study of drainage evolution as from an examination of the planation surfaces or deposits left by the sea.

This chapter will be divided into two main sections. The first will be concerned only with the *form* of drainage systems, and in particular with attempts to describe and classify them both in subjective and objective ways. The second will deal mainly with their *origin and development through time*. Of course, it is not always easy in practice to make a rigid distinction between the 'descriptive' and 'genetic' approaches, and some overlap between the two sections has proved inevitable—and indeed in many ways this is desirable.

DESCRIPTIVE STUDIES OF DRAINAGE SYSTEMS

There is an almost infinite variety in the patterns formed by drainage systems in areas of different rock-type, geological structure, climatic régime and erosional history. However, it is useful, if only for the purpose of simple description, to make a classification of some of the more obvious patterns—though all the time it should be remembered that, in nature, 'perfect' examples of these patterns are not readily encountered (Fig. 94).

A *Subparallel patterns.* These are perhaps the most simple patterns of all, and comprise a series of streams which run approximately parallel to each other. Subparallel patterns are especially characteristic of areas of

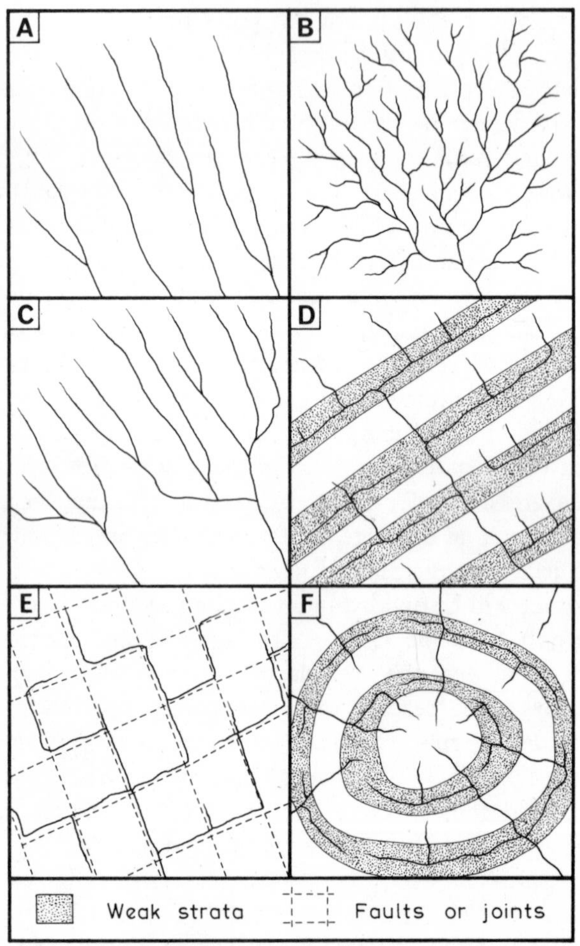

94 Types of drainage pattern: subparallel (A); dendritic (B); semi-dendritic (C); trellised (D); rectangular (E); radial-annular (F)

uniformly dipping rocks (such as the dip-slopes of cuestas) or areas which have recently been exposed by regression of the sea. In both cases, geological conditions and/or the time factor have not yet permitted the development of a more complex pattern by the process of adjustment to structure (p. 233). A subparallel pattern is therefore essentially an 'initial' drainage pattern.

B *Dendritic patterns*. These are very common patterns, and are mostly associated with areas of uniform lithology, horizontal or very gently dipping strata, and low relief (as on an extensive clay plain). They comprise a multitude of small branch streams which join each other, usually at fairly acute angles, to nourish a large trunk stream. The actual closeness of the pattern will vary a great deal, depending on the permeability of the underlying rock and the amount and nature of the precipitation, both of which will influence the intensity of surface run-off. The time-factor may also be of considerable importance, for it has been suggested that the initial dendritic pattern will be 'open', but that it will be gradually rendered more complex as new tributaries are added by the process of headward erosion. Later still, however, the pattern will again be simplified as 'surface abstraction' (p. 242) comes into play, causing a reduction in the number of its constituent streams.

C *Trellised patterns*. These are again common, particularly in areas of well-developed cuestas where the main dip-slope streams run broadly at right angles across alternately resistant and unresistant rock outcrops. The weaker strata are gradually eroded to form strike vales, separated by cuestas associated with the more resistant rocks, and are occupied by tributary streams which, with the passage of time, form an increasingly dominant component of the trellised pattern. The latter is, therefore, essentially a 'developed' drainage pattern, in which the process of adjustment to structure has played a very important part. Sometimes a pattern that has close affinities with a trellised pattern is found in areas affected by sub-parallel Jura-type folding. In this case the main stream elements are either synclinal or occupy vales formed by the breaching of anticlines (pp. 94–102). Secondary elements, at right angles to these, comprise small streams draining anticlinal flanks or larger streams crossing from one synclinal axis to another by way of structural saddles.

D *Rectangular patterns*. These again show some resemblance to trellised patterns. However, whereas in the latter the main and tributary streams join approximately at right angles, in a rectangular pattern the individual streams may themselves show marked angularities of course. These are invariably the result of geological controls, this time not the outcrops of weak strata but well-defined lines of weakness such as faults or joints, along which the streams have extended their courses by headward erosion and spring sapping. The process of adjustment to structure is thus again involved in this pattern.

E *Radial patterns*. These comprise streams diverging from a present or former high point, and are most usually associated with dome structures (sometimes involving igneous intrusion, as in batholiths and laccoliths) and large volcanic cones. The initial pattern is a simple one, but the gradual exposure of less resistant rocks within the core of the dome or cone will lead to the modification of the radial streams. In the case of a dome of sedimentary rocks, prolonged denudation may reveal a series of concentric rock outcrops, the weaker of which will favour the growth of tributaries. Thus a 'ring-like' element will be added to the initial pattern, and in time what has been referred to as an 'annular' pattern will come into existence.

F *Centripetal patterns*. These are formed by a series of streams converging on a central lowland from surrounding highlands, and are characteristic of many desert areas, where downfaulted blocks give rise to basins of internal drainage.

G *Deranged patterns*. These are essentially 'initial' patterns, despite the apparent implications of the name, and occur where there has been insufficient time for drainage integration. They are typical of areas of lowland glaciation, where as a result of glacial scour and/or the deposition of hummocky sheets of boulder-clay, sand and gravel the former drainage is completely or very nearly effaced. On the irregular surface exposed by deglaciation, a new pattern, characterised by small streams following irregular courses, boggy depressions and numerous small lakes, will be formed.

It is necessary to emphasise at this point that inexperienced students of geomorphology often attempt to add two further categories to those given above; these are so-called 'superimposed' and 'antecedent' patterns. However, this is not justified, as these terms are not at all descriptive of the drainage pattern as such, but relate only to its mode of origin and history (pp. 250–61); in other words, they are genetic terms. Furthermore, the process of superimposition may result directly in the formation of several of the patterns which have been listed.

During the last twenty years, this subjective (and at times naïvely pictorial) approach to the description of drainage patterns has been largely supplanted by more objective—and far more useful—techniques of study. These form the basis of what is known as drainage morphometry, which is an important branch of quantitative geomorphology (p. 4). It was to all intents and purposes initiated by Horton in 1945 and has been developed more recently by many other American research workers, notably Strahler (1950) and Schumm (1956). The prime objective of drainage basin morphometry is to gather accurate data of the measurable features of stream networks and drainage basins. This data is then subjected to careful statistical analysis, and can be used for comparing, in a precise and meaningful way, the properties of individual drainage basins (which may vary greatly in scale and form), and also for the establishment of certain basic laws of stream behaviour. For purposes such as these the classification given above would be totally useless.

The first step in the morphometric analysis of drainage basins is to apply the technique known as 'order designation'. On a detailed topographical map, or by surveying in the field if this is possible, the very smallest headwater tributaries of the basin are identified and designated as 1st order streams. Where two such streams join, a 2nd order segment is formed; where two 2nd order streams unite, a 3rd order segment results; and so on, in the manner depicted in Fig. 95. It follows that the trunk stream of the basin, through which all the discharge of the basin finds its outlet, is the stream segment of the highest order. The drainage basin itself is designated after the highest order stream segment that it contains; thus a basin containing a 4th order stream, plus numerous 3rd, 2nd and 1st order segments, is referred to as a 4th order drainage basin. It should be added that this method of order designation, devised by Strahler, is not the only one available, though it is the most commonly used at the present time. Horton used methods which in principle are similar to those described, but went on to redesignate his streams in such a way that, say, a 3rd order segment included the longest 2nd and 1st order segments feeding into it. In other words what was considered to be the trunk stream of the system was identified as far as its source, and not merely to the junction point of two streams of the next lowest order.

On completion of order designation, the task of 'order analysis' must next be undertaken. Basically this involves the counting of the number of stream segments of each order, for this may be an important determinant of the form of the drainage system. Thus in 2 5th order basins the distribution of stream segments might be as follows: 1 5th order segment in each case; 2 and 3 4th order segments respectively; 5 and 11 3rd order segments; 13 and 46 2nd order segments; and 30 and 139 1st order streams. In actual appearance the two basins will be different, for the simple reason that in the first there are, on average, two and a half times as many streams of one order as of the next highest order, whereas in the second this figure is approximately three and a half. These are in fact known as the 'bifurcation ratios' of the two systems.

The measurement of the total stream lengths of all orders, of the drainage basin area, and of the length of the basin perimeter (watershed) are other basic morphometric techniques. The data so obtained can be used for several purposes, such as the computation of 'drainage density' and 'texture ratio'. The former is simply the sum of the stream channel lengths divided by the total basin area; the latter involves recognition,

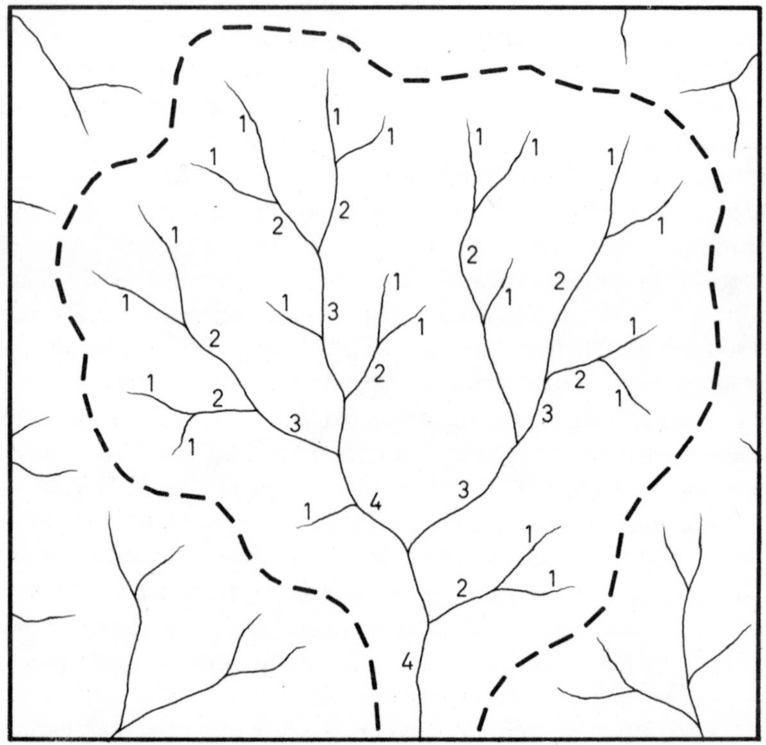

95 **Stream orders**

from a map, of the basin contour with the greatest number of crenulations (indentations marking valley courses), and dividing the number of crenulations by the length of the basin perimeter. As would be expected, both drainage density and texture ratio correlate closely with each other.

Some actual results gained by the application of these simple formulae will give an immediate impression of their value in making precise comparison of drainage characteristics. In the Chalk country of southern England the drainage network, as represented by the present-day dry valley systems, is very open. For instance, in the South Downs there are 2·8 miles of dry channel per square mile of basin area, in the Salisbury Plain area 2·9 miles, and in the North Downs 3·4 miles. The texture ratios for the three areas vary between 1 and 2 ('very coarse'). These figures must be contrasted with those for the resistant Carboniferous sandstones of the Appalachian Plateau of the U.S.A., where the stream channels are a little less widespaced. In this area the drainage density is between 3 and 8, and the texture ratio approaches 4 ('coarse'). In the deeply weathered igneous and metamorphic rocks of the California coast ranges, where the valleys are more numerous, the drainage density is 15–25, and the texture ratio 4–15 ('coarse'). In the Pleistocene sediments of southern California the corresponding figures are 25–40 and 10–25 ('fine'), and in the badlands of Arizona and New Jersey, where the stream network is exceptionally close, 200–900 and 100–300 ('ultra-fine').

The data used in drainage density calculation can also be used in computing what Schumm has termed the 'constant of channel maintenance'. This may be simply defined as the area of basin surface needed to sustain a unit length of stream channel (and is thus the inverse of drainage density). In an area of close dissection the ratio will be extremely low. For example, in the Perth Amboy badlands of New Jersey only 8·7 sq. ft of surface are required to support each foot of channel length. In the Chileno Canyon of the San Gabriel Mountains, California, the constant of channel maintenance rises to 316, and in the Chalk country of southern England ranges between 1500 (in the North Downs) and 1900 (in the South Downs). It will be clear that the constant is to some extent a function of rock-type and permeability, climatic régime, vegetation cover and relief, but other less obvious factors, such as duration of erosion (which can lead to integration of channels) and climatic history (for instance, recent periglaciation in the English Chalk), must also be considered in attempts to explain regional or even purely local differences.

Only a few of the more fundamental methods of drainage network analysis have been given here. Other morphometric techniques, some of them highly sophisticated, may be applied to the analysis of stream orders, basin relief, maximum and mean valley-slope angles, orientation of valleys and so on. In these ways a factual and accurate picture of all the morphological aspects of an area may be obtained. This is not, of course, to be regarded as an end in itself, but as an aid to the understanding of stream erosion and slope development.

GENETIC STUDIES OF DRAINAGE SYSTEMS

Broadly speaking, the initiation and subsequent evolution of any drainage system are determined by two main factors. Firstly, there is the nature of the surface on which the streams begin to flow. These will naturally follow the lines of steepest gradient; in other words, their courses will be consequent upon the form of the land surface, and such streams are accordingly referred to as 'consequents'. Clearly the initial consequent pattern will vary greatly in its complexity, depending on the degree of irregularity of the 'initial surface'. If this is produced by the uplift and gentle tilting of, say, a plain of marine erosion, the result will probably be a series of nearly parallel streams. If, on the other hand, the land surface is produced in the first instance by more complicated earth-movements, leading to the formation of a folded geological structure, the drainage pattern will tend to develop along different lines. The largest streams will flow along the synclinal axes, following the direction of pitch of their floors; such streams are referred to as 'primary' or 'longitudinal' consequents. Smaller tributary streams, known as 'secondary' or 'transverse' consequents, will drain the flanks of the anticlines and join the primary consequents to give a 'fish-bone' pattern.

The second main factor in drainage initiation and development is geological structure in the widest sense of that term (that is, including folds, faults, joints, angles of dip and lithology). As indicated already, the form of the initial surface may be directly determined by the underlying structure. Thus, in the second example given, the secondary consequents are at the same time true 'dip' streams, for the flanks of anticlines are no more than the dip-slopes of the youngest rocks involved in the folding movements. In other words, consequent streams can exhibit a close relationship both to the initial form of the land surface *and* to geological structure, where these happen to coincide.

However, this is by no means always the case. In the first example discussed above, the geological structure planed off by the sea may have included numerous folds, yet these would exert little or no influence on the courses of the consequent streams. In this case the drainage pattern would be 'discordant' to structure, whereas in the other example it would be 'accordant'.

Thus geological structure may or may not influence *initial* drainage patterns. What it certainly will do is to exert a close control over the later development of the rivers. The most significant feature in this context will be the appearance and growth of 'subsequent' streams, which by the process of headward erosion will extend along lines of geological weakness such as clay and sand outcrops, fault-lines, major joints and anticlinal axes. In this way, structurally guided streams will be continually added to the initial consequent pattern. Whether discordant or accordant, this will therefore be characterised by an ever-increasing degree of 'adjustment to structure'. It follows that the closeness of the relationship between stream courses and lines of geological weakness may afford some measure of the antiquity of a drainage system. For instance, if there is little adjustment of the stream pattern to structure the drainage may be extremely 'youthful'; but if the adjustment is very marked, it may be inferred that drainage development has proceeded for a long period, perhaps involving more than one cycle of erosion. A good example of well-adjusted drainage is afforded by the river systems of lowland England, where many of the larger streams follow the outcrops of weaker rocks in part at least of their courses. For example, the Severn in its lower section crosses easily eroded Triassic and Liassic marls and clays; the Trent for much of its middle course is guided by the unresistant Keuper Marls of the Midlands; the headwater tributaries of the Thames above Oxford follow the outcrop of the Oxford Clay; and within the Weald a whole host of smaller streams are orientated along the outcrops of the Weald Clay, the weaker divisions of the Lower Greensand, and the Gault Clay. Such evidence has led some writers, including Davis, to infer that the present drainage of lowland England began to develop early in the Tertiary era, after the great marine incursion responsible for the deposition of the Chalk had been terminated by a large-scale uplift of the land. However, conclusions such as this must not be drawn too readily. Very little is known about the rates at which subsequent streams can extend their courses and so achieve an advanced degree of adjustment to structure. Much will depend on the amount of uplift initiating the stream system, the precise resistance of the rocks involved, and the prevailing climatic conditions. There are, in fact, good reasons to believe that some of the well-known Wealden subsequents, far from being 'two-cycle' streams, have actually developed wholly within the last million or so years, since the regression from the area of the Calabrian sea (p. 279).

An important result of the growth of subsequent streams is the widespread disruption, through the process of river capture (p. 236), of the initial consequents. Thus, with the passage of time, the subsequents begin to dominate increasingly the drainage pattern as a whole, and the fragmented consequents become more and more difficult to distinguish. Yet if the history of drainage evolution is to be reconstructed, the initial drainage lines must be correctly determined. There are two principal ways in which this may be done. Firstly, consequent streams themselves display certain characteristics which are often easily recognisable. Secondly, river captures are among the most important events in drainage history, and their successful identification and interpretation is a very useful tool in the unravelling of drainage problems.

The characteristics of consequent streams

It has been stated already that the only factor determining the course of a consequent stream is the slope of the land surface on which that stream develops. In some circumstances it may be possible to reconstruct the form of this surface, and to demonstrate the close relationship between the direction of its maximum slope and that of a present day stream. For example, in the New Forest of Hampshire, the Tertiary sands and clays are overlain at many points by thick plateau-gravels of Quaternary age. These are evidently resting on a series of planation surfaces which decline in elevation from above 400 ft O.D. in the north to near sea-level in the south (p. 256). The main rivers of the Forest, including the Lymington and Beaulieu rivers, follow closely the slope of these gravels, although now incised into the underlying Tertiary rocks. It is logical to assume that these streams formed in the first instance on the surface of the gravels, and that they are in fact consequent streams of Quaternary date.

Needless to say, the identification and dating of consequents is not always as easy as this, particularly where geological conditions are more complex or where fragmentation by river capture is at an advanced stage. In an area where the structure comprises a series of anticlines and synclines, one is tempted at first sight to regard all synclinal streams as longitudinal consequents and all anticlinal streams as subsequents. In reality, it is quite possible for the synclinal streams to be subsequents. In a folded structure planed off by erosion, the synclines may be marked by the outcrops of unresistant rocks which promote the headward extension of tributary streams. In much of the Chalk country of Hampshire and Wiltshire synclinal streams are common (for example, the Ebble, upper Test and upper Itchen). In some instances these have been attributed, with evident justice, to growth along the axes of Chalk synclines once infilled by weak Tertiary sands and clays. Yet again, synclinal streams may even be 'resequents', which are traditionally regarded as the eventual outcome of drainage development in areas of folded rocks (p. 91). In general it may be said that, except where the folds involved are of very recent date, a subsequent or resequent origin must always be regarded as a possibility in the case of synclinal streams.

The problem is not only confined to synclinal streams. As Sparks (1953) has shown, anticlinal streams—which are normally identified with absolute confidence as subsequents—may in rather exceptional circumstances be actually of consequent origin. An example quoted by Sparks is the river Béthune, which closely follows the axis of the Bray anticline in Normandy. This stream has apparently been superimposed from a series of planation surfaces which, quite by coincidence, slope gently north-westwards in precisely the same direction as the fold. In England, the rivers Wylye and Nadder, draining the anticlinal vales of Warminster and Wardour (west of Salisbury) show certain similarities to the Béthune, and may conceivably have arisen in much the same way (p. 101).

One attribute that is normally taken to indicate a consequent origin for a stream is discordance with geological structure. As we have seen, subsequents grow along lines of weakness, and so promote a high degree of accordance between stream courses and the pattern of rock outcrops, folds, faults and joints. It is therefore very difficult to imagine how streams that transect, say, a series of important anticlines and synclines can possibly be of subsequent origin. Rivers such as the Taff, Rhymney, Ebbw and Lwyd, which cut discordantly across the faulted synclinorium of the south Wales coalfield, can hardly be regarded as other than excellent examples of consequent streams. Many of the rivers of the Weald show similar features. South of Pulborough, in Sussex, the Arun cuts directly through the Greenhurst anticline, whilst at Lewes the Ouse crosses in quick succession the Mt Caburn syncline, the Kingston-Beddingham anticline, and the Falmer syncline. Even small sections of streams may be identified as of consequent origin on grounds such as these. Although the river Dee in Cheshire is, in its lower course, a clear subsequent development, related to the weak Triassic rocks of the Cheshire plain, in its upper part it seems to be almost certainly consequent, for it flows from west to east across the complex structures of the Palaeozoic rocks around Corwen and Llangollen. The reasons why consequent streams are commonly discordant will be made more apparent on pp. 254–61, where superimposed and antecedent drainage are fully considered.

Finally, a word of caution is necessary. From what has been written the impression may have been gained that, while discordant streams are normally to be taken as consequents, the reverse is also the case. In fact, where initial surface form and geological structure exactly coincide, the consequent drainage will be accordant from the outset and will remain so despite fragmentation by river capture.

RIVER CAPTURE

In its simplest form, river capture is a well-known and easily understood phenomenon. The most straightforward type of capture is that shown in Fig. 96. A consequent stream (A) flows across bands of resistant and unresistant rock. A subsequent (C), tributary to a more powerful neighbouring consequent (B), erodes headwards along one of the unresistant outcrops, and in time approaches consequent A at point X. The weaker consequent is then diverted into the subsequent, a characteristic elbow of capture is formed, and the lower part of the dismembered stream, deprived of its upper catchment area, becomes a much diminished or 'misfit' stream occupying a valley that is manifestly too large for it.

Such a simple process seems hardly open to misinterpretation. However, in reality river piracy is often a more complicated affair, and many supposed captures have given rise to dispute among geomorphologists. Furthermore, there are some reasons to believe that captures of the type described may be less common than is generally believed.

In the first place, it needs to be emphasised that certain rigorous conditions must obtain before the diversion of one river by another can be effected. Most important, the pirate stream *must* be incised to a level substantially lower than its victim. It is not always easy to visualise how this can happen, at least in the 'classic' case shown in Fig. 96. All rivers possess an appreciable gradient downstream towards the sea, which acts as a common base-level for fluvial erosion. In streams flowing over similar rocks, and having similar catchment areas, this gradient may not

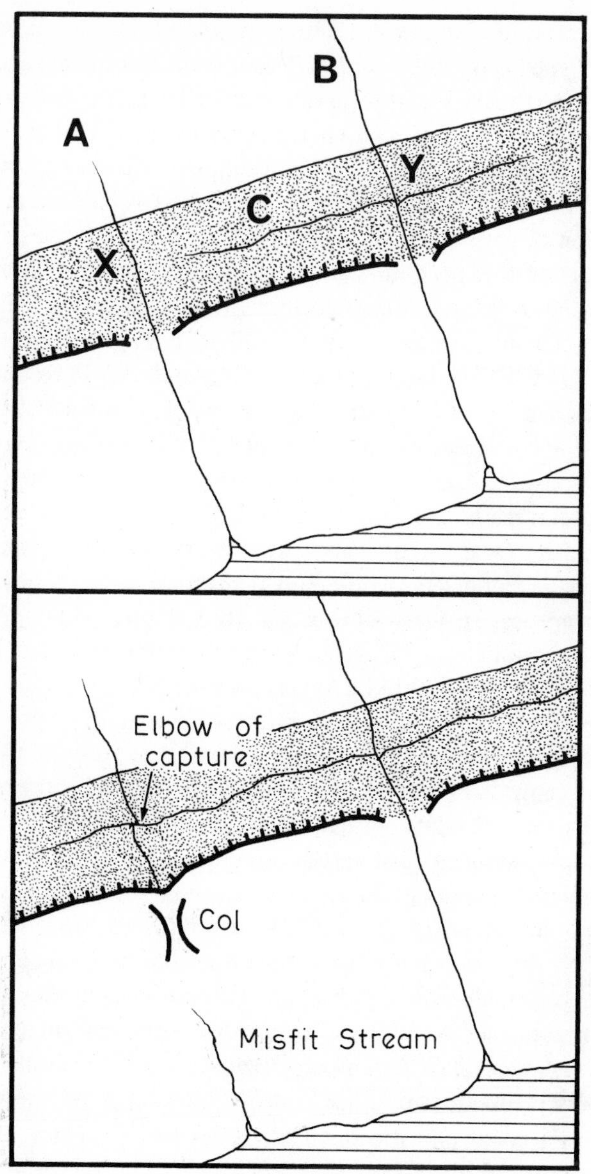

96 A simple river capture

vary greatly from one stream to another. Thus, the height of consequent B at point Y should not be greatly different from that of consequent A at point X (where the capture supposedly takes place). Yet the subsequent flowing from X to Y must also have a downstream gradient, so that at point X the uppermost part of the subsequent will be well above the level of the expected victim stream A. In these circumstances, capture and diversion of consequent A by consequent B is an impossibility.

The reality of this problem can be brought home by reference to an actual case. The two most important streams draining southwards through the Chalk country of central southern England are the rivers Avon and Test. The differences between these rivers, in terms of basin area and discharge, are not striking. In the area to the south-east of Salisbury, ideal conditions for the capture of the Avon by the Test apparently exist. A pronounced downfold in the Chalk (the Alderbury syncline) preserves here a tongue of weak Eocene rocks, along which a subsequent, the Dean Brook, has grown westwards towards the Avon. From the map alone, one would judge that the latter is in danger of imminent capture. In fact, there is not the remotest chance of such a diversion occurring in the foreseeable future, for the Dean Brook actually rises at a little under 250 ft O.D., whereas at the 'danger point' the Avon is at rather over 100 ft O.D. It is interesting to note in passing that, despite difficulties such as this, current theories of the evolution of the drainage of central southern England are based on the assumption that several captures have taken place in situations almost identical to this one.

It cannot be sufficiently stressed, therefore, that for river capture to occur one stream must obtain a *very great* erosional advantage over its neighbour. This advantage can only be derived from particular relief or geological conditions that are by no means encountered in all areas. For instance, some small-scale captures in southern England are associated with escarpment faces. The victim streams rise high up on the dip-slopes of, say, the Chalk or Upper Greensand, and in these situations their powers of downcutting are restricted owing to the resistance of the underlying rocks and the small discharge (itself further reduced by the permeability of the Chalk or Greensand). The pirate streams, by way of contrast, flow over weak clay outcrops, the Chalk Marl or Gault Clay, at the base of the scarp slopes, and thus have a height advantage of perhaps 200 to 300 ft. By headward erosion these scarp-foot streams are sometimes able to penetrate the escarpment and divert the upper courses of dip-slope streams. A clear example of such a capture, on a very local scale, is shown in Fig. 97. The deep valley incised into the Upper

Greensand near Binstead, Surrey, was eroded in the first place by a small dip-slope stream running north-westwards to the river Wey below Alton. It has subsequently been captured and overdeepened by a stream rising at the foot of the scarp and flowing southwards over the Gault Clay plain.

An example in which geological conditions have permitted capture on a much larger scale may be found on the northern margins of the south Wales coalfield. Here the generally north–south flowing consequent drainage has suffered much disruption by the rivers Neath and Tawe, which have eroded back along sharply defined lines of geological weakness ('shatter-belts') orientated from north-east to south-west. Other factors too have been favourable to the success of these two streams. Most important, they have had the overwhelming advantage of eroding headwards not from a consequent stream, but from the sea coast itself. The distances between the points of capture and the river mouths is comparatively small, and the gradient of the pirate streams has not had to be gentle for their headwaters to be at an elevation

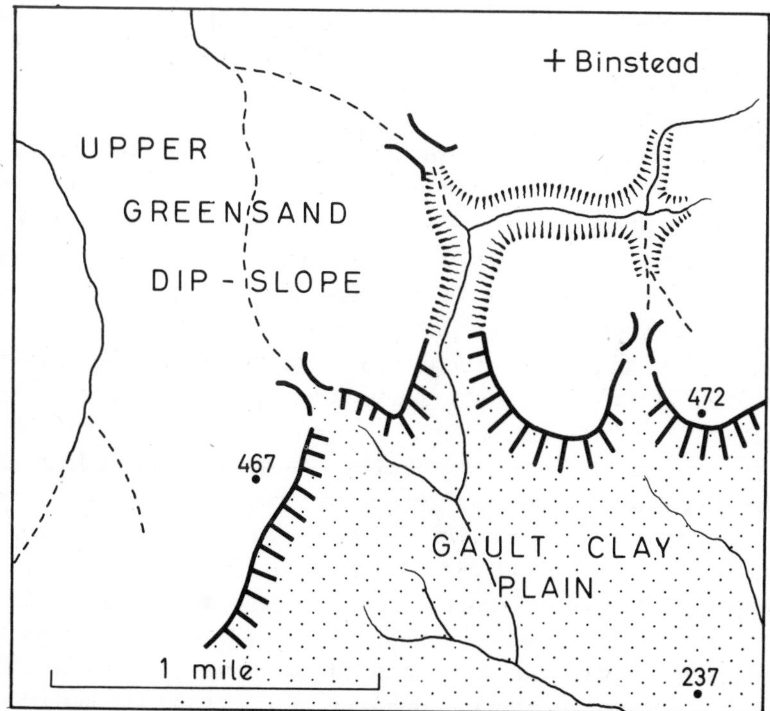

97 A river capture near Binstead, Surrey

substantially below that of the consequents (for example, the Mellte and Hepste) they were to divert. Put in another way, this means that the Neath and Tawe were able to possess quite steep courses, which in turn gave them additional powers of rapid headward erosion. Another significant factor here is that the lines of weakness sought out by the Neath and Tawe happened to cross the courses of the consequents in their uppermost parts, where they lay almost 1000 ft above sea-level and were thus unusually vulnerable to attack by deeply cutting subsequents.

It must be apparent from this discussion that the phenomenon of river capture cannot be 'taken on trust'. Good reason for its occurrence, as well as sound supporting evidence, must always be demonstrated. This is especially true in areas of geological homogeneity, where the possession by one stream of a striking erosional advantage over another must be extremely rare. Indeed, in a landscape underlain by one type of rock, be it chalk, limestone, sandstone or clay, and undisturbed by important folding and faulting, drainage may undergo comparatively little change with the passage of time. Furthermore, in areas of permeable rock, where headward erosion may be inhibited by falls in the water-table and the desiccation of the upper parts of valleys, river capture is not to be regarded as a 'normal incident in a veritable struggle for existence between rivers' (Wooldridge and Morgan), but as a decidely freak occurrence.

The study of river capture is also complicated by the varying nature of the actual mechanism of diversion. Broadly speaking, four main types of capture, using the term in its broadest sense, may be recognised. Firstly, there is the simple surface diversion, occurring in an area of impermeable rocks or where the water-table is high, in which the victim stream is able to maintain its flow until the moment of diversion. Secondly, in an area of permeable rocks the surface diversion may be preceded by underground abstraction, which will usually be so marked that at or below the point of incipient diversion the victim stream may disappear beneath the surface. Since it is working to a lower level, the capturing stream can tap water percolating through the bed of the higher stream, thus increasing its own discharge and powers of erosion whilst weakening its victim (Fig. 98A). Thirdly, part of the drainage basin of one stream may be incorporated with that of another stream without any surface diversion as such occurring. As a result there are none of the usual symptoms of capture, such as elbows or misfits. This process, which is achieved through the 'migration of divides', is common where drainage systems flow away from each other in opposite directions. It

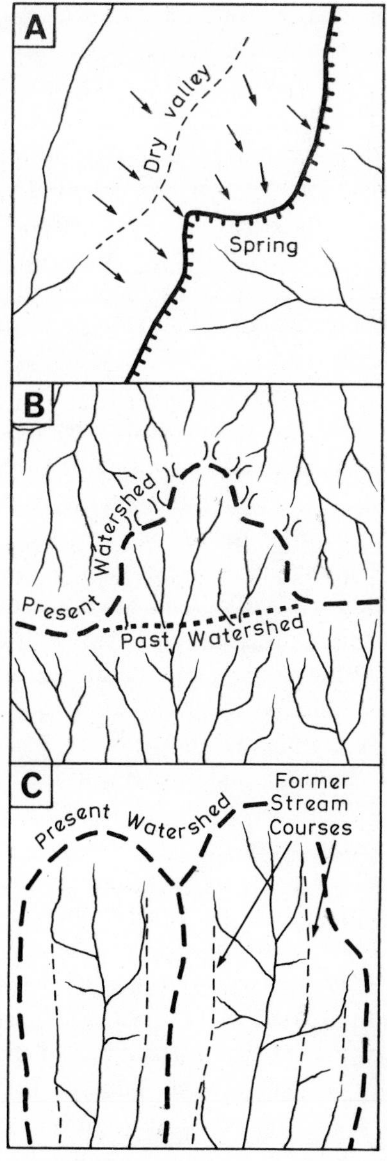

98 Types of capture: underground abstraction (A); divide migration (B); valley abstraction (C)

can also take place in areas of homogeneous rock, providing one stream system possesses steeper gradients and greater erosional energy than its neighbours (Fig. 98B). Fourthly, surface abstraction can occur where closely spaced streams flow roughly parallel to each other. If any individual stream can erode more quickly than immediately adjacent streams, perhaps because it has succeeded in penetrating to a weak stratum or follows a joint or fault-line, it will be able to enlarge its valley laterally and incorporate the catchments of the less successful streams (Fig. 98C).

It is now necessary to consider in more detail the morphological features arising from capture, and to discuss problems of interpretation posed by these features.

Evidence of river capture

A *Elbows of capture*. As stated already, one of the best-known results of river capture is the so-called 'elbow'—indeed, inexperienced students of geomorphology are tempted to attribute any sharp change in a river's course to a history of diversion. A case in point is the right-angled bend of the eastern Yar, in the Isle of Wight, where the stream after flowing generally from west to east turns sharply northwards to pass through the Chalk ridge by way of the Brading gap. At first glance it seems obvious that the Yar once ran on eastwards into the sea near Sandown, and that it has recently been diverted northwards by the 'Brading stream'. Detailed study shows that this explanation is in fact unlikely. The west–east section of the Yar is evidently of subsequent origin, for it is developed along the outcrop of the weak Lower Greensand. Indeed it seems to have arisen as a tributary to a consequent south–north river, responsible for the cutting of the Brading gap in the first instance, which has been destroyed in its upper part with the formation by marine erosion of Sandown Bay (Fig. 99). The present elbow of the Yar is thus quite fortuitous. Admittedly this is a rather unusual example, but it serves to underline the point that alternative explanations of suspected elbows of capture are always worth considering. It can often be shown that changes in a river's course reflect the influence of geological factors such as rectangular faulting or jointing, to which the stream has become adjusted. In conclusion, it is probably fair to say that elbows of capture are the most suspect evidence for river piracy.

B *Misfit streams*. Like elbows, misfit streams have for long been regarded as a classic symptom of capture. It is certainly true that river capture must result in the formation of misfits, but it must be realised that such features can also result from other causes. As G. H. Dury has demonstrated, it is common to find virtually *all* the streams of an area, whether affected by capture or not, showing evidence of greatly reduced discharge. Typically such streams, which are now more usually referred to as 'underfit', are marked by small amplitude meanders, which are contained within larger valley meanders, presumably cut under past conditions of much enhanced surface run-off. One can only assume that the present-day streams have all been affected by a climatic change involving a reduction in precipitation and a relative increase in importance of evapo-transpiration and percolation. It is worthy of note that streams which have clearly effected captures, and which have thus received increments of discharge, themselves often display underfit characteristics.

C *Cols or wind-gaps.* An important feature of the physical landscape, not previously mentioned but normally resulting from river capture, is

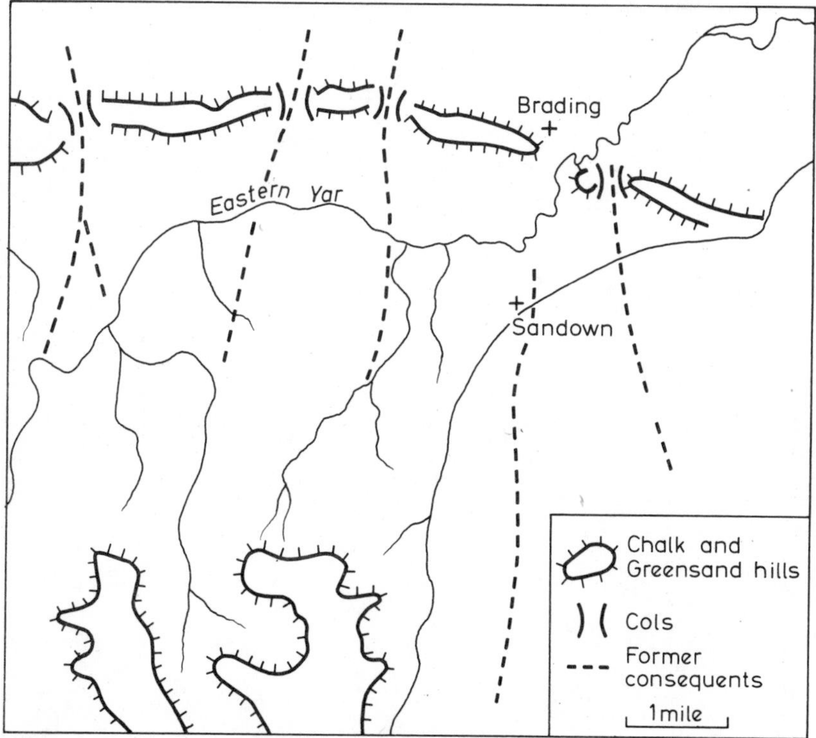

99 The drainage of the Vale of Sandown, Isle of Wight

the col. This is the section of the captured valley, immediately adjacent to the point of capture, which is left dry after diversion has taken place (Fig. 96). The existence of such a col is often taken as certain evidence of capture, though as will be shown below difficulties may in fact arise in the interpretation of cols. Furthermore, cols may be especially valuable aids in reconstructing drainage history in areas where numerous captures have occurred. As a general rule geomorphologists assume that, if there are no reasons to believe that a col has been modified by more recent erosion (associated, say, with the extension into the col of an obsequent stream from the point of capture), its height will give some measure of its antiquity. Fig. 100 shows the development of a hypothetical drainage system, consisting initially of four consequent streams, three of which have been later disrupted by an important subsequent. Three cols (at 500, 400 and 300 ft) mark the sites of the captures, and reveal that the captures occurred in the order A, B, C. Thus the first capture (A) was of consequent I by a subsequent tributary of consequent II. Later, a tributary of consequent IV captured consequent III(B), and then extended itself to capture consequent II(C).

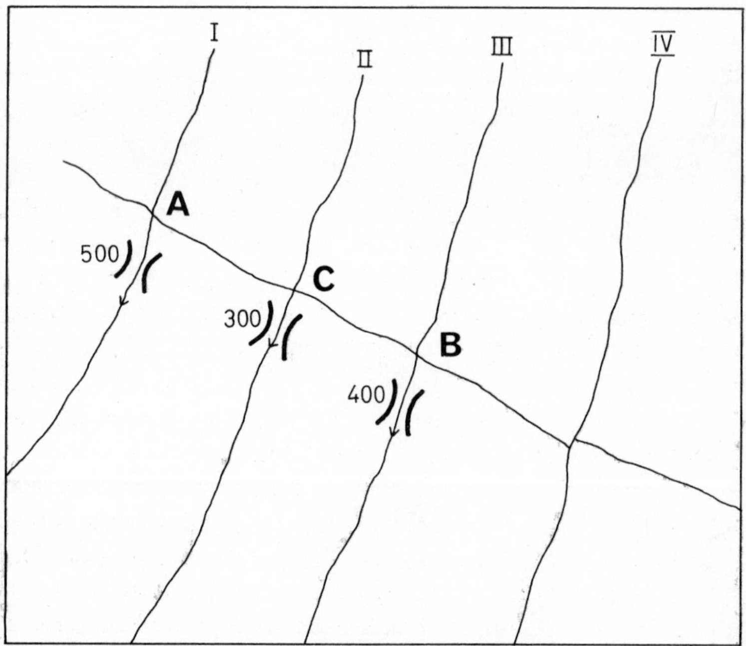

100 **The use of cols in the interpretation of river captures**

This is clearly a very straightforward example, and in reality the situation is rarely open to so easy an interpretation. The height differences between cols may be very much less, and application of the axiom 'greater height means greater age' then becomes dangerous, for the height of a col is determined not only by the date of the capture, but also by the size of the stream that once occupied it. In other words, if two streams are captured simultaneously and one is much larger than the other (and thus capable of grading its course to base-level more perfectly), it is most unlikely that the cols will be left at precisely the same elevations. Theoretically at least, the later capture of a very small stream could result in a col substantially higher than that formed by the earlier capture of a much larger stream, which had succeeded in cutting a deeper valley.

Another important point is that cols are subject to wasting by weathering, soil creep and rainwash. For this reason too, the present-day heights of cols may be misleading, particularly in areas where, owing to inequalities of rock resistance, differential wasting can occur. In short, cols are ephemeral features; when youthful they are well defined, and in cross-section show the usual characteristics of a valley profile; when ancient they are much blunted by prolonged denudation, and commonly take on the form of gentle and ill-defined 'sags' in a crestline. Ultimately, as the cycle of erosion enters the stage of old age, all cols marking captures are effaced by general landscape lowering. In fact, Davis long ago emphasised that in a two-cycle landscape the captures of the first cycle can be differentiated from those of the second or current cycle by the presence or absence of cols.

Finally, it must be added that, although cols are a normal result of the capture process, they can result too from other causes (for example, recession of escarpments (p. 85) or watershed breaching by ice (p. 373)). Therefore the existence of a col is not by itself certain proof of capture. Other lines of evidence, such as suspected elbows and underfits, must also be taken into account—unless, by a stroke of good fortune, fluviatile gravels are found to occupy the floor of the col, in which case its former occupation by a stream can hardly be disputed.

The problems that are encountered in col interpretation are neatly illustrated by reference to the western part of the South Downs, between Petersfield and the Arun gap. At many points the crestline of the Chalk escarpment is interrupted either by deep valley-like gaps, with floors lying some 200–300 ft below adjacent summits, or very subdued depressions at much higher elevations. Either type of feature could mark the former passages of north–south streams through the Downs. Such streams might well have been beheaded by the West Sussex Rother, a long subsequent running parallel to and north of the present escarpment. However, alternative explanations are at least as likely. Many of the high-level and gentle cols occur at points where scarp-slope and dip-slope streams once flowed away from each other. Soil creep and rain-wash into the heads of the valleys eroded by these streams could well be responsible for the local lowering of the Chalk divide, to give col-like forms. Most of the deeper cols, on the other hand, seem to be due to the recession of the Chalk escarpment, which has caused the beheading of dip-slope valleys. During the whole process the upper parts of these valleys have probably been dry, and the formation of the recession-cols (p. 85) has not been associated with any surface diversions of streams.

D *River profiles.* The study of stream long-profiles can, in favourable circumstances, afford useful evidence of river capture, and can also help in the more detailed reconstruction of drainage history. As stated above, the success of river capture depends on the ability of one stream to cut to a level appreciably below that of another. At the moment of capture, therefore, the captured stream is subjected to what is in effect a sudden fall of base-level. The result is similar to that produced by a drop of sea-level. A knickpoint is formed in the profile of the diverted stream, and this migrates steadily by headward erosion towards the source of the stream. Above this knickpoint, the profile remains as it was prior to capture—in brief, it is the equivalent of a graded reach (p. 61)—and

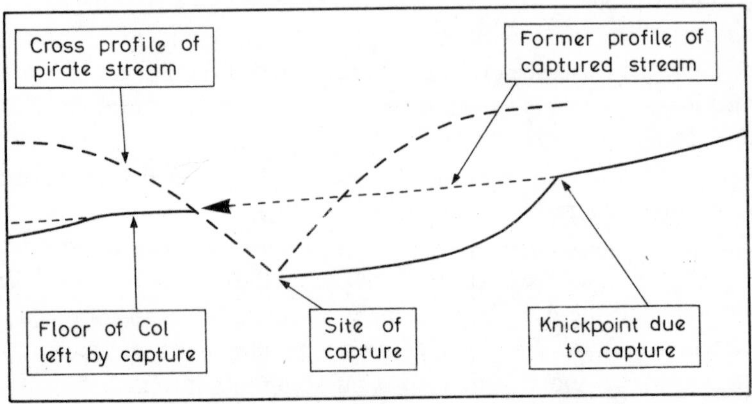

101 The effect of river capture on the profile of the captured stream

sometimes it is possible to extrapolate this upper course to 'line up' with the col over which the stream formerly ran (Fig. 101).

The use of this method of drainage interpretation is illustrated hypothetically in Fig. 102. This depicts a stream pattern of three initial consequents which have been affected by two captures, leaving cols at 300 and 150 ft respectively. The most obvious inference is that the subsequent stream worked back as a tributary to consequent I, and successively captured consequent II (leaving col A) and consequent III (leaving col B). However likely this interpretation, it is not the only possible one. It is also conceivable that the earliest capture was of consequent II by a subsequent tributary of consequent III (leaving col A), and that later a tributary of consequent I rediverted consequent II (so that X marks the site of a double capture) and then went on to capture consequent III (leaving col B). If this more complicated sequence of events were true, the upper course of consequent II should display two knickpoints, with two graded reaches above them (Fig. 102). The higher graded reach should line up with the col at 300 ft (A), formed before the disruption of consequent II, whereas the lower reach

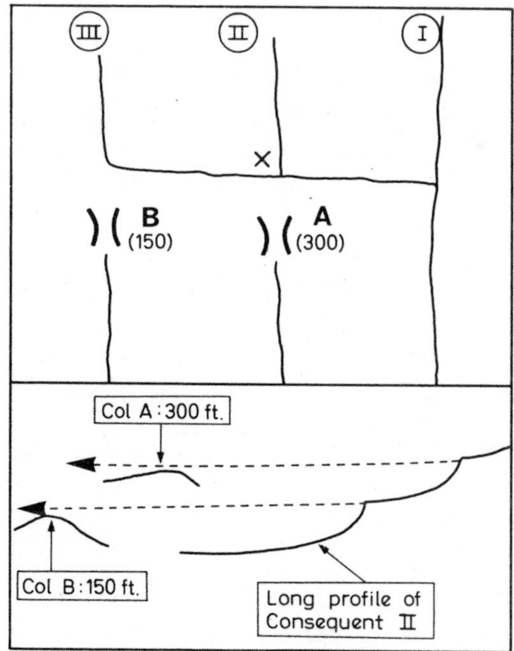

102 The use of profile extrapolation in the interpretation of river captures

should be associated with the col at 150 ft (B), indicating the passage of the stream through that col after the first capture and prior to the second capture at point X. If, in fact, the upper course of consequent II displays only one knickpoint, with a graded reach above it lining up with the col at 300 ft (A), the more simple sequence of captures first postulated is supported.

E *Gravels.* The examination of gravel deposits can provide further evidence in support of river capture. A case in point is the western part of the North Downs, where gravel deposits on the Chalk contain a good deal of hard chert (a flint-like concretion) which can only have been transported from the Hythe Beds division of the Lower Greensand within the Weald. At present, these deposits are separated from their source in the greensand hills by a major east–west vale eroded by subsequent streams. One can safely infer the former existence of chert-carrying south–north streams which have been broken up by the growth of these subsequents.

Within this area is found one of the most famous of all river captures, denoted by the deep gap at Farnham and by the sharp change of course of the river Wey at the point where it is apparently about to enter the gap (Fig. 103). The upper part of the Wey, running north-eastwards from Alton over Chalk and Upper Greensand, nowhere exposes Hythe Beds. Yet chert-bearing river gravels are found within the Farnham gap. This leads to the supposition that originally the gap was formed by a south–north river, draining from the Hythe Beds hill-country around Hindhead. The present Alton branch of the Wey can only have been a left-bank tributary of this river. Prior to the famous Farnham capture, the south–north stream must itself have been diverted by the lower Wey, working back westwards from the Guildford area, in the vicinity of Tilford village. As a result, the Alton branch of the Wey was left flowing through the Farnham gap into the present-day Blackwater. Later, however, the Farnham capture was effected by an obsequent branch of the Guildford Wey, working back northwards along the line of the old south–north stream from the site of the first capture at Tilford. This is therefore an interesting example of how the study of gravels can reveal a capture (that at Tilford) which might not have been suspected at all from other evidence.

The interpretation of gravel evidence is, on the other hand, sometimes beset by problems. In areas of glaciation or where there has been a history of marine transgression in comparatively recent times, the large-scale movement of gravel from one area to another can be effected by glaciers or longshore drift. Subsequently such material may be moved downhill to become incorporated with true riverine gravels, and a false notion of the source of the river laying down those gravels might be

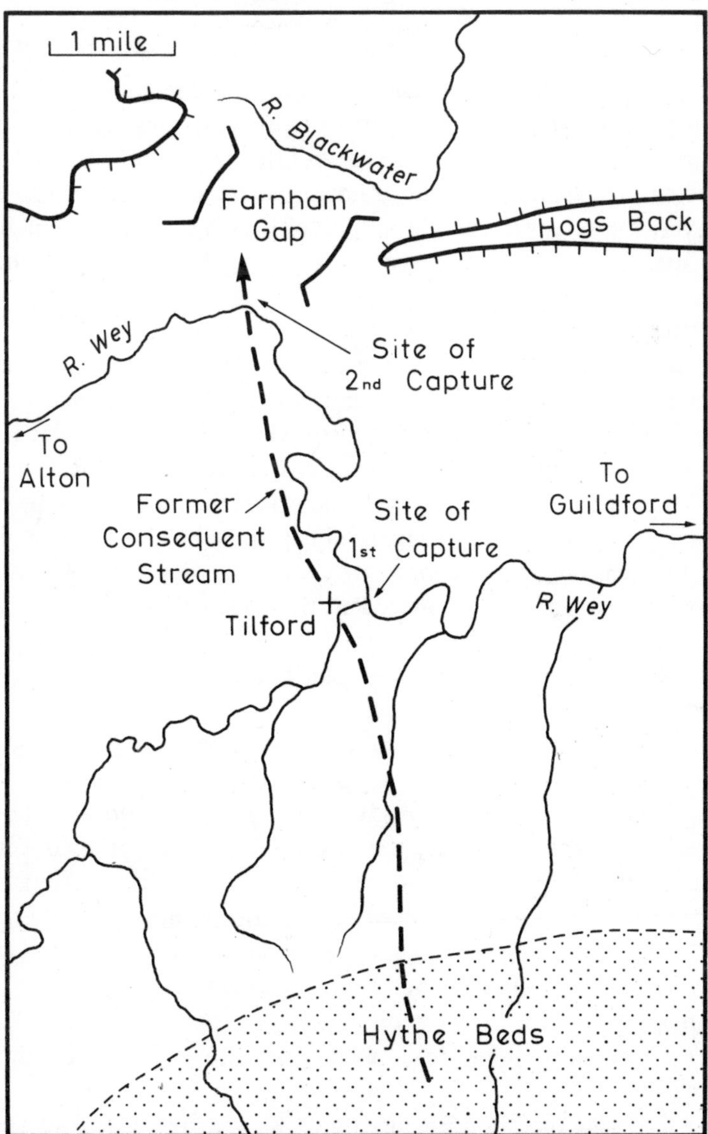

103 The capture of the river Wey near Farnham

gained. Some slight doubt might even obtain in the case of the Farnham gap gravels, for the Calabrian sea (p. 278) spread across the outcrops of the Chalk and Hythe Beds, together with intervening strata, in this western Wealden area, and some marine transportation of chert is feasible.

ANTECEDENT AND SUPERIMPOSED DRAINAGE

The main themes of this chapter have been (i) the relationship between stream patterns and the initial form of the land surface, and (ii) the changes which affect stream patterns with the passage of time, including the increasing adjustment of the drainage pattern to geological structure. However, the relatively simple outline of drainage development so far given may be rendered more complex in either of the two following ways:

(i) After the initiation of a consequent stream pattern, earth-movements may lead to a substantial alteration of the original geological structure nourishing that drainage. If rapid and violent, these movements might succeed in entirely disrupting the existing river system, and give rise to a wholly new consequent pattern, intimately related to the form and orientation of the new structures. If, however, the movements are more protracted and gentle, the original drainage may be able to maintain its form by incision into the new structures as they develop. Such a phenomenon is known as 'antecedent drainage'. It is a common error to suppose that antecedent drainage may simply result from a *general* uplift of the land; in fact, this only leads to rejuvenation of the drainage system (p. 60).

(ii) As consequent streams initiated on a certain geological formation or structure vertically corrade their courses, they may in time erode through an unconformity and encounter an older and substantially different structure. However, the streams cannot adapt their courses immediately to conform with the new geological conditions. Adjustment to structure will certainly proceed in due course, but in the meantime the old consequent directions will persist and, even after the rocks of the overlying structure have been entirely removed, will often be easily recognisable because of their discordant relationship with the newly exposed structure. This phenomenon is known as 'superimposed drainage'.

It will be seen that the basic difference between antecedent and superimposed drainage is that in the former the rivers are actually older than the structures they cross, whereas in the latter the rivers are by definition a good deal younger than the underlying folds and faults. However, a point of resemblance is that both processes may be associated with the development of discordant patterns, though this is not always the case with superimposition.

Antecedent drainage

Theoretically at least, examples of antecedent drainage should occur frequently, for earth-movements affecting land-areas do not always operate with great rapidity, whereas under favourable circumstances the rates of downward corrasion by rivers are often astonishingly high. Thus, the first requisite of antecedence, the existence of a river with a potential rate of downcutting in excess of the actual rate of uplift of a newly forming structure (such as an anticline or horst), should not be regarded as in any way unusual. However, although some examples of antecedence have been recognised (such as the gorge section of the Rhine, the crossing of the Cascade Range by the Columbia river, and the great cleft of the river Arun incised through the Himalayan folds between Everest and Kangchenjunga), geomorphologists are for the most part reluctant to invoke it as an explanation of stream courses that ignore geological structures. Indeed, antecedence is usually considered as a possibility only when superimposition, which also can lead to discordance, can be firmly ruled out. Thus Sparks writes: 'antecedence, as a hypothesis, should be the last resort of a geomorphologist seeking to explain an insequent pattern of drainage, as it is, except in ideal cases, undemonstrable'. However, the antipathy towards antecedence seems to go deeper, for geomorphologists are sometimes willing to propose theories of superimposition on the very slenderest of evidence. Furthermore, one must be careful not to assume too readily that because a process cannot be easily demonstrated, it cannot actually occur. As will be shown, antecedence by its very nature leaves only slight and impermanent evidence, and is hardly ever provable.

None the less, it must be admitted that in some instances the impossibility of antecedence, as a mechanism to account for a discordant river pattern, *can* be easily shown. A good example is afforded by the rivers of south Wales, which cross discordantly the Hercynian fold- and fault-lines of the coalfield synclinorium. One of the most obvious features of this river pattern is its patent youthfulness. The consequent streams have largely remained intact, though geological conditions, involving west–east outcrops of weak formations such as the

Lower and Upper Coal Measures, would seem to favour subsequent stream development and numerous river captures. Yet if the drainage is antecedent it must date from the Carboniferous period or earlier (that is, some 250 million years ago)—an obvious impossibility. In Carboniferous times first marine and then swamp-like conditions existed in the area, precluding the possibility of a normal drainage system. Even if such a system began to form as the earliest Hercynian folds were initiated, it would have later been effaced by aridity in Permo-Triassic times or by marine inundation in the Jurassic and Cretaceous periods. It is quite certain therefore that the discordant rivers of south Wales must be explained either in terms of superimposition (perhaps from a layer of Chalk resting unconformably on the Palaeozoic rocks of the coalfield) or as resulting from headward erosion by streams working inland from the coast. What has been said of the south Wales coalfield drainage must apply equally well to all other areas of Britain where the drainage is discordant to geological structures of Hercynian or greater age.

The situation is rather different, of course, when the drainage of an area where the structures are much younger is being considered. In the Weald, for example, there are many discordant rivers, notably those which cut directly across the Alpine folds of the South Downs and the area lying immediately to the north. It is worth considering whether a case could be made here for antecedence on general grounds. Firstly, it is possible to work out an average rate of growth for the Alpine folds, which probably began to form in the Eocene period and were 'completed' in the Oligocene and Miocene periods. The amplitude of folds crossed by rivers such as the Arun, Adur, Ouse and Cuckmere is rarely more than 1000 ft. Distributed evenly over the time-interval involved (some 35 million years), the average rate of growth of the folds is easily calculated as 1 ft per 35,000 years. Obviously this figure would have been greatly exceeded at times, notably during the culminating movements of the Miocene, yet it offers some kind of yardstick of the slowness of fold growth. Secondly, some idea of the potential speed of downward river erosion in the Weald may be gained from the fact that, during the Quaternary, the rivers have been able to grade their profiles to a base-level that has fallen by approximately 600 ft. Assuming for the purposes of argument that the duration of the Quaternary was 1 million years, it can be seen that the average rate of downcutting of the Wealden rivers must have been in the order of 1 ft for every 1700 years—or twenty times the average rate of fold growth. Furthermore, this average rate of erosion must have been greatly exceeded during times of rapid sea-level fall (that is, at the onset of each glacial period). Clearly these figures rest on too many assumptions, and in no way prove that the Wealden rivers are antecedent; they merely suggest that the possibility is worth investigating. For proof, one must seek actual field evidence.

Briefly, the one piece of evidence that could clinch the case for antecedence is the occurrence in a valley of warped river terraces, with the warpings related to some structural axis (a fault or anticline) along which active movements have recently taken place (Fig. 104). Obviously, since the terraces are the remains of former valley bottoms, their possession of up-and-down gradients is open to no other explanation than localised uplift at a rate not exceeding that of river erosion. If the latter condition had not obtained, upstream from the axis of warping the true valley bottom, floored by solid rock, would have become buried beneath a considerable thickness of alluvium laid down by the river as it was impeded by the structural barrier rising across its path. Even in these circumstances, the river might not be diverted along a new course, providing that the rate of alluviation can keep pace with the rise of the barrier. In this case it is the evidence of the alluvial infilling rather than warped terraces which is important (Fig. 104).

Unfortunately, evidence such as this is rarely found. River terraces and alluvial accumulations are recent features of most landscapes—indeed, virtually all the terraces found in the British Isles date from the Quaternary and are related to its numerous changes of base-level. Hence, it is hopeless to try to prove that a river crossing an Alpine (Miocene) fold in Sussex was antecedent to that fold, for terraces developed at the time have long since disappeared. It is not surprising to find that the examples of antecedence which are fully substantiated by field evidence are, in geological terms, extremely youthful. For instance, Coleman (1958) has shown that an uplift of 120 ft across the course of the river Salzach in the Alps has occurred during the last 9000 years. Similarly, the famous instances of antecedent drainage in the Himalayas, involving the river Arun and its tributaries, are testified to by river gravels of recent date. These indicate an uplift of 100 ft or so across the course of the Dzakar Chu, a feeder of the Arun. This uplift is undoubtedly due to the continuing isostatic recovery of the Himalayan range. The possibility that over a much longer period the uplift has been greater is indicated by the depth of the gorges cut by the Arun and other rivers across the structural grain of the Himalayas. Certainly the discordance of these rivers could not be explained in terms of superimposition. There is no possibility whatsoever of either an undisturbed layer of sediments or a widespread planation surface, supporting a river system with north–south consequent elements, being developed above the present Himalayan summits in the short interval of time since the Miocene earth-movements. However, care is needed in interpreting the evidence here. The courses of the Himalayan rivers could be the result of headward extension by rivers which derived great erosive powers from the steep gradients over which they flowed down to the Gangetic Plain. Thus, rivers such as the Arun may have worked back comparatively recently through the uplifted and folded Himalayan block, and were *then* able to maintain their courses despite further uplift and even slight folding of the block. In short, to show that a river has successfully combated recent limited uplift across its path is by no means the same as proving that the river pre-dates entirely all the geological structures that it transects.

Superimposed drainage

The basic mechanism of this process has been described on p. 250. In scale and complexity the phenomenon may vary a great deal, as will be shown in the following pages. What must be emphasised is that superimposition of drainage does not merely involve the cutting down of rivers initiated on one *stratum* into an underlying *stratum*, the two forming part of a conformable sequence. The rocks exposed by drainage incision must be components of an altogether different sequence, normally of different structure, from those of the overlying cover.

A *Superimposition from marine benches.* During periods of stillstand of sea-level, wave action will produce erosional platforms of up to a mile or more in width. Such benches will frequently transgress geological structures, will possess a gentle gradient away from the shoreline, and will be covered to a greater or lesser extent by deposits of shingle, sand and mud. If these marine platforms are subsequently exposed by a fall of sea-level, any existing streams will extend their courses over the newly emerged sea-floor towards the displaced strandline. In a short time they will succeed in cutting through the veneer of marine deposits, and will be superimposed on to the underlying rocks (Fig. 105).

There is much evidence to show that superimposition of this type has occurred in many parts of Britain. At the onset of the Quaternary, the sea stood some 600 ft higher than at present, and large areas of what is now lowland Britain were submerged. During the Quaternary, the sea-level has fallen spasmodically, and numerous small marine benches

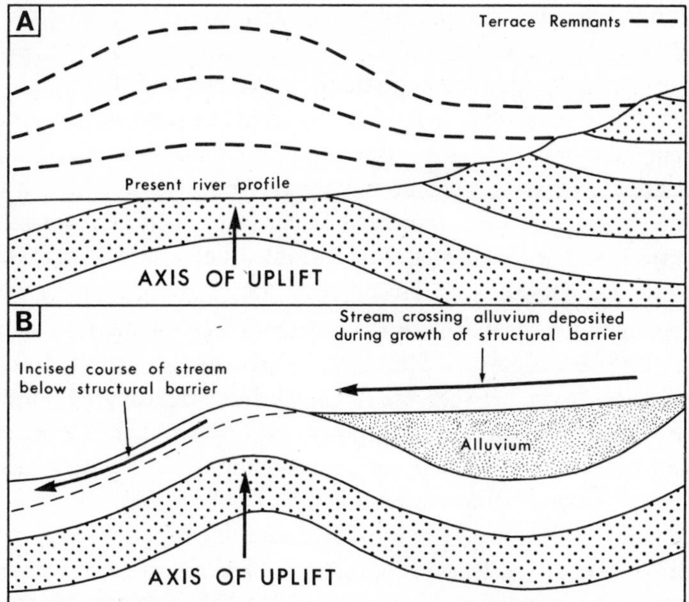

104 **Evidence of antecedence: warped terraces (A) and alluviation (B)**

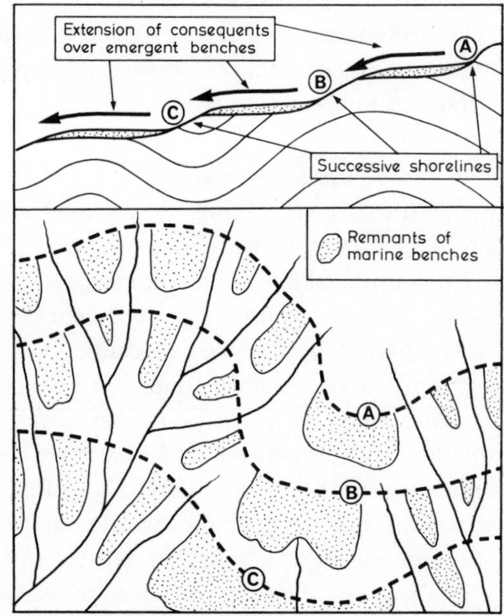

105 **The superimposition of rivers from emergent marine benches**

were formed during stillstands of the sea at 475, 430, 380, 330, 290, 230 and 180 ft O.D., as well as at even lower elevations. In some areas, the benches form a distinctive morphological staircase, and the streams here have gradually extended their channels over the successively revealed wave-cut platforms. At the same time, they have been repeatedly rejuvenated, and have become deeply incised into the higher older benches. In some instances, marked discordances between streams and structures have been produced. For example, the lower Meon of Hampshire, which crosses the western end of the anticlinal ridge of Portsdown, has demonstrably been superimposed from late-Quaternary benches covered by plateau-gravels. However, in areas such as the central part of the South Downs between the Arun gap and Brighton, the old shorelines are aligned parallel to the strike of the Chalk and the slopes of the emerged benches roughly followed the direction of dip; hence a basically accordant pattern of superimposed valleys has been formed.

In detail, the drainage patterns produced by superimposition from marine benches reveal features of interest. Where the old shorelines are straight in plan and the various benches slope in a constant direction, a simple pattern of subparallel consequents will tend to develop. However, where the shorelines are more irregular, with pronounced headlands and embayments, a semi-dendritic pattern will result. The reason is that marine platforms normally possess gradients approximately at right angles to the shore, and in embayments there will be a tendency for convergence of drainage to occur (Fig. 105).

B *Superimposition from extensive marine surfaces.* Over longer periods of geological time, particularly if the sea is slowly transgressing previous land-areas, much more extensive surfaces of marine planation may, in the view of some authorities, be formed. An example is the Calabrian surface of south-east England, whose form and extent have been traced by Wooldridge and Linton (1955). In some localities, such as the Chalk country of Hampshire and Wiltshire, the marine plain may have approached a width of 20 miles. It has been suggested that during the Calabrian transgression all previous relief features were obliterated, and following the regression of the sea a wholly new pattern of drainage, controlled by the form and slope of the marine surface, was superimposed on to the underlying structures. At present there is a fundamental contrast between, say, the river pattern west of Salisbury, where streams such as the Wylye, Nadder and Ebble are closely guided by the Alpine structures (p. 101) and where there is no evidence of extensive

Calabrian planation, and that of the Chalk area to the east. Here morphological evidence of the marine transgression is quite strong, and the discordance of streams such as the Avon, Test and Itchen, which run broadly from north to south across the structural grain, can be explained by superimposition from a southward-sloping Calabrian sea-floor and from the series of marine benches below 475 ft O.D. left by the regressing sea (Fig. 106).

c *Superimposition from fluvio-marine deposits.* In certain circumstances, associated with very active fluviatile erosion and transportation and very weak and ineffective wave attack, the zone immediately offshore may be the scene not of marine planation but of large-scale alluvial sedimentation. The result will be a 'fluvio-marine' plain, gradually built out seawards from the land and crossed by the lower courses of the rivers which provided the material for the plain. A subsequent fall in sea-level will cause the superimposition of these rivers, from their own alluvial deposits, on to the older rocks of the original sea-floor. This process, referred to as 'auto-consequence', was at one time thought to be the

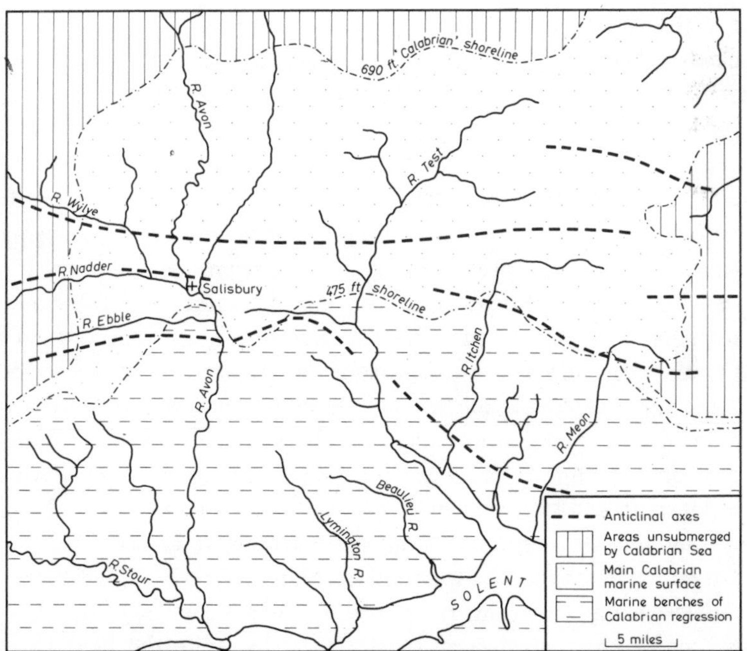

106 The drainage pattern of central southern England

cause of the discordance of the Hampshire rivers. More recently, it has been revived by the French geomorphologist, Pinchemel (1954), who postulates that the drainage system of much of south-east England has been superimposed from an aggradational plain of early-Pliocene date, uplifted and warped along axes passing through the Weald and the Vale of Pewsey. Such a feature is well attested in Belgium and northern France, but the evidence in England is by no means clear and the whole theory must be regarded as highly controversial.

D *Superimposition from major geological formations.* Over extremely long periods of geological time, ancient structures may become masked by accumulations of vast thicknesses of younger sediments. For example, during late-Cretaceous times much of Britain was affected by a major marine transgression (the so-called 'Cenomanian' transgression). All of what is now lowland England was inundated, and some authorities believe that many parts of highland Britain (including the uplands of the South-west peninsula, the Welsh Massif, the Lake District, the Pennines and the Southern Uplands of Scotland) were at least partially submerged. The Cenomanian sea seems to have spread over a land-area previously reduced by erosion to a peneplain. It was therefore associated not with planation, but with the accumulation of Upper Cretaceous marine rocks, of which the Chalk is easily the most important single formation.

At the end of the Mesozoic era a major uplift, centred on the old Caledonian-Hercynian massifs of western and northern Britain, expelled the sea and gave to the country as a whole, with its covering layer of Chalk, a broad easterly tilt. On the new land-area, a consequent drainage system, comprising many large streams draining eastwards towards the present North Sea, must have been formed. In the view of many geomorphologists, prolonged erosion during the Tertiary era has led to the stripping of the Chalk from the greater part of the country, and the superimposition of the Chalk drainage on to the older Mesozoic and Palaeozoic rocks. Subsequently, adjustment to the rock outcrops and structures thus exposed has reached an advanced stage. Even so, many instances of streams which disregard structural trends, and which may therefore be attributed to superimposition from the Chalk, may be identified in areas to the west and north of the present Chalk margins. In Wales, many important rivers contain long stretches in which they run broadly from north-west to south-east, despite the fact that the Caledonian and Hercynian structures are aligned from north-east to south-west (the central Wales synclinorium and Teifi anticlinorium) or from east to west (the south Wales coalfield synclinorium). However, superimposition from a Chalk cover is not the only available explanation of such discordances between streams and structures. In some areas, the drainage system inherited from the Chalk may itself have been greatly modified or even destroyed altogether during later erosional episodes. Sissons (1954) has shown that certain rivers of south-west Yorkshire (including the Aire, Calder and Don), long regarded as 'Chalk consequents', may in fact have been superimposed from a staircase of marine benches of Tertiary age, in the manner described on p. 256.

Among the geomorphologists who accept the theory of superimposition from the Chalk there are naturally some divergences of view. This is largely inescapable, for the reconstruction of old drainage patterns and their subsequent histories must remain speculative when so much of the evidence has long since vanished. One of the earliest proponents of the superimposition theory, Davis (1895), postulated that over central and northern England, away from the structural complexities of the Weald and London and Hampshire Basins, the Chalk was uniclinal in structure, and nourished a series of subparallel consequent streams (represented today, for instance, by the rivers draining the eastern flanks of the Pennines). Linton (1951), on the other hand, has suggested that the surface of the Chalk was interrupted by east-west synclines, forming in effect northerly counterparts of the London and Hampshire Basins. These downfolds would have been occupied by longitudinal consequent streams, similar to but longer than the present-day Kennet-Thames, which would have been joined from north and south by transverse consequents draining intervening anticlinal uplands. Following their superimposition on to older rocks, both sets of consequents have been largely disrupted. Even so, Linton was able to make out a case for the former existence of one longitudinal consequent, the 'Proto-Trent', which he envisaged as developing in a Chalk syncline crossing north Wales and the English Midlands. Today this great river is represented only by the upper course of the Welsh Dee and the upper and middle reaches of the modern Trent. Among the transverse consequents which joined it from the north may be the rivers Dove and Derwent, trenching the southern borders of the Pennines. These streams show a clear disregard for the ancient geological structures of the area, and are perhaps most satisfactorily explained in terms of superimposition from a stratum or surface which no longer exists. In southern Scotland, where the greater overall resistance of the old Palaeozoic rocks has militated against the large-scale development of subsequent streams, the river Tweed may represent another synclinal consequent which in its middle and lower course is virtually intact. Only the upper part of the stream, west of Biggar, has been diverted northwards to form part of the Clyde system.

Several writers have adopted the theory of superimposition from a Chalk cover to explain the Welsh drainage system. The Chalk here is generally assumed to have had a tilt towards the south-east. Jones (1951) has suggested that it in fact formed part of a dome-like structure, centred over Snowdonia and giving rise to a radial pattern of consequent streams. The existing consequent elements in Wales certainly have a tendency to fan out south-eastwards in precisely the way that such a theory would demand (Fig. 107). The theory is undoubtedly attractive, but some features of the pattern indicate that in detail the process of superimposition has been more complicated than would appear at first sight. Thus, although many consequent elements are still easily identifiable, the original system has been much broken up by subsequent stream growth. The activity of those working back from the western and south-western coasts, taking advantage of faults, shatter-belts and the like, is especially noteworthy. This high degree of adjustment to structure of the Welsh drainage points to the rapid removal of the Chalk cover early in the Tertiary. The late Tertiary has seen the virtually uninterrupted denudation of the Palaeozoic rocks, encompassing in the opinion of Brown (1957) at least three cycles of sub-aerial erosion (p. 285), during each of which the adaptation of the streams to lines of weakness would proceed a stage further. However, there are also in Wales stream elements which, to judge from their orientation, are of subsequent origin, but which are not related to known geological weaknesses. It is just conceivable that these are examples of 'superimposed subsequents'.

In most theories of superimposition, it is conveniently assumed that the initial consequent system is lowered in an unchanged form on to an older system of rocks, and that the formation of subsequents, the fragmentation of the consequents, and the whole process of adjustment to structure all take place only *after* the event of superimposition. When marine benches or surfaces veneered by deposits are involved this is quite reasonable, for the conditions do not favour the appearance of any subsequents prior to superimposition. However, when rivers are let down from a great thickness of younger rocks, the situation is totally different. In the example of the Welsh drainage, it cannot even be assumed that superimposition from the Chalk alone has occurred. It is

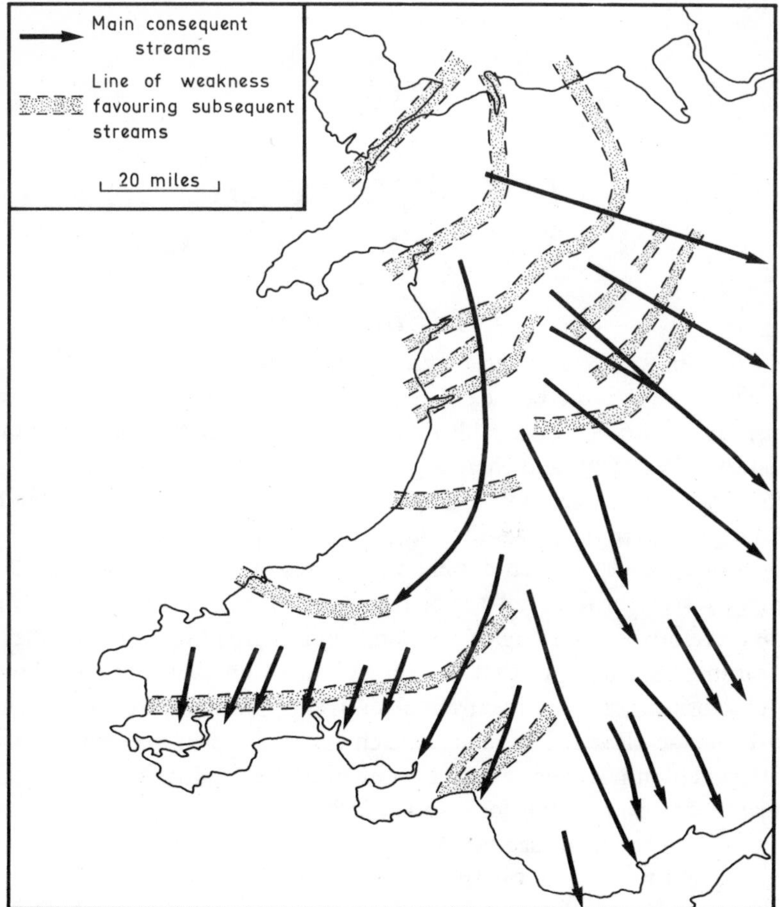

107 Factors in the drainage evolution of Wales

probable that in many localities Jurassic and/or Triassic rocks were sandwiched between the Chalk and the underlying Palaeozoics. If so, it is hardly likely that consequent rivers, cutting vertically through a thousand or more feet of varied Mesozoic sediments, would remain wholly unaffected by the development of subsequent streams. Thus the great complexity of the Welsh drainage may stem not only from the rather distant date (early Tertiary) of the actual superimposition, but also from the fact that the superimposed river system had itself undergone considerable modification.

8

THE STUDY OF PLANATION SURFACES

INTRODUCTION

It has been widely accepted by geomorphologists that so-called 'erosion surfaces' form important elements in the landscape of the earth. Such surfaces afford an invaluable means of reconstructing the erosional history of a region, and their study may throw light on other geomorphological problems, for example those posed by the evolution of drainage patterns (pp. 232–61). Furthermore, in some areas they actually form prominent, or even dominant, features of the physical landscape (as on Dartmoor (pp. 128–30) or the Grands Causses (pp. 141–2)), and are therefore as deserving of geomophological attention as river valleys, escarpments, slopes, coastal landforms and so on. More usually however, erosion surfaces are far from obvious as physical features, for reasons to be made clear below, and in such instances their recognition and interpretation pose very difficult problems. Often the very existence of a surface may be a matter for dispute; and even a surface which is readily apparent in the field may be interpreted, in terms of its origin, age and subsequent history, in a number of ways.

It is perhaps unfortunate that by constant usage geomorphologists have given the term 'erosion surface' a very limited and specialised meaning. It is, in fact, almost universally used to describe only flat or near-flat erosional plains, formed very close to base-level and resulting from cycles of erosion that have reached well beyond the stage of youth and in many instances into the stage of old age. Thus peneplains, panplains, pediplains and planes of marine erosion are all erosion surfaces in the accepted sense of the term. In addition to these major features there are, of course, near-level surfaces that are much more limited in extent and local in distribution. Within an individual river valley that has experienced a series of base-level changes, remnants of former valley floors may occur as pronounced 'levels' or spur-flattenings on the present valley-side slopes. Geomorphologists commonly refer to such features as 'benches' or 'valley-side benches'; and where they have not been greatly dissected by more recent erosion, so preserving some continuity along the valley edge, and are overlain by river gravels or other fluviatile deposits, the levels are known as 'river terraces'. Again, in areas formerly submerged by the sea and affected at the higher levels of the sea by wave erosion, bench-like forms (representing old wave-cut platforms) backed by comparatively steep bluffs (old degraded sea-cliffs) may be identified in the present landscape. These features have been variously referred to as 'marine benches', 'marine flats', 'marine terraces' and 'marine platforms'. Where the benches are occupied by marine sands and shingle they are known as 'raised beaches'.

Strictly speaking, of course, all parts of the land surface that are not directly depositional in origin are surfaces of erosion. Thus the dip- and scarp-slopes of a cuesta, although owing their form in large measure to the influence of geological structure, are 'surfaces' produced by weathering and erosion. Again, many geomorphologists have recognised the existence of 'structural surfaces' (p. 69), which are carefully distinguished from 'erosion surfaces'. Yet these are formed where weak rocks (for example, clays and sands) have been stripped away neatly from underlying more resistant strata (sandstone, limestone or chalk) in such a way that the topmost surfaces of the latter are undissected and remain as striking 'plane' elements in the landscape; in short, the structural surface has itself arisen only as a result of denudational processes. Even where deposition has recently been active, as in the boulder-clay plains of eastern England or the peats and clays of the Fenland, some erosional modification of the surface forms has usually occurred.

A good case could be made out for abandoning the term 'erosion surface', and for simplifying the confusing terminology which has been referred to above. Indeed, it is the intention in this book to use as far as possible only two terms for the description of these forms. The major surfaces, known hitherto as erosion surfaces, will be referred to as 'planation surfaces', and the more limited surfaces will be described simply as 'partial planation surfaces'. It will be convenient, however, to retain 'river terrace' and 'raised beach', since these have a geomorphological and also geological meaning; furthermore, they are necessary for use in describing fluviatile and marine deposits which do not rest on planed surfaces but are wholly aggradational in origin. It is considered that the terms 'planation surface' and 'partial planation surface' are more appropriate than 'erosion surface' for two reasons. They enable one to make an immediate distinction between surfaces cut across differing rock-types and varied geological structures (planation surfaces) and those aligned parallel to the dip of the underlying rocks (structural surfaces), without the implication that formation of the latter owes nothing to erosional processes (Fig. 108). Further, the term 'planation surface' is sufficiently non-specific to be used in describing erosional plains irrespective of their mode of origin (whether they are peneplains, panplains, pediplains or the result of marine planation).

THE RECOGNITION AND INTERPRETATION OF PLANATION SURFACES

The most usual ways in which planation surfaces are formed are fully discussed in chapter 6. The aim of the following discussion is to illustrate the methods used and the problems encountered by the geomorphologist in his study of particular surfaces. By way of introduction certain general considerations must be outlined.

Perhaps the first, if not the most obvious point, is that there is little or no reason for seeking planation surfaces only at or near the present base-level of erosion. Indeed, if surfaces are found close to the existing sea-level, the chances are that this is purely coincidental and that they have, in fact, been raised or lowered from their original altitudinal positions by more recent crustal movements. A great deal of evidence exists to show that the sea has stood at its present level for only a very short period, geomorphologically speaking. In the British Isles, and indeed elsewhere, the base-level of erosion has changed many times during the Quaternary. These 'shifts' have been caused in part by periodic depression and recovery of the land isostatically as the ice-sheets have waxed and waned, and in part by actual eustatic rises and falls of sea-level caused by changes in the volume of water locked up in the ice. Another important factor, sometimes overlooked, has been a more continuous decline in sea-level, quite independent of those due to glacial causes, which has been going on since mid-Tertiary times and perhaps earlier. At the onset of the Quaternary era, the sea-level was some 600 ft higher than at present; and some authorities have even suggested that at some stage in the Tertiary it may have stood at a height of nearly 2000 ft O.D. There is indeed a case for regarding most landforms below 600 ft as of wholly Quaternary origin, and those above 600 ft as of Tertiary age—though some glacial and periglacial modification of these must of course be envisaged. Furthermore, since the duration of the Quaternary was in the order of only 1 million years, there has been insufficient time for the development of other than very limited partial planation surfaces, both sub-aerial and marine, related to the spasmodically falling base-level below 600 ft.

It follows from what has been written that all true planation surfaces, since they must be at least of Tertiary age and related generally to base-levels above 600 ft O.D., will have undergone a greater or lesser amount of subsequent modification by river erosion, weathering and other processes. As a result, their former characteristics may no longer be recognisable, and it may therefore be impossible to decide easily whether the surfaces are peneplains (comprising gently undulating convexo-concave slopes) or marine planes (almost level or with a continuous but barely perceptible slope away from the ancient coastline). The actual degree to which the surfaces will have been modified or dissected will depend on several factors. The older the surface, the greater will be its fragmentation, unless special circumstances, such as the existence of overlying deposits, have conspired towards its preservation. The resistance of the underlying rocks will obviously be a factor of paramount importance. In an area of weak sedimentary rocks, in which a later erosion cycle is able to progress rapidly, the planation surface will soon be greatly modified or disappear altogether in the general lowering of the land surface; conversely, in an area of resistant metamorphic and igneous rocks, its existence will be far more protracted. One might go on to argue that if two planation surfaces of the same state of preservation are found in adjacent areas of resistant and unresistant rocks, then that in

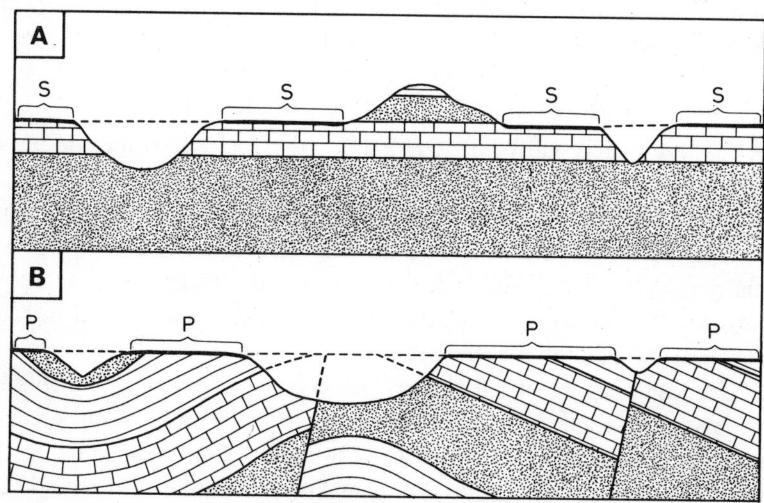

108 Structural surfaces (A) and planation surfaces (B)

the former must be considerably the older. Another factor influencing the preservation of planation surfaces will be the closeness of the drainage network, which in turn will be influenced by geological and climatic controls. Thus the greater the drainage density, the greater the likelihood of early destruction of the surface by interfluve wasting.

It will be clear that in the present landscape planation surfaces may exist in a variety of forms. If of comparatively recent date (say late-Tertiary) or for a long period protected by overlying rocks (p. 268), if developed in very resistant rocks, and if dissected by an open network of river valleys, the surface will comprise extensive plateaus and flat-topped divides, all occurring at approximately the same elevation (Fig. 109). If geological conditions are less favourable, or the surface of much greater age (early Tertiary or Mesozoic), it may be represented only by a general accordance in height of the main summits of an area. If conditions are decidedly adverse, or the surface extremely ancient, it will almost certainly have long since disappeared as an identifiable element in the landscape.

Obviously the argument has been oversimplified so far, in that any one planation surface will transect a variety of rock-types and will be drained by variably spaced stream networks. When subjected to rejuvenation it will therefore be wasted and modified more in some localities than in others. For example, a particular surface may be exceptionally well preserved in a region of very resistant rocks, like the Palaeozoics of Wales or the south-west peninsula of England, but destroyed altogether in an intervening zone of weak rocks, such as the Triassic and Liassic formations of the lower Severn–Bristol Channel depression.

The correlation and dating of planation surfaces

For reasons which will become apparent, the correlation and dating of planation surfaces is beset with difficulties, and has given rise to some of the most enduring controversies in geomorphology. It must be emphasised that correlation and dating do not form distinct topics, but are usually closely related—indeed, the only way of determining the age of a surface may be to relate it, on morphological grounds, to a surface elsewhere the date of which is already known.

The most simple method of correlation available to the geomorphologist involves the equating of surfaces which occur in different areas at the same elevation above sea-level. The assumption is made that the surfaces were once continuous, but that a concentration of erosion in intervening parts has led to the geographical fragmentation of the original

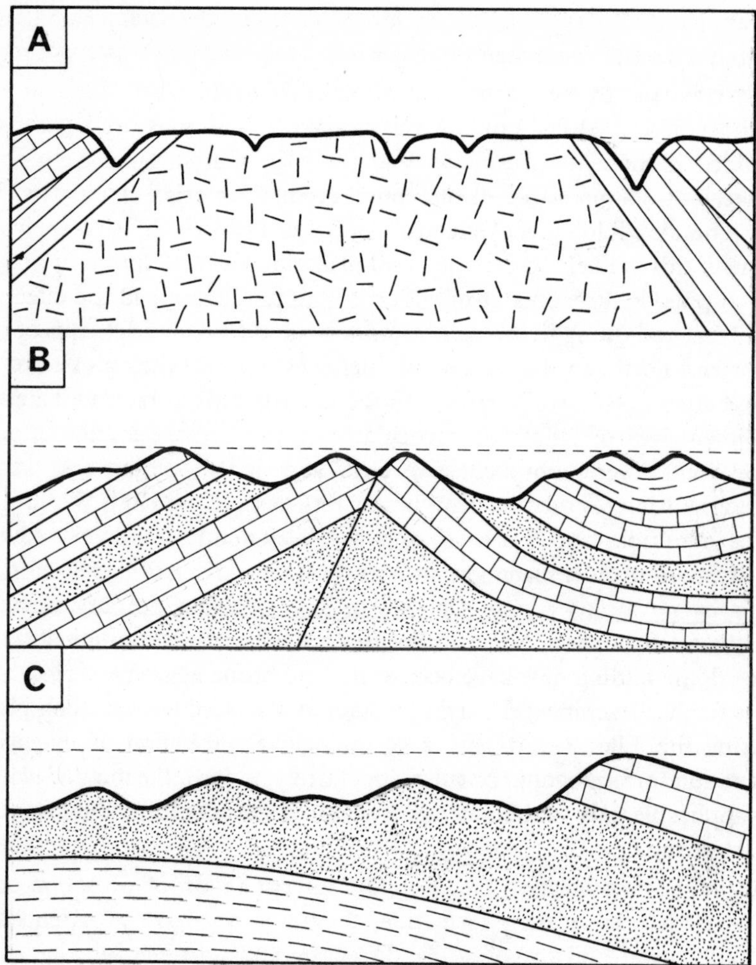

109 **Different degrees of dissection of upraised peneplains**

feature (much as in the example referred to above). Thus it is possible to make a 'height correlation' between a hill-top planation surface at 700–900 ft O.D. in the Chalk and Lower Greensand country of south-east England and marked erosional bevellings at 950–1250 ft on the Carboniferous Limestone and Millstone Grit of the southern Pennines. The small difference in height of the two may be ascribed to the fact that in the former area the rocks are only moderately resistant, whereas in the Pennines they are very resistant and have not subsequently been wasted to the same extent. In the intervening area of the Midlands, however, the

Jurassic and Triassic formations are weak, and the surface has disappeared with the fashioning of major vales by rivers such as the Trent.

It need hardly be stressed, however, that height correlation in the manner described can only be very tentative, and must be rigorously avoided if there is a suspicion that the planation surfaces have been subjected to appreciable displacement from their original 'heights' by warping and folding movements. So far as the British Isles are concerned, this is likely to be true of all planation surfaces formed prior to the Alpine disturbances of mid-Tertiary times. These had the effect of uplifting and tilting the ancient Caledonian and Hercynian massifs of the west and north, so that large-scale deformation of existing surfaces in those areas must have occurred. In the less rigid Mesozoic and Tertiary sedimentaries of England, particularly in the south-east, the Alpine movements were more effective still, and sharp folding and some strike-faulting resulted in many parts, for example the High Weald, the Isles of Wight and Purbeck, and the Weymouth area. Very severe 'distortion' of older planation surfaces must have arisen in such localities. It is by no means certain whether or not even late-Tertiary surfaces escaped deformation. Some authorities consider that the culmination of the Alpine folding may have been at the end of the Miocene period, and that further warpings, the dying echoes of the earth-storm, took place during the Pliocene. If this were so, height correlation of extensive planation surfaces would be ruled out entirely; indeed, the method would be applicable only to the much more limited surfaces formed during the Quaternary.

Another complicating factor in the interpretation of planation surfaces is that, since their formation, they may not have experienced a straightforward history of uplift, dissection and warping, but that over long periods of geological time they have been 'fossilised' in an intact state by overlying younger deposits. Such surfaces do not, of course, always remain as geological unconformities, for during a later period of erosion the cover rocks may be removed and the buried plane 'resurrected' or 'exhumed' as an important feature of the landscape. In the process of revival a great deal will depend on the precise nature and degree of resistance of the rocks which lie above and below the unconformity. Probably the most important feature of this type in southern England is the so-called 'sub-Cenomanian' (or, more loosely, 'sub-Chalk') unconformity. This represents a major eroded surface, of very faint relief, which is developed across tilted and folded rocks ranging in age from Permian to Lower Cretaceous. It has been preserved for a very long time by overlying Upper Greensand and Chalk, which are marine deposits laid down during a great transgression of the sea in Upper Cretaceous times (Fig. 110). In much of Devon and Dorset, these younger rocks are being dissected and eroded away (pp. 69–72), but the sub-Cenomanian surface is not being resurrected as an important element in the present landscape, for the simple reason that the underlying rocks (for example, the Keuper Marls of east Devon) are less resistant than the Greensand and Chalk. Another notable unconformity in southern England separates the lowermost Eocene rocks of the London and Hampshire Basins from the eroded upper surface of the Chalk. This is known as the 'sub-Eocene' surface, and was formed in very early-Tertiary times by planation of the recently uplifted and exposed Chalk. For most of the Tertiary era the surface has remained fossilised by overlying sands and clays, but on the margins of the two sedimentary basins these weak formations are being removed to reveal the sub-Eocene surface as a striking element in the landscape of some areas. The surface is no longer horizontal, however, since it has been affected by the more recent Alpine movements. In general it takes the form of a 'plane' surface tilted at about 2° and forming the lower parts of the Chalk dip-slope (Fig. 111).

On general grounds one might assume that such exhumed surfaces would be particularly evident in the west and north of Great Britain, where the old Palaeozoic rocks are for the most part very resistant to erosion. Furthermore, it has been postulated that these Caledonian-Hercynian uplands have been buried, at least in part, beneath younger sedimentary formations of Mesozoic age. Thus much of Wales may have been 'submerged' beneath Triassic and Liassic rocks, and most of highland Britain may have been inundated by the Chalk sea. In the

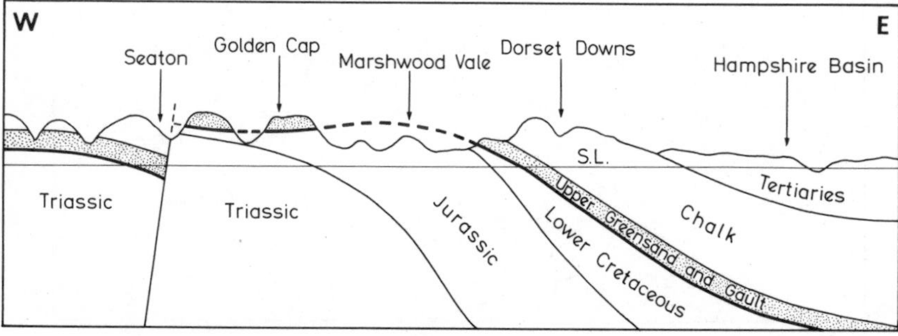

110 The sub-Chalk surface of southern England

light of these possibilities the existence of exhumed surfaces in the old massifs may be suspected with even more confidence.

The definite identification of such surfaces is, however, not an easy task. In the case of the sub-Eocene surface there is little difficulty, for a direct link between the surface and the unconformity (outcropping at the base of the Eocene scarp-face) can be readily established. More commonly, however, the overlying strata will have been completely removed, at least in the area of study, so that interpretation again becomes very tentative. This problem of the disappearance of deposits is not confined to exhumed surfaces. In theory at least, all planation surfaces should bear an overlying cover of detritus (marine shingle, fluviatile gravels, or material weathered *in situ*) which can be used as a guide to their mode of origin and possibly assist in the dating of the surfaces. In reality, such deposits rarely remain on uplifted and well-dissected surfaces, or if they do still exist have been so modified by subsequent disturbance, weathering and the incorporation of fresh material as to be virtually indecipherable. Probably the best-known planation surface in Great Britain is the so-called Welsh 'tableland' (pp. 281–6). As will be shown, this has been interpreted in a variety of ways, but no final explanation and dating can be forthcoming since all the deposits that must once have lain on the surface have been removed, mainly by glaciation and allied processes in the Quaternary. Another well-known surface is the hill-top peneplain of south-eastern England, now represented by numerous summits on the Chalk and other rocks of the area at a height of 700–900 ft O.D. (pp. 276–8). In certain places, as to the west of Salisbury, the planation surface is associated with thick deposits of

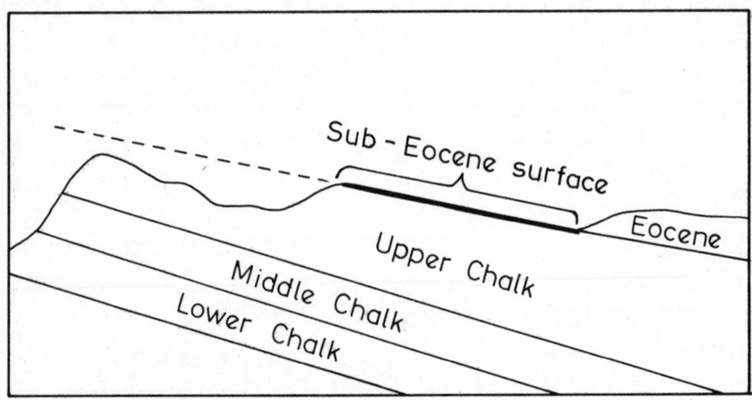

III The sub-Eocene surface of southern England

'clay-with-flints', the study of which ought, theoretically, to throw light on the mode of origin and age of the surface. However, close examination reveals that the clay-with-flints is a complex deposit, containing various elements of widely differing age (for example, rounded flints, red clay and sands from a former Eocene cover, clay derived from the weathering of the Chalk in a 'tropical' climate in the Tertiary, and solifluxion debris and loess of Quaternary date). Moreover, the clay-with-flints occurs elsewhere on the Chalk, at elevations below 700 ft and both on the 'sub-Eocene' and on Quaternary partial planation surfaces of sub-aerial and marine origin.

Even where surfaces are mantled by deposits that are manifestly less complicated in origin, a great deal of care in interpretation is needed. In many areas which were subjected to cold climatic conditions in the Quaternary, the planation surfaces have a veneer of frost-shattered and soliflucted debris, but it would be wrong to assume that they have been more than modified by cryoplanation processes (pp. 180–2). Again a surface with an extensive cover of gravel containing obvious marine elements, such as rounded beach cobbles, is not necessarily the product of wave action. The marine material may just have been lowered from a higher marine surface on to a sub-aerial 'flat' by agencies such as soil creep, slumping and solifluxion. Conversely, planation surfaces of marine origin may bear fluviatile gravel, for such surfaces, when exposed by retreat of the sea, usually have very gentle gradients, so that streams extending from the old land area across them will flow slowly. As a result their powers of both erosion and transportation are restricted, and they meander and anastomose freely, laying down extensive spreads of riverine gravel.

The pitfalls to be avoided in the interpretation of deposits on planation surfaces have been demonstrated by King (1950) in his study of the great African pediplains (pp. 170–4). King considers that these surfaces have been wrongly interpreted and dated by other geomorphologists, who have failed to realise that datable deposits give (i) only an *end-date* for the formation of a surface, and (ii) only an indication of the *final* processes responsible for the fashioning of a surface. Thus a pediplain formed in its essentials during Jurassic and earlier times may have deposits resting on it that have accumulated at any time since, and that have resulted merely from the retrimming of the already existing plain. The most important planation surface in Africa, termed by King the 'Gondwana Pediplain', was in fact developed prior to the break-up of Gondwanaland and the roughing-out of the African continent by continental drift during Cretaceous times. Yet in some localities it is overlain by Miocene deposits, and has accordingly been assigned, in King's view mistakenly, to a mid-Tertiary erosion cycle.

If planation surfaces cannot in many instances be interpreted by study of deposits resting upon them, there is at least the possibility that they may be dated—and indeed their actual mode of origin determined—by an examination of deposits found elsewhere. As an uplifted land-mass is progressively reduced to a near-level plain during a major cycle of erosion, so in adjacent downwarped areas (either marine basins or enclosed terrestrial basins) large-scale accumulation of the detrital material must occur. Thus the formation of considerable thicknesses of sands and clays in the 'Wealden Lake' in early-Cretaceous times was undoubtedly related to sub-aerial erosion in the more northerly and westerly parts of Britain. This erosion may indeed have led to the fashioning of the planation surface later transgressed and fossilised by the Chalk sea, as described above. Sometimes an actual 'cycle of deposition' may be recognised. In this the basal deposits will comprise very coarse materials (conglomerates), which give way vertically to medium-grained sandstones and finally to fine clays and marls (Fig. 112). This particular sequence may be taken to indicate a gradual decline in the relief and angles of slope in the land-mass providing the detritus, or in other words a steady evolution towards a planation form. Such a sequence occurs, in a very general form, in the Triassic rocks of parts of Great Britain. Thus, within Devon and Somerset, the Lower Triassic (Bunter) rocks are sandstones and conglomerates, and these are overlain by Lower Keuper Sandstones and Upper Keuper Marls. The manner in which the uppermost of these deposits transgress near-by 'bevelled' areas of Palaeozoic rocks, and the fact that the Triassic rocks appear to have accumulated in desert basins, both indicate the probable existence of a pediplain surface in Britain by the close of the Triassic period.

Finally, information about the mode of origin and date of a surface can sometimes be derived from study of the drainage patterns developed upon it. The method involves basically the recognition of the consequent elements in the drainage, and the reconstruction of the conditions that would have promoted such a consequent pattern. For example, it may be possible to show that the original drainage lines of an area where a planation surface is well developed were markedly discordant to the geological structure. Providing that the possibility of antecedence can be ruled out, it may be inferred that superimposition of drainage has

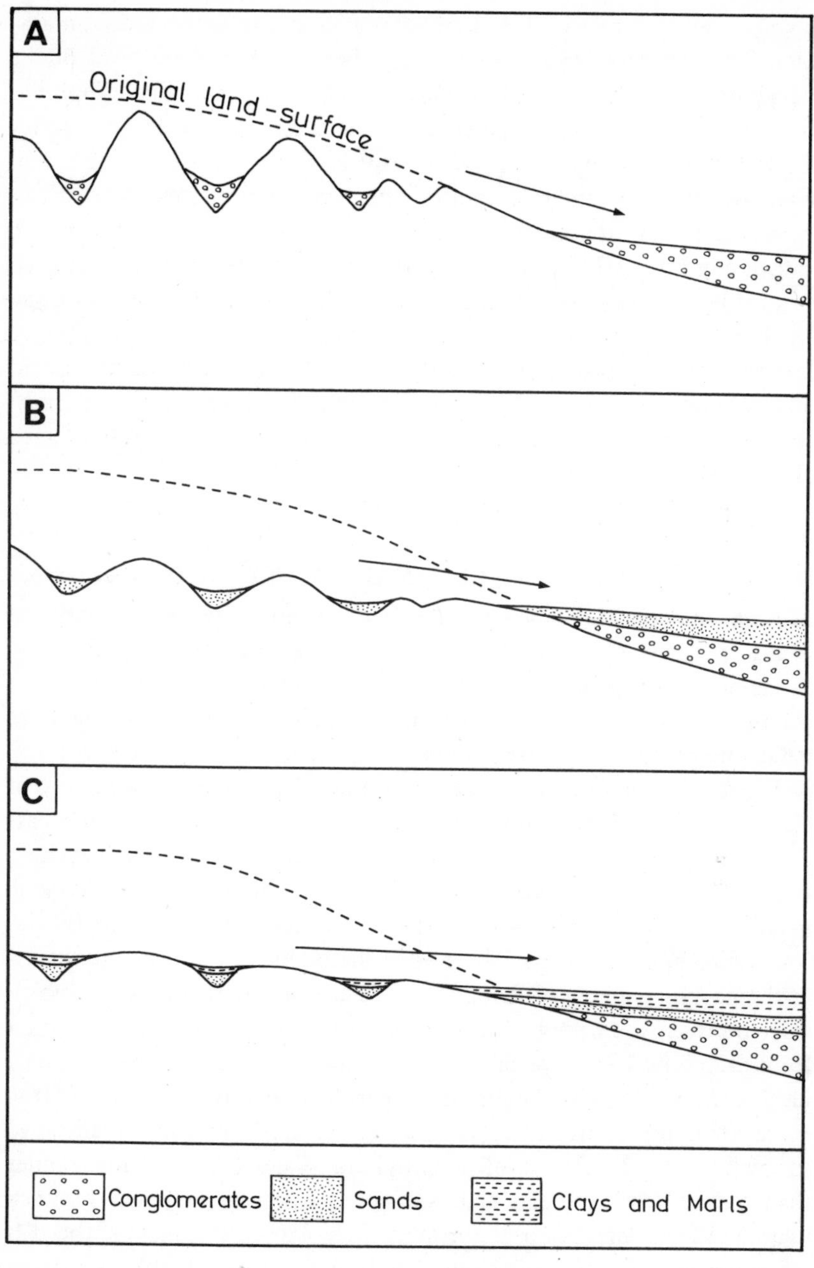

112 The idealised relationship between erosion and sedimentation

taken place (pp. 254–61). The drainage system of the British Isles contains many discordant elements, which have been explained in terms of the development of the initial drainage on a former layer of Chalk, uplifted from beneath the sea and given a general west–east tilt in very early-Tertiary times. The subsequent exposure of the sub-Chalk surface in areas of very resistant rock is a possible corollary of this thesis. Elsewhere in Britain there are planation surfaces drained by rivers which show a high degree of accordance with geological structure, and in such cases it is difficult to escape the conclusion that sub-aerial erosion alone has been involved, and that there have been no marine episodes (since these give rise to planation of rocks irrespective of their resistance, and new drainage systems which ignore in many instances structural lines) or fossilisation and exhumation (which also tends to produce discordance of drainage).

PLANATION SURFACES IN ENGLAND AND WALES

Most geomorphologists are agreed that the landscape of England and Wales contains clear evidence of several episodes of past planation. Some of this evidence, together with the interpretations that have been put upon it, will now be discussed, with the aim of exemplifying in more detail the problems outlined above.

The planation surfaces of England and Wales occur at a variety of elevations. In the scarpland country of the Midlands and the south-east they are represented by the cuesta summits, sometimes clearly bevelled, at between 500 and 1000 ft O.D. However, in the older Palaeozoic massifs of the west and north, surfaces between 1000 and 2000 ft (and sometimes higher still) have been identified. One of the major problems in the geomorphology of Britain is to decide on the relationships which exist between these higher and lower surfaces. Are they quite distinct from each other, in terms of age and/or mode of origin, or are the high-level surfaces merely the upraised parts (relatively speaking) of the more low-lying plains? If the former is the case, then it is realistic to suppose that the surfaces of, say, upland Wales are very much older than those of, say, the Chalklands of south-east England, and that they may date from late-Mesozoic or even early-Mesozoic times. If the two sets of surfaces are of similar age, then it ought to be possible to account for the differences in elevation in terms of the effects of known earth-movements (such as those of the Alpine period). Furthermore, the surfaces should not be horizontal, but should have perceptible gradients which can be extrapolated from one surface fragment to another.

In order to illustrate the complexity of the problem, it will be sufficient to discuss the form, possible mode of origin and likely age of the surfaces in two areas only: central southern England, as a representative part of lowland Britain, and the Welsh massif, as a representative of highland Britain.

The planation surfaces of central southern England

These are, in order of decreasing age, as follows:

A *The sub-Cenomanian, or sub-Chalk, surface.* This takes the form of a major unconformity which is exposed most clearly in the coastal sections of southern England between Weymouth and Exeter. The surface is eroded across previously deposited and disturbed Triassic and Jurassic sediments, and is protected today by an overlying cover of Upper Greensand and Chalk (see also pp. 69–72). At present the unconformity lies at a depth of over 2000 ft below sea-level in southern Hampshire, but it rises westwards towards Dorset and Devon, though not steadily since it has been disrupted to some extent by folding and faulting (as at Beer) in early-Tertiary and mid-Tertiary times (Fig. 110). In parts of Dorset (for example, on the coast near Weymouth) the unconformity is carried well above sea-level by local folding, but elsewhere (as near Dorchester) it still lies well below base-level. Over much of eastern Devon, on the other hand, the sub-Chalk surface is above sea-level, and is displayed by many sea-cliffs, where Chalk and Upper Greensand are seen resting on the eroded Jurassic and Triassic rocks. The general rise is continued to the west of the Exe valley, and in the Haldon Hills, which comprise Upper Greensand resting on Permian rocks, the unconformity reaches a height of 700–800 ft O.D. Further still to the west any Cretaceous strata that once existed have been destroyed.

The two most noteworthy attributes of the sub-Chalk surface are (a) that it is not an important feature in the landscape of central southern England (for the reasons given on pp. 268–9), and (b) that there is a rise in the unconformity when traced from east to west. It seems possible, therefore, that remnants of the surface may exist, as planational elements in the present landscape, on the old hard Palaeozoic rocks of western Devon and Cornwall.

B *The sub-Eocene surface.* This occurs both as an important unconformity beneath the lowermost Tertiary rocks of the Hampshire Basin, and as an exposed and easily recognisable surface feature of the Chalk dip-slope adjacent to the margins of the Tertiaries. At present the surface has an average gradient of about 2°, though much higher angles are encountered in areas where the Alpine movements (which, of course, post-date the surface, and thus led to its deformation) were more pronounced, as in the Isles of Wight and Purbeck. The difference in angle of dip between the Eocene rocks and the Chalk is generally slight, but is sufficient to cause an 'overstep' of the tilted zones of the Chalk by the younger rocks. This overstep is orientated broadly from south to north, and as a result in the London Basin the highest zones of the Upper Chalk are absent. From this valuable piece of evidence it may be inferred that, to the north and north-west of the Chilterns, Berkshire Downs and Marlborough Downs, the sub-Eocene surface must at one time have passed on to rocks older than the Chalk and exposed during the sub-Eocene cycle of erosion. Remnants of the surface may still exist in such areas, but owing to the effective removal of all Tertiary rocks these can no longer be recognised with any certainty. There is also the possibility that within the Chalklands themselves the sub-Eocene surface is not confined to localities close to the present Tertiary margins, but is represented by some at least of the higher summits of the Chalk many miles from the nearest Eocene outcrop (in other words, at elevations up to 975 ft O.D.).

It remains to be added that the term 'sub-Eocene' may be something of a misnomer. The surface may not in fact have been completed prior to the deposition of the earliest Eocene rocks (the Reading Beds in the case of the Hampshire Basin), but away from the Tertiary basins its erosional development may have continued throughout Eocene times and even into the Oligocene and Miocene periods. If so, it would be more accurately described as the 'early-Tertiary' surface.

C *The Pliocene peneplain.* Many geomorphologists have commented on the fact that the higher summits of central southern and south-eastern England show a marked degree of accordance at about 700–900 ft O.D., and have inferred the existence of an uplifted and dissected peneplain of great importance. In the last century the geologist Topley recognised the 'summit-plain' in the Weald, and ascribed it, in the fashion then current, to marine erosion. The surface was later studied by Bury (1910), who attributed it again to marine action, associated with an early-Pliocene transgression of the Weald. Bury further suggested that after the withdrawal of this sea and the warping of the exposed surface the present drainage system, in places highly discordant to structure, was superimposed from a thin cover of Pliocene sediments, remnants of which cap the North Downs at certain points (for example, Netley Heath in Surrey and near Lenham in Kent).

In the 1930s a reappraisal of the hill-top surface was made by Wooldridge and Linton, who revived a suggestion made in 1895 by Davis that the surface was in fact a sub-aerial peneplain. Their interpretation was based on the following lines of argument. Firstly, the peneplain is associated in many areas with clay-with-flints, believed by them to be a sub-aerial deposit that had accumulated at least in part during a prolonged period of weathering (but see p. 271). Secondly, the surface appears to be gently undulating, with barely perceptible slopes leading down towards the shoulders of more deeply incised valleys formed as a result of a later fall of base-level (Fig. 113A). Thirdly, in the areas where the hill-top plain is most in evidence (for instance, to the west of Salisbury), a remarkable accordance exists between the main

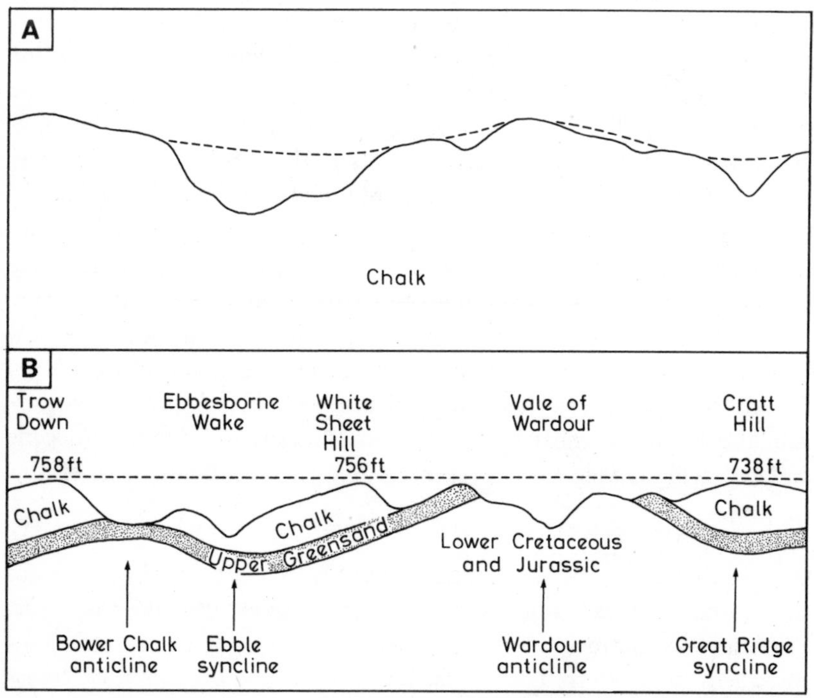

113 The Pliocene peneplain of southern England: form (A); relationship to Alpine folds (B)

structural lineaments and the principal rivers, which follow either synclinal or anticlinal axes. Wooldridge and Linton considered that, in south-east England as a whole, the peneplain is essentially horizontal (except for the minor local gradients referred to), and has not been affected by any subsequent warpings, except near the coast of East Anglia where the eroded top of the Chalk declines beneath Pliocene and Pleistocene formations. This horizontality of the surface is important in helping to demonstrate that it is not merely the dissected sub-Eocene surface, since the latter must, as we have seen, been everywhere deformed by the Alpine earth-movements. The hill-top peneplain in fact appears to transect the Alpine folds, the crests of which were evidently eroded during the cycle of erosion that produced the surface (Fig. 113B). Wooldridge and Linton consider that such evidence proves conclusively the 'post-Alpine' age of the peneplain. (They do not discuss the possibility of erosion actually accompanying the folding movements, in which case the structures and surface would be contemporaneous.) They postulate that the Alpine folding occurred in late-Oligocene and early-Miocene times, and that the peneplain was formed during a period of earth stability in late-Miocene and early-Pliocene times.

It seems reasonable to suppose that such a major episode of erosion would have left its mark elsewhere in Britain, both on the Mesozoic rocks of the English lowlands and on the Palaeozoic rocks of the highlands. For example, the highest summits of the Cotswolds, at just over 1000 ft O.D., might well preserve remnants of the 'Mio-Pliocene' surface. Clayton (1953) has suggested that the '1000-foot' plain of the southern Pennines (actually at 950–1250 ft) may date from the same period, as may the Lower Surface of Dartmoor (at 750–950 ft).

Finally, it must be pointed out that some geomorphologists now regard the Alpine movements as having reached a climax at the end of the Miocene period, so that if the hill-top peneplain does entirely post-date the folding it must be re-dated as Pliocene. The term 'Pliocene surface' will in fact be used throughout the remaining part of this chapter.

D *The Calabrian marine surface.* There is incontrovertible evidence that in late-Pliocene times much of central southern and south-east England was submerged by the waters of a transgressing sea. This episode is now referred to as the Plio-Pleistocene or Calabrian transgression. The evidence for the submergence takes a number of forms. Firstly, there are particularly in the Chalkland areas numerous bevellings at elevations of between 540 and 690 ft O.D. These are backed by steep slopes (apparently much degraded cliff-lines) which lead up to the higher ground where the Pliocene surface is preserved (Fig. 114). Secondly, certain of these flats are allied to shingle and sand deposits, preserved by piping into the Chalk. Formerly these deposits were dated as Pliocene (hence the original use of the term 'Pliocene transgression'), but they are now known to be of early-Pleistocene age. Thirdly, the drainage pattern of those parts of southern England which were inundated by the sea shows a marked disregard for the Alpine structures. This latter feature has led Wooldridge and Linton to suggest that, although the remaining marine surfaces are very limited in extent, the Calabrian sea led to widespread erosional bevelling at about 600 ft and the total obliteration in some areas of the pre-existing Pliocene relief and drainage. Thus the greater part of the Chalk country of Hampshire may have been base-levelled at this time by wave action, and rivers such as the Test, Itchen and Meon are regarded as superimposed elements of post-Calabrian age (pp. 256–7). The spasmodic retreat of the Calabrian sea during the Quaternary is well evidenced by a number of small marine levels, at elevations below 475 ft O.D., on the dip-slope of the South Downs and on the northern margins of the Hampshire Basin.

In general, then, the main stages in the evolution of the landscape of central southern England seem to be well understood, and only details of chronology are in serious dispute among geomorphologists. The reasons for this comparative lack of contention lie in the existence here of datable deposits of late-Cretaceous and Tertiary age, and in the very

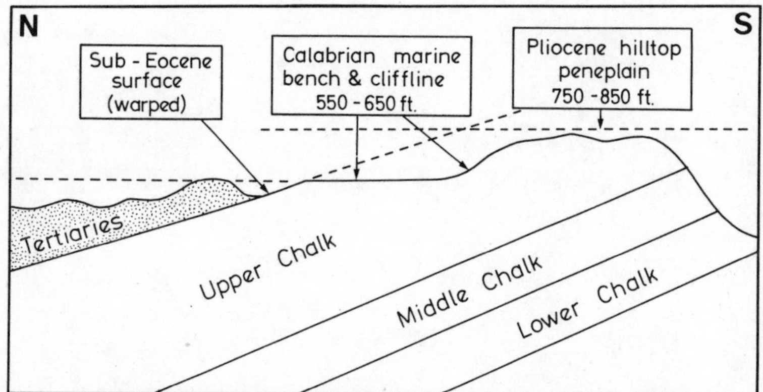

114 The relationship between the sub-Eocene surface, the Pliocene peneplain and the Calabrian bench in the central North Downs

valuable 'datum-line' afforded by the Alpine movements. In the Welsh massif, the situation is quite different. Over most of Wales there are no post-Palaeozoic rocks, and the major geological structures date back to Hercynian and Caledonian times. There is thus a great gap in the geological history of some 200 million years or more. It is hardly surprising that the surfaces of Wales, which must have been developed at some time during this interval, have given rise to immense controversy.

The planation surfaces of the Welsh massif

Planation surfaces in Wales have been recognised at a variety of heights and in many different localities since the last century. For example, in north Wales three 'coastal platforms' have been identified at elevations of 500–600 ft (represented by the hill-tops of Anglesey), at 400 ft (on the flanks of Snowdonia), and at an average of 270 ft (the so-called 'Menaian Platform' of Anglesey and the Lleyn peninsula). In south Wales, a 600 ft surface has been widely traced, and is supposedly represented by the summits of the Old Red Sandstone Hills of Gower

115 A planed-off surface of steeply dipping Carboniferous Limestone at approximately 200 ft O.D. near Paviland, Gower, south Wales. [Eric Kay]

(pp. 92–3). A 400 ft platform is well developed in the Vale of Glamorgan, in Pembrokeshire, and along the coast of Cardigan Bay. Finally, a 200 ft surface is again much in evidence in Gower, where it is eroded across steeply dipping Carboniferous Limestone, and in south Pembrokeshire (Fig. 115). Most of these platforms have been attributed to marine action, and are related by most authors to a progressive fall of base-level affecting the Welsh massif during Pliocene times.

Of much greater importance than these partial planation surfaces is the so-called Welsh 'tableland'. That the upland plateau surfaces of Wales, which reach generally to a height of 1500–2000 ft, bear the imprint of a major episode of planation has been recognised for a century. However, the only thing that can be said with absolute certainty about this surface is that it post-dates the Hercynian earth-movements, since it cuts across the Hercynian folds and faults and in some places has been associated with the removal of 10,000 ft or more of rock from the crests of the folds.

We shall describe four of the theories that have been put forward to explain the origin of the tableland.

(i) The geologist Jones (1931) has postulated that the Welsh surface may represent an ancient Triassic peneplain which owes its remarkable state of preservation in part to the great resistance of the underlying Palaeozoic rocks, but even more to its protection over a very long period of geological time by overlying Mesozoic strata, of which the Chalk was the youngest and probably the most important. According to this thesis, the 'sub-Triassic' surface has been exhumed as a result of a major uplift and prolonged erosion during the Tertiary era. That a sedimentary cover once existed over much of Wales is indicated by the discordant nature of much of the consequent drainage, which bears all the marks of a history of superimposition (pp. 260–1). The post-Cretaceous uplift, which initiated the erosional stripping of this cover, may have been greatest in the Snowdonia area, thus producing a dome with flanks declining eastwards and south-eastwards over much of present-day Wales. Elements of the resulting consequent drainage, forming part of an overall radial pattern, are still recognisable today.

The evidence favouring Jones's interpretation can only be stated very briefly here. Firstly, there are stratigraphical reasons for inferring the planation of some Hercynian areas by the close of Triassic times (p. 272). In some areas the youngest Triassic rocks and oldest Jurassic strata rest on folded Palaeozoic rocks (notably the Old Red Sandstone and Carboniferous Limestone) which have been effectively planed across by erosional processes. This unconformity is particularly apparent to the north of the Bristol Channel, in Glamorgan and on the south-eastern margins of the Forest of Dean, where it occurs usually at less than 600 ft O.D. Secondly, there are good reasons for supposing that the present-day scarp-face on the south side of the Welsh coalfield upland is, at least locally, an exhumed Triassic feature (as are similar scarps in the Mendips to the east). Jones argues that, although there is no direct confirmatory evidence, the coalfield scarps on the eastern side (between Newport and Abergavenny) and to the north (in the Brecon Beacons and the Black Mountain) may be of the same origin and age. Thirdly, on the basis of a detailed analysis of topographical maps, Jones concludes that the Welsh tableland is not horizontal, as so many have supposed, but declines in elevation southwards from about 2000 ft in the Plynlimon area to only a little over 1000 ft at the foot of the northern escarpment of the coalfield. That the surface itself and the great scarp-face overlooking it are of the same age becomes a distinct possibility. Furthermore, the southward decline of the tableland, which can be explained most satisfactorily in terms of a major post-Triassic downwarping centred on the Bristol Channel, permits the establishment of a direct link between the 2000 ft surface of north Wales and the low-lying sub-Triassic surfaces south of the coalfield. Prior to the southern downwarping the coalfield plateau of south Wales apparently stood as a great residual upland mass above the Triassic plain, in much the same way as do the mountains of Snowdonia today (Fig. 116).

(ii) It is possible that the Welsh tableland represents not a sub-Triassic surface but a sub-Chalk surface, formed as a result of late-Jurassic and early-Cretaceous uplift and erosion, and fossilised by the deposits of the transgressing Chalk sea (p. 275). If this is so, the exhumation of the surface and the superimposition on to it of the Chalk drainage pattern must have taken place wholly during the Tertiary era, after the retreat of the Chalk sea had been initiated by an uplift of the Welsh massif. Thus for the first time in this discussion a somewhat tenuous link may be established between a major planation surface in upland Britain and one, admittedly represented only by a major unconformity, at a much lower elevation in southern England. The manner in which the sub-Chalk surface 'climbs' westwards along the south coast of England has already been noted. At first sight it would appear that the rate of this ascent would be inadequate to take the sub-Chalk plain up to the required elevation in Wales. However, it seems that the gradient of the surface towards the north-west (that is, in the direction of Wales) was

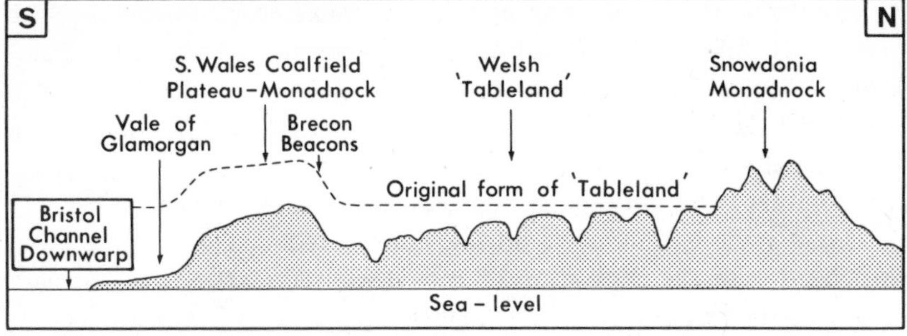

116 The southern downwarp of the Welsh tableland

rather greater. For instance, along the Chalk scarp of Wiltshire the unconformity outcrops at about 400 ft O.D. on average; yet it obviously cleared the top of the Cotswolds at over 1000 ft some 20 miles away (Fig. 117). Thus the mean gradient of the surface must have been at least 30 ft per mile, and perhaps a little more since no Cretaceous deposits rest now on the Cotswold crests. The nearest part of the Welsh plateau (at 1834 ft O.D.) occurs on the eastern margins of the Welsh coalfield, some 30 miles beyond the Cotswolds. A simple calculation, based on the assumption of a continued gradient of 30 ft per mile, will demonstrate that the sub-Chalk surface in the eastern coalfield should be at 1900 ft O.D. There is therefore a remarkable measure of agreement between the actual height of the land and the theoretical height of the sub-Chalk surface, so much so that the hypothesis stated above is given added support (Fig. 117).

This method of extrapolating a surface at a given angle of slope clearly has dangers (one does not know whether the surface originally had constant gradient, or whether it had been differentially warped), yet

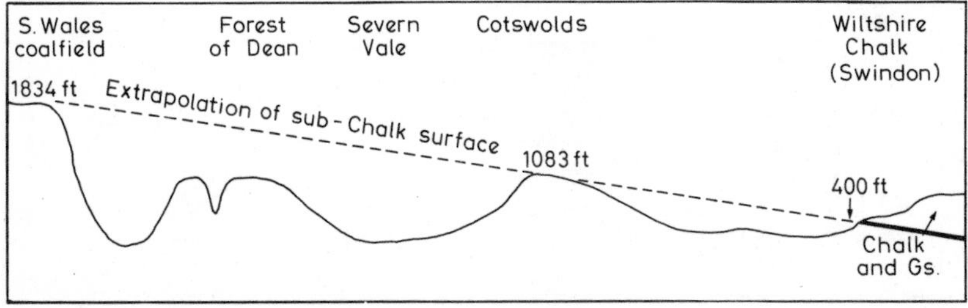

117 The possible relationship between the sub-Chalk surface of central southern England and the Welsh tableland of the south Wales coalfield

it is sometimes the only way in which possible relationships between surfaces at different elevations in widely separated areas can be established. Ideally one needs further evidence to support the correlation, such as Chalk outliers or deposits of flints high up on the Welsh plateau. Unfortunately these do not exist.

(iii) Another possibility is that the Welsh tableland is an early-Tertiary planation surface, developed during Eocene and Oligocene times, and associated with the deposition of considerable thicknesses of Tertiary sediments in eastern and south-eastern England. As has been demonstrated above, a sub-Eocene or early-Tertiary surface is at present being exposed around the margins of the Hampshire and London Basins. The form of this surface is complicated, owing to its distortion by the Alpine earth-movements, but like the sub-Chalk surface it has a tendency to climb both westwards and north-westwards. Thus, beneath the main axis of the Hampshire Basin the surface lies at a depth of several hundreds of feet, whereas its existence has been shown in east Devon at heights of up to 1035 ft O.D. A correlation with the Welsh tableland must again be regarded as a possibility, though no further evidence in support of the hypothesis is available.

This theory of early-Tertiary planation can still embrace the view that the Welsh massif had previously been covered by the Chalk. Indeed Linton has drawn attention to the possibility that the sub-Chalk surface is represented in the Welsh landscape, not by the main tableland but by the summits of the high residual masses, such as the Snowdon and Berwyn ranges. These certainly show some tendency to decline in

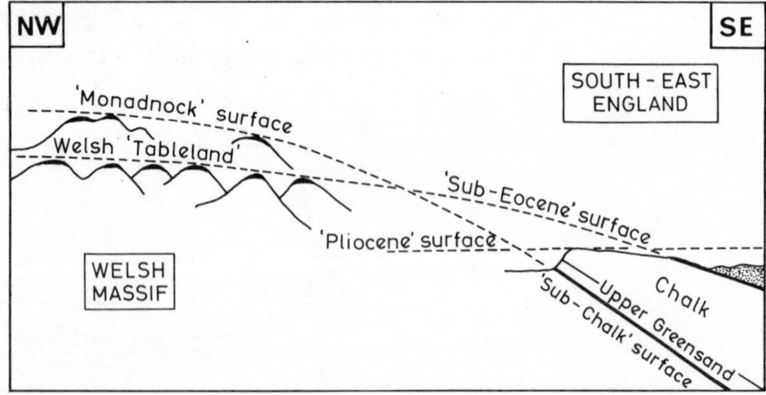

118 The possible relationship between the sub-Chalk, sub-Eocene, Pliocene and Welsh surfaces

height to the east and south-east towards the sub-Chalk unconformity of England. The main Welsh surface, at a lower level and thus post-dating the 'monadnock' surface, has been fashioned from the Palaeozoic rocks below the former base of the Chalk in Tertiary times. The suggested relationship between upland and lowland surfaces in England and Wales is depicted in Fig. 118. The manner in which the two main surfaces intersect can be attributed to a double tilting of the sub-Chalk surface (in pre-Tertiary and Alpine times), and a single tilting of the early-Tertiary surface as a result of the Alpine uplift.

(iv) Finally, the Welsh surface may be regarded as of still more recent date. This is basically the view taken by Brown (1957), who after a detailed field study of the area has concluded that there are in fact three main planation surfaces (in addition to a possible sub-Chalk surface preserved by the monadnock summits). These are: the 'High Plateau' (the tableland proper) at 1700–2000 ft O.D.; the 'Middle Peneplain' at 1200–1600 ft; and the 'Low Peneplain' at 800–1000 ft. Brown considers, from the evidence of the form of these surfaces and their distribution in relation to existing drainage lines, that all are of sub-aerial origin. They date from Miocene and Pliocene times, since they are evidently quite unwarped (although the mid-Tertiary Alpine movements would have had a major impact in the Welsh massif). The surfaces appear to have resulted from a series of base-level falls, which may have been caused by eustatic shifts of sea-level—though how the sea can undergo an actual fall of some 1500 ft is by no means clear. The two higher surfaces can have no remaining counterparts in central southern England, but a correlation can be satisfactorily made between the Low Peneplain and the Pliocene hill-top plain. The latter can be traced from the Marlborough Downs (maximum height 887 ft O.D.), through the Cotswold summits (maximum 1083 ft at Cleeve Hill) to the bevelled surface of the Forest of Dean plateau (1005 ft at Beacon Hill), which forms part of the

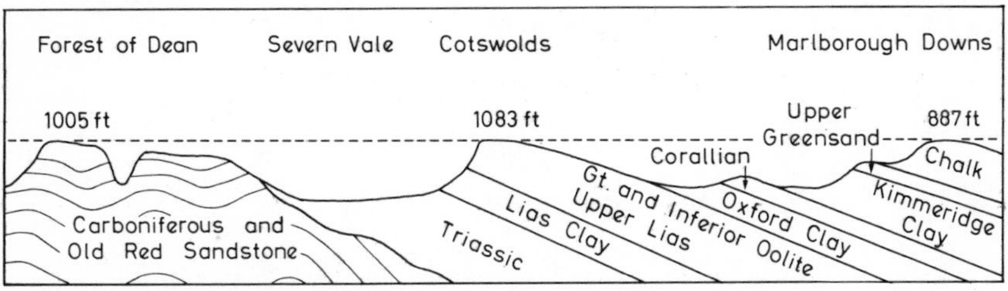

119 The possible height correlation of surfaces in the Marlborough Downs, Cotswolds and the Forest of Dean

Low Peneplain as mapped by Brown (Fig. 119). In addition to the three Mio-Pliocene peneplains, a 600-ft coastal plateau is recognised by Brown at many points around the margins of Wales. This bears again no overlying deposits to aid interpretation, but the coastal situation and constant height of the surface both point to a marine origin. A tentative correlation, on the basis of height alone, may be made between the 600-ft surface and the Calabrian marine platform of south-eastern England.

Brown's researches have undoubtedly contributed much to our knowledge and understanding of the landscape of Wales, but in view of the long-continued controversy over the origin of the Welsh tableland it would be optimistic to assume that all problems have at last been solved. In the first place there is the great difficulty of explaining the very large late-Tertiary falls of base-level. (This problem does not exist if the tableland is, in fact, either a sub-Chalk or sub-Triassic surface, since the height of the surface can then be attributed to bodily uplift by pre-Tertiary and Alpine earth-movements). Secondly, it is perhaps strange that the three sub-aerial cycles were not associated with considerable sedimentation in, say, lowland England, much of which must have lain beneath the sea while the two upper surfaces were being formed. To suggest that the third and most recent cycle, which has left an erosional imprint on lowland England in the form of the Pliocene peneplain, led to the complete destruction of all previously formed late-Tertiary sediments is a possible but not wholly satisfactory solution.

9

ARID AND SEMI-ARID LANDFORMS

INTRODUCTION

It has been estimated that deserts and semi-deserts occupy about one-third of the earth's land-area. For the reason of coverage alone any student of geomorphology must seek a detailed knowledge and understanding of the landforms of these areas. However, there is another and equally important reason: arid lands are commonly supposed to exhibit morphological characteristics quite different from those of more humid regions. A. A. Miller, for instance, has stated that in deserts the processes of weathering and erosion, greatly influenced by the prevailing conditions of temperature and humidity, 'produce a hard clear profile and jagged lines in exposed rocks'. He has also drawn attention to the absence of 'the graded line, the steady fall of the ground towards base-level, the ineluctable waste of land and the transport of the debris to the sea'. In the words of Balchin and Pye, 'there . . . exists a fundamental difference between the smooth graded profiles associated with the humid cycle, and the sectioned profiles, with angular junctions, typical of the arid cycle' (1955).

These statements are, of course, generalisations, and indeed some of the descriptive phrases used could well be applied in, say, areas of mountain glaciation or of strong karstic development. Furthermore, the landforms of all deserts cannot be adequately summarised in so simple a way, for arid landscapes vary enormously both in broad outline and in detailed configuration. Among the chief landscape-types five are readily recognisable.

(i) Mountain-girt basins (typical of the arid south-west of the U.S.A.), which are associated with centripetal drainage patterns, deep dissection of adjacent plateau-blocks, steep boulder-strewn slopes, gentle rock pediments and wide spreads of alluvium.

(ii) The structurally simple sandstone plateaus of Libya and western Egypt, incised by long ramifying wadi-systems, with steep or near-vertical valley-side slopes, and flat valley floors choked by alluvial infillings.

(iii) The great and almost perfectly levelled erosional plains (pediplains) of the African and Great Australian Deserts, with their barely perceptible concave slopes, resulting from the coalescence of innumerable rock pediments, and their isolated steep-sided hills or inselbergs.

(iv) The great accumulations of sand and sand-dunes, forming the ergs of the Sahara (for example, Erg de Fachl, Erg Chech and Erg Rebiana, whose continuity is broken only by the great volcanic massifs of Ahaggar and Tibesti) and the Great Sandy Desert of north-western Australia.

(v) The broad areas of rocky or stony desert (hammada or reg), from which all fine material has been swept by running water or wind deflation into adjacent depressions occupied by alluvial spreads or ergs (Fig. 120). Clearly such landscapes can form component parts of the more extensive pediplains referred to above.

These geomorphological contrasts within deserts are sometimes explicable in terms of geological structure or denudational history. For example, the complex relief patterns of the basin-and-range country of the U.S.A. are in large measure the result of block-faulting during the Tertiary era. Many of the steep 'mountain fronts' here are deeply scarred fault-scarps, and the much aggraded lowlands ('bolsons') encircled or bounded by these scarps are structurally determined depressions. The greater extent of plain-like country in Africa, on the other hand, arises largely from the structural stability of the continent (outside of the areas affected by recent earth-movements, such as the Rift Valley and the Atlas Mountains.) In fact, the underlying rocks of much of Africa form a 'basal-complex' of great antiquity, and in many places have evidently been subjected to almost continuous erosion, encompassing several major cycles of erosion, since the beginning of Mesozoic times. As a result landforms either produced by or related to initial structural features have long since been obliterated by erosion. Such evidence supports the notion that desert landscapes owe much to stage of development (pp. 165–7); indeed it is worth noting that even the most extreme opponents of the cycle concept are prepared to accept the existence of widespread planation surfaces in arid and semi-arid regions of Africa.

Different desert landscapes may be the result not only of the varying influence of structure and stage, but also reflect climatic factors, which obviously determine the main denudational processes at work in arid regions. An obvious initial distinction must be made between the 'hot' deserts (for example, the Sahara, the Arabian Desert, the Atacama Desert, the Kalahari and the Great Australian Desert), in which there is no real cold season at all, and the so-called 'cold' deserts (for instance, the Great Basin of the U.S.A., the deserts of the Asian interior, and the Patagonian Desert), in which winter conditions may be severe and mean January (July) temperatures approach freezing-point. In cold deserts frost weathering may be a process of considerable importance, but its

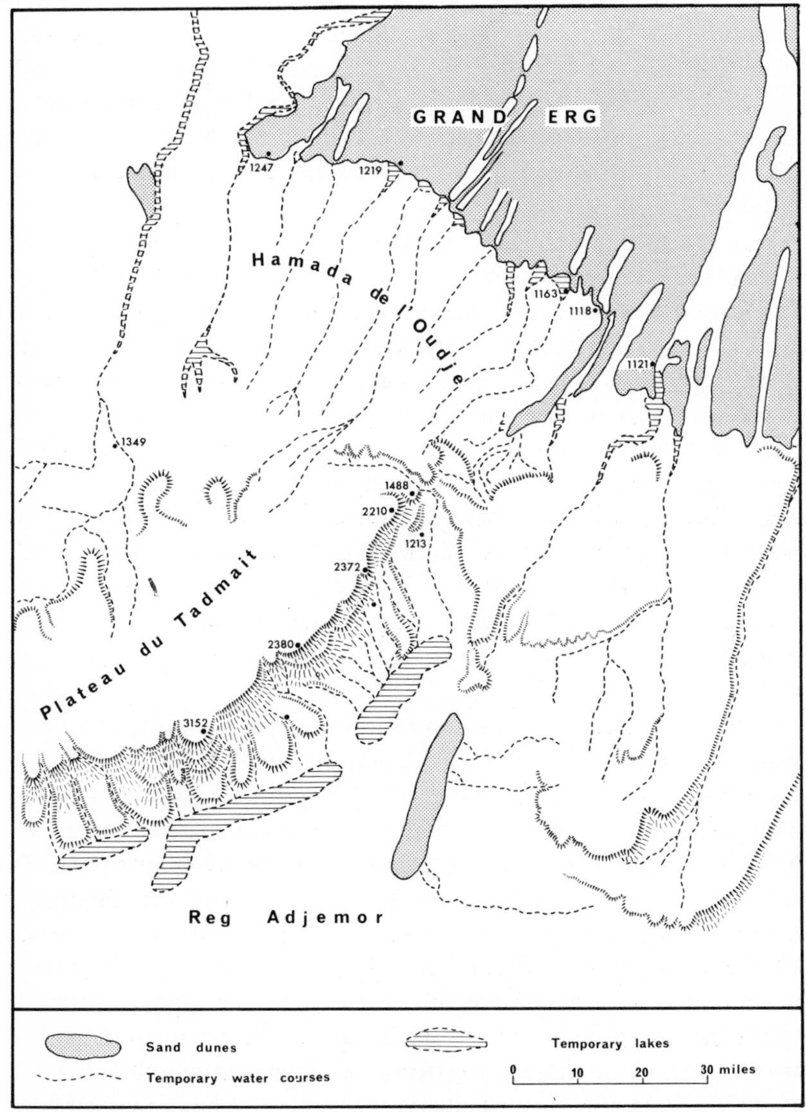

120 The geomorphological features of the Sahara Desert, southern Algeria

effects in hot deserts—in spite of the low nocturnal temperatures and occasional frosts—are not great. Another major division of deserts is into 'continental' and 'maritime' types. Even when occurring in low latitudes, the latter are surprisingly cool; indeed, the mean temperature of the hottest month may not exceed 70 °F. Again, because of the regulating influence of the near-by sea annual temperature ranges are low in maritime deserts, by comparison with those of continental interiors, and even diurnal contrasts are far less sharp, so that insolation weathering will be less pronounced.

However, whilst these variations of climate do exist, it is still possible to recognise some characteristics which are common to most deserts, and which help to account for the differences between arid and humid landscapes.

From the geomorphological point of view, a fundamental feature of the great inland deserts, whether hot or cold, is the wide diurnal range of temperature. This is most marked in the tropical continental deserts, where the high angle of the sun, the low relative humidity and the general absence of cloud cover combine to cause rapid heating of the ground-surface. This heat is in turn transmitted by conduction to the air above, and afternoon temperatures frequently exceed 100 °F, and may reach as high as 120 °F or more. However, at night the heat is lost owing to unimpeded radiation, and air temperatures fall by 30 °F or more; in exceptional circumstances the diurnal range may exceed 60 °F. Such extremes are, as stated above, not encountered in the more restricted coastal deserts, where daytime temperatures even in summer are generally in the order of 80 °F or less, and the diurnal range below 20 °F.

Another basic feature of deserts is, of course, their aridity. This is a result both of the low annual rainfall and the high rates of evaporation, especially from bare rock surfaces affected by the hot sun and strong winds. In tropical deserts the rainfall may approach or exceed 10 inches per year, but this is so ineffective that only xerophilous plants can be sustained. Needless to say, in many desert areas the total precipitation is in fact minute. Thus in Egypt the annual rainfall decreases rapidly inland from 8 inches at Alexandria to 3 inches in the middle Nile delta area, and at Aswan is considerably less than 1 inch—indeed in many years there is literally no rain at all. In these drier parts the rainfall is often, though by no means invariably, convective in type, and although of brief duration may be very intense. Sutton (1949) has described how, in April 1909, thunderstorms gave nearly 2 inches of rain in the Eastern

Desert of Egypt during a period of only 24 hours. Such violent rainfall is accompanied, in areas of bare rock and sparse vegetation, by a very high percentage of run-off. Sutton records how a sudden storm in 1902 transformed the Wadi Alaqui, which joins the Nile about 50 miles south of Aswan. The normally dry stream course near Um Geriat was quickly occupied by a mighty stream nearly 300 yards wide and from 3 to 8 ft in depth; water continued in fact to flow in the wadi for about three days. However, in spite of its great initial volume this stream failed to reach the Nile. Instead, after running for some 40 miles, it entered a wide depression where it formed a large pool, which in turn was rapidly depleted by evaporation and percolation into the desert sand.

A final consideration in the study of desert landforms concerns recent climatic changes and their effects. There is abundant evidence to show that during the Quaternary the growth of the continental ice-sheets had serious repercussions on the climates of the extra-glacial areas. Some authorities have detected a shifting of the main climatic belts towards the Equator, so that mid-latitude cyclonic rainfall may have penetrated, for instance, into the North African Deserts. Furthermore, the southern margins of the Sahara seem at times to have encroached on to adjacent areas of savanna. Probably this took place during the glacial periods, whereas a return to more humid conditions occurred during the interglacials. These climatic migrations have been recorded by Lake Chad, the size of which fluctuated greatly during the Quaternary.

The existence of past 'pluvial' periods, affecting particularly the poleward margins of present-day tropical deserts but also to some extent the remote desert interiors, is almost certainly a factor of geomorphological significance. As will be shown, many desert landforms bear the clear imprint of water action and may therefore have been fashioned at a time when annual rainfall and surface run-off was much in excess of that of today. For example, the great wadis of the North African and Arabian deserts may well have been eroded under pluvial conditions. Peel (1941) has ascribed such valleys cut into the weak sandstones of the central Libyan desert to the action of both surface streams and springs, largely during a past period of greater precipitation which is attested to by dry lake beds, archaeological evidence, and legends telling of abundant rains and plentiful pasturage.

It is necessary to add, however, that increased precipitation does not always produce an accentuation of erosive processes—indeed the reverse may sometimes be the case. A small and uncertain rainfall, taking the form of sharp convectional downpours of limited duration, will tend to produce short-lived but powerful surface flow on an unvegetated or sparsely vegetated surface. Such flash floods can produce quite startling results in terms of debris transportation and possibly even erosion. On the other hand, a much greater and seasonally well-distributed rainfall will favour a denser vegetation cover, which will in turn retard surface run-off and promote large losses of water through evapo-transpiration and percolation. Such conditions, typical of humid-temperate areas at the present day, cause 'stagnation' of erosional processes. It is quite possible that, by comparison, all but the most arid deserts are even now, despite their desiccation in post-glacial times, quite active geomorphologically speaking. The study of deserts thus emphasises once again that the greatest problem confronting geomorphologists is probably that of distinguishing between the influences of 'current' and 'historical' processes on present-day landforms.

DESERT PROCESSES

Weathering

For many years it has been assumed that weathering in deserts is overwhelmingly of a mechanical kind, and is essentially the result of the high diurnal ranges of temperature described above. The process is in fact sometimes referred to as 'insolation weathering'. During the daytime intense heating by the sun's rays will cause surface layers of a rock to expand. Most rock minerals are very poor conductors of heat, so that the pressures set up by this expansion are strongly localised in a very shallow surface zone. It was originally believed that in this way fracturing parallel to the surface of the rock would be produced, and platy, often curvilinear fragments of rock ('spalls') would thus be loosened. Such 'onion weathering' or exfoliation does seem to be common in areas of little rainfall and high temperatures, and is especially effective in certain intrusive igneous rocks such as granite and metamorphic rocks like gneiss. Exfoliation on a large scale plays a major part in the development of 'exfoliating half-domes' (rounded outcrops of granite projecting through steep debris-mantled slopes) and of the larger dome-like inselbergs (often referred to as 'bornhardts', after the German geologist W. Bornhardt, who was one of the earliest students of these remarkable landforms) found in many parts of Africa. It must be stressed, however, that such features are characteristic not only of arid areas, but occur very widely in seasonally wet savanna climates. Again, the famous 'sugarloaves' of eastern Brazil (those at Rio de Janeiro experience an average annual rainfall of 40–50 inches) probably owe their main outlines to the process of exfoliation.

Other important weathering processes that are usually attributed to insolation effects are the breakdown of well-jointed rocks into boulders ('block disintegration') and the fragmentation of crystalline rocks into small particles ('granular disintegration'). The latter process results from the different colours (and thus different capacities for absorbing heat) and the varying coefficients of expansion of the constituent minerals of the rock. Many rocks are affected simultaneously by block and granular disintegration. The former may lead to the development of a near-vertical face, from which the released boulders undergo free fall and accumulate as talus at the slope base. Alternatively, on steep slopes of 25–45° the joint-bounded blocks, after separation by weathering, remain *in situ* or experience very slow downslope movement. On both the talus and boulder-covered slopes finer material results from granular decay, and is easily transported to the foot of the slope by running water, giving fans and spreads of alluvium.

Most modern geomorphologists consider that these three processes of exfoliation, block weathering and granular disintegration, though leading to a physical breakdown of the solid rock, are not solely the result of mechanical processes. In their view, the importance of the aridity factor (which seems to preclude chemical reactions, since these usually depend on the presence of moisture) and of diurnal temperature changes has been overstressed. For one thing laboratory experiments carried out by Blackwelder (1933) and Griggs (1936) have shown that most common rocks are remarkably resistant to mechanical breakdown, even when temperature changes are exaggerated in amount and rapidity. For another, the layers of rock detached in nature by exfoliation are often too thick (up to several feet) to be affected by temperature changes alone, and it is reasonable to suppose that the rock is in fact being broken along pre-existing joints which run parallel to the rock surface. Such joints, which give to igneous rocks a stratified appearance and are accordingly referred to as 'pseudo-bedding planes', may result from the process known as 'pressure release' or 'dilation'. This affects mainly plutonic rocks, which tend to recoil as the great overburden is gradually worn away, giving rise to a system of curvilinear sheet jointing at the surface. Of course, the problem remains of deciding which weathering agent takes advantage of these joints. In view of the limited penetration of the rock by surface temperature changes, the possibility of chemical decay by percolating or even rising water must be seriously considered.

Here it must be emphasised again that water is not totally absent from desert areas. Moreover, detailed study of weathered rocks reveals quite clearly that some chemical breakdown of constituent minerals *has* occurred under arid conditions, probably in the presence of very small quantities of moisture. This may have been derived from episodic rainfall, or it may have been drawn up by capillary action from a zone of saturation lying at some depth. In the latter case, the rising water could well have contained dissolved salts, and in passing through pores and along joints have attacked and loosened the rock. Possibly the process of exfoliation may be initiated in this way. On a small scale capillary rise is associated with the chemical breakdown of the interior of large boulders and the deposition of a hard crust of 'desert varnish' (comprising oxides of iron and manganese) on their surface. Breaching of this outer shell may be followed by removal of the rotted core, giving the 'hollow blocks' found particularly on desert margins. Cavity weathering, leading to the formation of rounded hollows ('tafoni') in granite, may also be due to chemical decay, since it occurs commonly in shaded areas where moisture is likely to linger. Many pedestal rocks in deserts once attributed to the action of sand blasting just above ground-level are now believed to be the result primarily of chemical weathering. A relevant factor in this context is the widespread occurrence in deserts of dew. Although the relative humidity is normally very low, and may fall during the daytime to 25% or less, nocturnal chilling may be so intense that the ground temperature falls below the dew-point. Copious condensation can result, and although the water droplets are quickly evaporated the following morning their effects can in the long run be important. Peel (1966) has suggested, in fact, that dew is probably responsible for the etching of solution patterns observed on limestone surfaces in deserts.

In summary, it may be said that there has been a tendency to overestimate mechanical weathering and to underestimate chemical decay in deserts. Needless to say, the former must not be dismissed; nor is the latter nearly as effective as in humid areas, where the large-scale breakdown of certain rock minerals into clay is achieved. Rather chemical weathering in deserts is highly selective in operation, attacking joints, shaded places, the bases of rock masses, and some rock minerals. The result is the actual physical breakdown of the rock into boulders, spalls and small grains. Thus the effects are basically similar to those produced in other climates (for example, periglacial) by true mechanical weathering. The surface of a typical desert is therefore mantled by predominantly coarse and often angular detritus rather than fine clays.

Running water

It is fashionable nowadays to attribute most major landforms of deserts to the action of running water, either during past pluvial periods or under existing conditions. Observations such as those of Sutton and many other writers show that it is no longer feasible to ignore erosion, transportation and deposition by surface water in favour of the action of wind, once thought to be the most important geomorphological agent in deserts. Furthermore, not only have floods actually been observed, but many of the characteristic landforms of deserts (such as wadis, rock pediments, and extensive spreads and fans of alluvium) bear the clear imprint of water action.

One of the most common forms of desert run-off is the so-called 'stream-flood' (a term coined by Davis). These occur usually in dissected upland areas, containing stream channels that are normally dry but which are occupied by spates after heavy local rainfall. The flood of the Wadi Alaqui, described above, falls obviously into this category. Not all stream-floods are quite on this large scale, though, and much smaller channels cut into the steep mountain-fronts (fault-scarps) and rock pediments of the basin-and-range country of the western U.S.A. are from time to time the scene of spasmodic and impetuous flows.

The geomorphological role of stream-floods is not wholly clear. They may on the one hand have been responsible for the carving of deep branching wadis, such as those of the Libyan and Arabian Deserts, or on the other they may merely follow the courses of former semi-permanent streams of Quaternary age. Some authorities have argued that the smooth and unbroken 'graded' profiles of many wadis point to their erosion by normal streams rather than by brief and irregular floods. At the present day, many wadi floors are occupied by thick and extensive alluvial deposits, and transportation of this material, largely weathered and washed from the valley sides, rather than attack on the underlying rock floor may be the chief effect of the stream-flood. Certainly desert streams suffer severe losses of water by percolation and evaporation, and after an initial period when a heavy load of alluvium is rapidly picked up this reduced discharge produces a condition of overloading. Hence deposition rather than erosion is inevitable in the middle and lower parts of wadi courses.

However, one erosional task that may be performed very successfully by stream-floods is lateral corrasion. The steep slopes of many desert valleys, which are often separated by a sharp angle from the flat valley bottoms, may be undercut by laterally swinging stream-floods. In this way interfluves may be gradually consumed, and a surface of lateral planation (similar in essentials to the 'panplane' of Crickmay) may be formed. Such a sequence of landform development has been postulated by Peel (1941) to explain the major features (erosional plains, residual hills resulting from the imperfect destruction of interfluves, and plateaus dissected by large wadis) of the southern Libyan Desert (Fig. 121). Peel also considers that in the permeable Nubian sandstones here spring action may have been a contributory factor in the undercutting and retreat of the valley-side slopes. In the deserts of the south-western U.S.A. lateral corrasion by stream-floods is, in the view of Johnson and other geomorphologists, particularly active at points where upland valleys debouch into areas of lowland (Fig. 122). Fan-shaped embayments in the mountain-front usually mark the exits of these valleys, and allow the stream-floods to swing laterally and undermine the steep slopes on either side. Retreat of the mountain-front, an important process leading to the reduction of upland masses in this area, may result primarily from the 'trimming back' of the spurs between adjacent embayments by stream-flood erosion. The insistence of many geomorphologists on the importance of lateral erosion in arid lands rests partly on observation and inference (undercut slopes and sharp 'knicks' separating valley slopes and floors), and partly on acceptance of a theory propounded by the American geologist, G. K. Gilbert. In this it is postulated that streams which are 'fully loaded'—a likely feature of desert flows because of the absence of impeding vegetation and the availability of abundant rock waste—are not able to corrade vertically but can erode laterally to great effect (p. 56).

The second main type of desert run-off is the 'sheet-flood' (a term suggested by the American McGee in 1897 and widely adopted by other writers). Sheet-floods, as the name indicates, comprise extensive

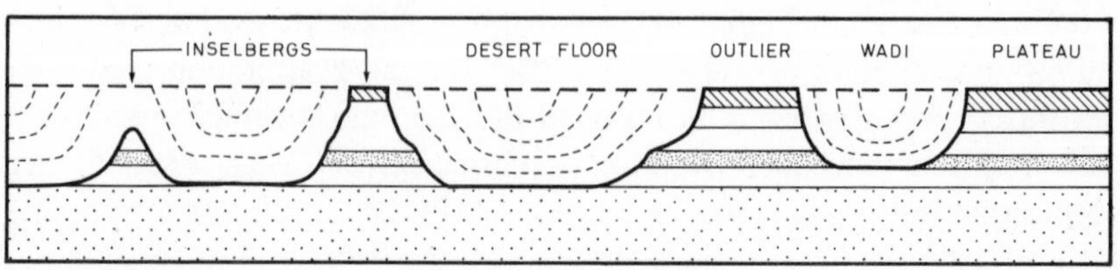

121 Relief features of the Libyan Desert (after R. F. Peel)

flows of water which are not confined to channels but are spread over virtually the whole ground surface. Such floods clearly cannot occur in well-dissected areas, but develop to best effect on undissected gentle slopes. The rock pediments and alluvial fans which fringe desert uplands provide ideal conditions for the formation of sheet-floods. In these localities sheet-floods may follow heavy local rain, or may result fom the transformation of a stream-flood emerging from a near-by canyon mouth. Sheet-floods do not normally comprise a uniformally deep layer of water flowing at a constant speed, but possess an extremely intricate network of high-velocity threads or 'rills'. The anastomosing pattern of these rills is often imprinted on the layer of detritus over which the sheet-flood flows.

The geomorphological role of sheet-floods has for long been disputed. Some early writers contended that sheet-floods were important eroding agents, lowering the surfaces of rock pediments and alluvial fans, and McGee himself argued that the widespread occurrence of near-level areas in deserts reflected the potency of sheet-flood action. However, it seems impossible to envisage the existence of sheet-floods except where the land had already been levelled, and it is more likely that sheet-floods *result* from pediments and similar features rather than fashioning them

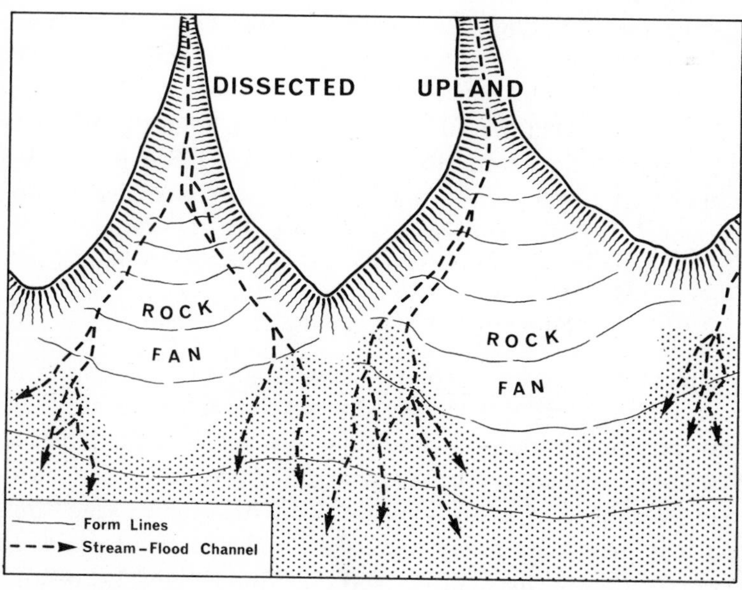

122 Rock fans

in the first instance. W. M. Davis and others have argued that sheet-floods act mainly as transporting agents, carrying finely comminuted detritus from the mountain-front above, across the pediment, to the zone of alluviation ('peripediment') below. The pediment itself is thus regarded as a 'slope of transportation' on which sheet-floods are the most active agent. Davis has further suggested that occasionally pediments are stripped entirely of their alluvial veneer, and that the sheet-floods are then replaced by stream-floods which deeply gully the solid rock. Eventually the interfluves between these trenches are removed by the laterally corrading stream-floods, and 'the restoration of sheet-floods on a smoothed rock surface' will ensue. Thus the pediment is lowered or regraded by alternate periods of sheet-flood and stream-flood action.

It is necessary to conclude this brief account of running water in deserts with some cautionary remarks. As will be evident from what has been written, much of the study of water action has been carried out by American workers, notably McGee, Lawson, Davis, Bryan and Johnson, in the arid south-west of the U.S.A. However, as Peel (1966) has pointed out, this area is 'not very typical of the World's dry lands'. Furthermore, 'the massive and extensive sheet-flood described many years ago by McGee . . . seems to have been witnessed by rather few field-workers since, even in so well-peopled and comparatively rainy an area'. Peel also states that the widely held belief that *most* desert rainfalls are of exceptional intensity is not sustained by actual observations—though these observations are themselves not yet as numerous as is desirable. Freak storms do occur, but more on the desert margins than in the truly arid interiors. It does seem that over great tracts of the Sahara, where conditions are 'hyper-arid', water action today is of 'rather minor importance'. It is presumably in such areas that the wetter conditions of the Quaternary pluvials must have had their greatest impact.

Wind

It is evident to any desert traveller that wind must be an agent of some importance in desert morphology. Indeed, the numerous small dust-storms, resulting from local convection currents, and the larger-scale *simoon* of the northern Sahara (described by A. A. Miller as a 'swirling rush of scorching air, laden with dense clouds of blistering sand through which it is impossible to see more than a few yards') seem to have impressed early students of desert landforms far more that the occasional flash-flood. Similarly the great sand seas, evidently the work of wind transport and deposition, attracted from the start a good deal of attention—hence the popular misconception of the typical desert as an area of unending sand-dunes. Certainly the individual ergs of the Sahara may be of enormous extent, covering areas as large as that of England, but in fact it has been estimated that they occupy only one-tenth of the total desert surface. The remainder comprises reg and hammada, in which sand may be present but in comparatively small quantities.

It is not surprising therefore to find some authorities regarding wind action as paramount, even in the formation of large-scale desert landforms. Thus in 1904 the German geologist Passarge postulated that the widespread erosional plains, surmounted by steep-sided residual inselbergs, of much of arid, semi-arid and savanna West Africa were the product of alternating humid and dry epochs in the past. Under humid conditions deep chemical weathering would produce a deep layer of decomposed rock, which during a subsequent phase of desiccation would be removed by wind action. Inselbergs would develop at points where the more resistant rock limited the depth to which weathering could penetrate, whereas the plains would mark areas where the detrital layer had been thick and evenly developed. This theory of Passarge's is in some ways remarkably similar to modern explanations of savanna landscapes (pp. 174–80), except that it is nowadays assumed that wet-season floods are responsible for removal of the weathered overlay to reveal the basal weathering surface and subsurface domes.

Probably no modern geomorphologist would accept that wind action has been of any real importance in the evolution of features such as inselbergs, pediments and wider erosional plains. For instance, L. C. King in his cycle of pediplanation (pp. 170–4) assumes that the initial dissection of the landscape, to produce the upland blocks which in time will be reduced by weathering and scarp retreat to inselbergs, koppies and tors, is the result of stream erosion during a period of falling base-level. Only one important type of desert landform is today considered to be the outcome of wind 'erosion'. This is the enclosed depression, a feature of all arid areas and one that varies in scale from the small 'buffalo wallows' to the large 'P'ang Kiang' hollows of the Mongolian Desert and the great Qattara depression of the Western Egyptian Desert. The last-named measures approximately 200 by 100 miles, attains a depth of 440 ft below sea-level, and has been formed, according to Peel, by the removal of some 800 cubic miles of rock material. It is known that some enclosed depressions are of structural origin, resulting from faulting or broad regional downwarps, but this cannot be the explanation of all such features. Possibly the majority of depressions were initiated in areas of unresistant rock, where moisture was comparatively abundant and chemical breakdown of the rock facilitated (Fig. 123). The material so loosened could then be blown away ('deflated' or 'exported'), and the formation of the hollow begun. The whole process would tend to be self-reinforcing, for the growing hollow would increase the accumulation of moisture and so accelerate chemical decay. Deflation would gradually deepen the hollow until the water-table was exposed, at which point oases formed or evaporation produced a protective cover of salt inhibiting further wind action. Deflation hollows seem to undergo both deepening and lateral extension, for some are bounded by cliffs showing evidence of active retreat. There are often signs of basal undercutting, attributed by some to wind eddying and scour due to the lee effect of the cliffs, and by others to chemical weathering at the level of the moist depression floor. However, the main

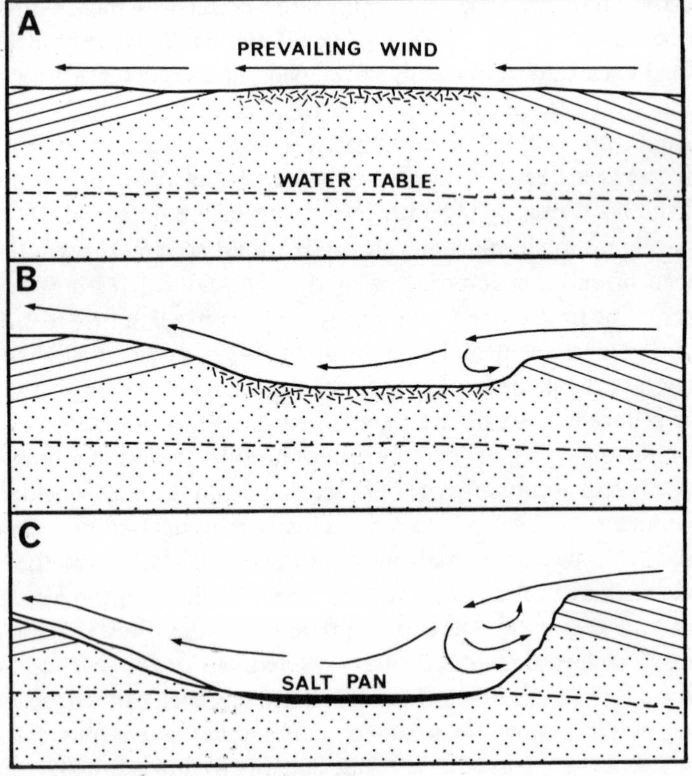

123 The formation of deflation hollows

process responsible for the recession of the cliffs may be simply gullying by surface run-off.

On an altogether smaller scale wind erosion leads to certain grotesque rock formations standing above some desert surfaces. Movement of sand particles by the wind takes the form of 'creep' along the surface or of saltation, a process involving the hopping of sand grains in a turbulent air flow. This moving sand, which is concentrated in the zone a foot or two above ground-level, produces a strong blasting or abrasive effect at the base of any rock masses lying in the path of the wind. Such abrasion is capable of polishing hard rocks, of undercutting soft rocks, and of shaping individual boulders and small stone into 'ventifacts', notably the well-known 'dreikanter' with their three faceted sides. It was formerly considered that the 'pedestal' or 'mushroom' rocks of some deserts were the result of basal corrasion by wind, but it is now believed that weathering near ground-level is probably a more important factor than sand blast. Pedestal rocks are in fact by no means confined to arid lands, but occur frequently in more humid climates (for instance, the oft-quoted Brimham Rocks of Yorkshire) where wind action cannot be invoked. Probably the only landforms of deserts that can be confidently ascribed to wind abrasion alone are the comparatively unimportant 'yardangs' and allied 'ridge-and-furrow' features. These comprise elongated depressions, developed by deflation along lines of weakness in the rock and commonly orientated in the direction of the prevailing winds, which are separated by upstanding or isolated masses of more resistant rock, the yardangs proper. These are rarely more than 25–30 ft in height, and frequently less, and show distinct signs of basal abrasion on the upwind side only.

The great sand accumulations are the features of wind deposition that obviously claim first consideration. In these, the sand may be spread over an area of thousands of square miles, and attain such a depth that the solid rock is nowhere visible. The surface forms of these sand seas vary immensely. Often the pattern of the individual dunes is highly complex. In parts of Australia the dune-lines form a curious anastomosing pattern, rather like that of a great braided river, but it is more usual to find numerous longitudinal sand-ridges, which are sometimes from 300 to 600 ft in height and extend for as much as 100 miles, or lower and broader elongated dunes referred to as 'whalebacks'. In much of the Sahara and the Arabian deserts, these major dune-lines trend broadly from north-east to south-west or from north to south. This seems to reflect the control exercised by the north-east trade winds.

The very existence of these great sandy areas would seem to provide a *prima facie* case for still regarding the wind, despite what has been said above, as a major erosive (that is, deflating and abrading) force in deserts. Briefly, the argument would be that what the wind deposits at one place it must have previously picked up elsewhere. In fact, some of the ergs of the Libyan Desert do lie to the leeward of large hollows that appear to have been overdeepened by deflation. However, it is highly significant that the great ergs of the Sahara occupy low-lying basins, surrounded by reg and hammada from which the surface sand has been removed by some agency (Fig. 120). This collection of the sand seems therefore to have resulted from some process that can operate centripetally and that is also influenced by the gradient of the land-surface. This can only have been water transport, involving stream- and sheet-floods under arid conditions or semi-permanent streams in a climate more humid than that of today. In some instances, the sand accumulation occurred well before Quaternary times. For example, W. D. Thornbury has pointed out that the Sand Hills region of Nebraska, covering approximately 24,000 miles, is developed primarily from the weathering of local unconsolidated Tertiary sandstones. Many of the greatest sand areas in deserts consist, then, of alluvial material laid down by fluvial action in low-lying basins, perhaps occupied at the time by playa lakes or even extensions of the sea. The role of the wind has been merely to rework this material *in situ*, and to shape it into the dunes of the present day.

It is appropriate to discuss briefly the origin and form of individual types of dune. There is, of course, an infinite variety in the size and precise shape of these, but Bagnold (1941), perhaps the foremost expert on the subject, has suggested that two major dune-types may be recognised. The first of these, the crescentic dune or 'barchan', is said to develop where the wind blows very largely from one direction only. The necessary prerequisite of barchan formation is a mound of sand. This may accumulate to the lee of and in the shelter of any available obstruction (for example, a rock or bush) to air flow and sand movement immediately above ground-level. Alternatively, it may become detached from the end of a sand-ridge which has formed, say, beyond a gap in a ridge or scarp through which sand drift has been funnelled. Once in existence, the mound will grow by trapping more sand and will also experience migration, for the wind will blow sand up the windward slope and cause it to accumulate on the sheltered upper part of the lee slope. In profile the mound will soon become asymmetrical. The leeward slope will steepen to an angle of approximately 34° (the maximum angle of rest of dry sand), after which shearing will take place, giving a slip face at rather less than 34°. In plan this face will be transverse to the wind direction, and will stand above slumped sand at the lower angle, so that the leeward slope as a whole will be concave in profile. This process will be repeated continually, and the dune will migrate downwind at a speed determined by the rate of sand supply and the velocity of the wind. The rate of movement will be slowest at the centre of the dune (its highest point), for more energy is needed here to move sand up the long windward slope, and greatest at its margins where the dune is less high. As a result 'horns' begin to develop, and the original sand mound is transformed into a typical migrating barchan.

The second main dune-type, the longitudinal dune or 'seif', is less well understood. Such dunes are parallel, instead of transverse, to the direction of the prevailing wind, and often reveal the influence of secondary winds in possessing an asymmetrical profile, comprising a gentle slope similar to the windward face of a barchan and a steep slip face. This evidence has naturally led Bagnold to the conclusion that seifs occur in areas where the winds are predominantly from one direction but where there are also occasional strong cross-winds from another. Indeed, Bagnold has shown that some longitudinal dunes result from the transformation of barchans. As the latter are affected by cross-winds, the windward of the two horns becomes greatly elongated, and the slip face, initially transverse to the prevailing winds, is gradually reorientated across the path of the new winds. In areas where dunes are numerous, such a process could lead to the widespread coalescence of individual barchans and the development of a series of semi-parallel longitudinal dunes.

THE DESERT PIEDMONT

The desert localities which have proved of greatest interest to the geomorphologist are those separating mountains, plateaus and residual upland masses from the broad plains of erosion and deposition that are marginal to or surround them. These are the so-called 'piedmont' or mountain-foot zones, and it is here that the most significant processes of desert erosion (the backwearing of slopes, leading to the destruction of upland blocks and their replacement by the gently sloping desert pediplain) take place (Fig. 124). Piedmont-zone landforms have been studied in detail in a wide variety of desert environments, and although precisely the same features are not found everywhere there are certain elements (particularly 'knicks' or 'piedmont angles' and rock pediments) which are almost invariably present.

The mountain-front

Sometimes referred to as the 'mountain slope' or 'scarp', this comprises a high and steep slope, ranging in angle from 25° to 90°, which rises abruptly from the gentler slopes beneath. In some instances, the mountain-front is capped by a near-vertical cliff, below which is developed a rectilinear denudational element at 25° to 35°; in others the whole slope is of rectilinear form, though some steeper sections may occur if there are outcrops of unusually resistant rock, and in granite and similar formations exfoliating half-domes may occur. In any one area (especially if it is underlain by uniform rocks), the rectilinear elements—whether developed on detached hill-masses, on the mountain-front proper, or on the sides of canyons incised into the mountain-front—are usually developed at a constant angle. From such evidence it is safe to infer that these slopes are retreating without loss of angle, for it would be too much of a coincidence if all were at exactly the same 'stage' of development. Several authorities have tried to explain such parallel retreat in terms of some form of basal undercutting, by agents such as stream-floods or sheet-floods, which effectively counteracts any tendency of the slope to decline in angle as it is weathered back. However, this theory cannot explain satisfactorily either the rectilinearity of many slopes or their occurrence at a constant angle in any one locality.

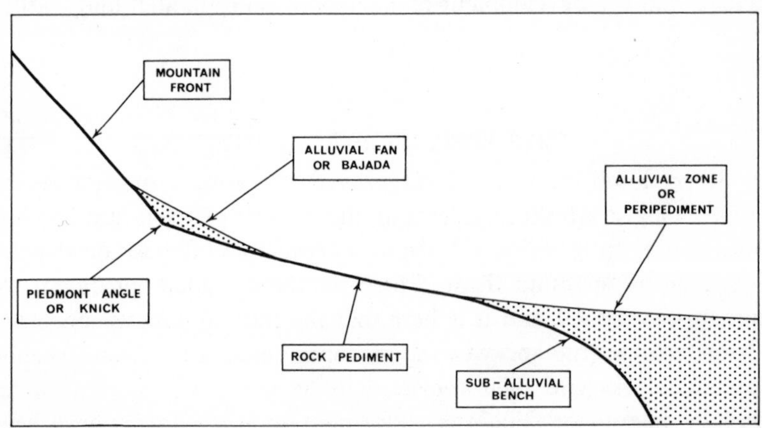

124 Idealised cross-section of the desert piedmont

Detailed study reveals that in fact the mountain-front is often a 'boulder-controlled' slope; in other words, although underlain by solid rock and therefore denudational in origin, the slope develops at the angle of repose of the detritus lying on its surface (Fig. 125). On a typical mountain-front this weathered material is usually of very coarse grade, and its main component may be large boulders which have been released by block disintegration. The size and shape of these boulders (which control their angle of rest, and thus in turn the angle of the slope on which they lie) is determined by the spacing of the joints either in the underlying rock or on the steep free face rising above the rectilinear slope. Providing the joint pattern remains constant, weathering back of the slope will continue to produce detritus of the same calibre, and no change in the inclination of the slope need occur. The boulders occupying the mountain-front either move very slowly downhill or decay, as a result of granular disintegration, *in situ*. Fine sandy material will thereby be released, and this can be washed very quickly to the foot of the steep slope by run-off during occasional desert rainstorms.

The knick

This very sharp break in profile, also named the 'piedmont angle', separates the steep mountain-front or the slopes of a residual hill from the rock pediment or alluvial fan below. Usually, there is a marked change in the character of the superficial weathered waste at the

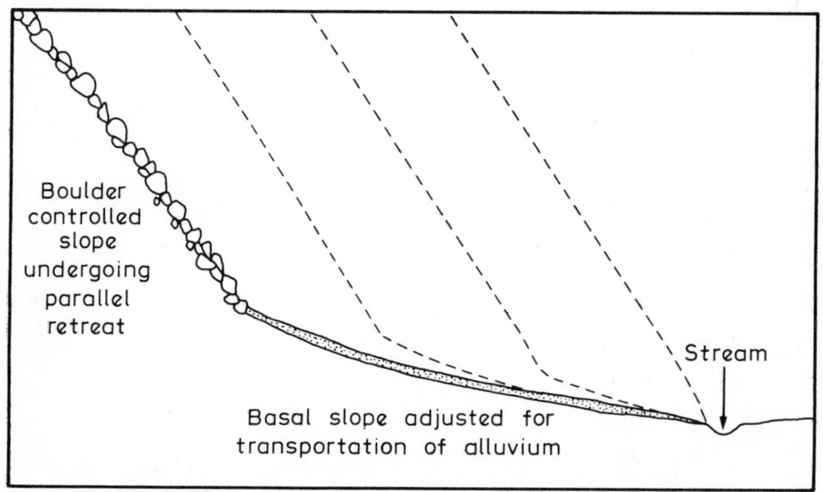

125 Parallel slope retreat and pediment formation

piedmont angle, from the coarse boulders which litter the slopes above to fine-grained alluvium resting on the pediment. On occasion the knick itself is actually buried beneath alluvial fans, but more commonly it is exposed. This suggests that transportational processes are so effective on the upper part of the pediment that detritus moved to the foot of the mountain-front can normally be washed away with ease, despite the low slope angles which prevail here.

The origin of the knick has long been disputed. Some geomorphologists have attributed it to lateral cutting by running water. Such a process is most feasible in 'pediment embayments' or 'pediment passes', where stream-floods issuing from canyon mouths or actually developing within the pass are able to swing against the base of the mountain-front on either side. However, where no such embayments or passes exist, sheet-floods are more likely to develop. Since these may comprise water running off the mountain-front on to the pediment (or in other words across the line of the piedmont angle) lateral corrasion as the cause of the knick is not easy to envisage. Other writers, notably King and Peel, regard the piedmont angle as essentially a weathering phenomenon. It is noteworthy that similar breaks of slope occur in savanna lands, where the rock at the head of the pediment may be so rotted that it is easily

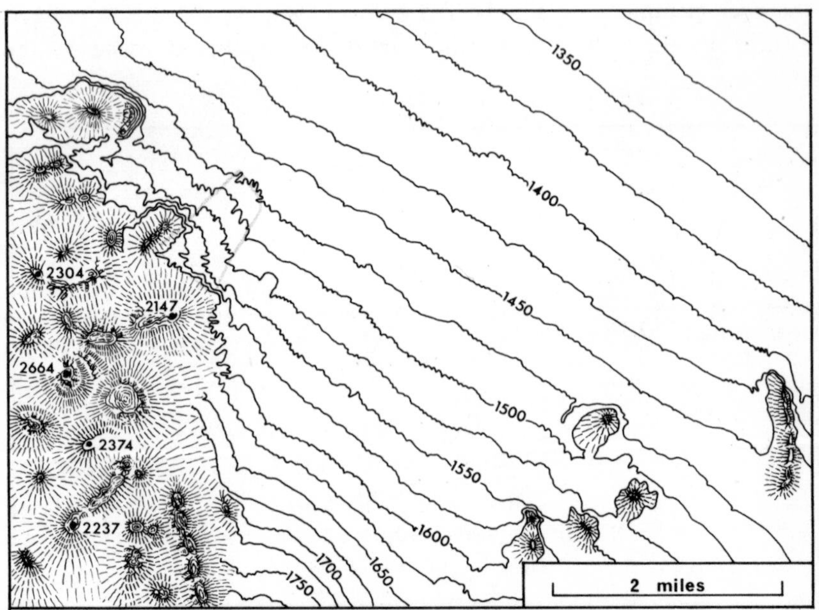

126 Rock pediment near Antelope Peak, Arizona

127 A pediment, separated from a small residual hill by a pronounced 'knick', in a semi-arid landscape, Uganda. [G. R. Siviour]

removed by springs and streams to give a well-defined trench ('linear depression') at the foot of the mountain slope or inselberg. In arid areas a similar if less marked concentration of weathering might well occur, for rainwater drains rapidly from the upper slopes to the lower, where it tends to soak in and persist, particularly if detrital fans have temporarily formed. Peel has noted that in the Sahara some residual pinnacles and inselbergs are surrounded by annular depressions, where moisture-assisted weathering has attacked the rock and paved the way for strongly localised wind deflation.

The rock pediment (Fig. 126)

This is perhaps the most problematical of all desert landforms, and has certainly given rise to the greatest controversy. However, it is a feature by no means exclusive to deserts, for it is widely developed in savanna areas, and in the opinion of some can be equated with the basal concave slopes formed under other humid conditions. Basically, the pediment is a gently inclined rock surface, leading up to the foot of the mountain-front, and planed across underlying rocks irrespective of their varying resistance to erosion (Fig. 127). In cross-profile it is concave, with slope angles ranging from 7° on the upper pediment to only about half a degree on its lower margins. This concavity immediately suggests that running water is an important agent in the formation of pediments. This seems to be confirmed by the existence on many pediment surfaces of an alluvial veneer which is usually patchy but may mask completely the solid rock. Downslope this veneer tends to become more continuous and thicker, and eventually the pediment passes beneath the alluvial zone or 'peripediment'. Borings for water have revealed that beneath the alluvium the pediment may form a convex profile, giving what Lawson referred to as the 'sub-alluvial convex bench'. Another feature of pediments is the manner in which they extend into canyon mouths, so forming 'pediment embayments'. Such embayments on either side of an upland mass may grow towards each other and coalesce to give a 'pediment pass'. The various theories which have been proposed to explain pediments are briefly reviewed below.

The bajada

This term is usually applied to a series of coalescent alluvial fans, laid down by ephemeral streams at their points of exit through the mountain-front into the piedmont zone, but is sometimes used to describe the peripediment. The marked reduction of gradient at the base of the mountain-front causes a diminution in stream energy, and hence carrying capacity, and the resultant deposition of boulders, gravel and sand causes the build-up of the fans. The angle of slope of these is sometimes as high as 20°, but is more often comparable with that of the rock pediment (7–1°)—indeed there has undoubtedly been confusion in the field between true bajadas and well-veneered pediments. The smaller and steeper bajadas tend to mask only the upper part of the rock pediment, burying the piedmont angle, but the larger and more gently sloping fans may be so extensive as to grade into the alluvial deposits of the peripediment; the resultant sub-alluvial rock surface is then referred to as a 'concealed pediment'.

The alluvial zone

Sometimes referred to as the peripediment, this is a broad area of fine material washed by stream- and sheet-floods across the pediment from near-by uplands. Often the alluvial layers have been steadily built upwards over a very long period. In the south-west U.S.A. the alluvium has infilled fault-troughs ('bolsons') of Tertiary age. As the marginal fault-scarps have been driven back to give the present-day mountain-fronts, so the products of their degradation have been transported into the structural basins, to accumulate to a depth of a thousand feet or more. Such alluviation has naturally been greatest in totally enclosed basins, which have nourished a centripetal drainage system and temporary lakes. The latter exist only after rainstorms, and their sites are marked by flat plains ('playas') covered by salt left after evaporation of the lake water. In some basins, however, through or outflowing drainage has developed, and in these continuous alluviation, leading to a rising local base-level, will not be encountered. Instead, the alluvium will be removed from the basin as rapidly as it is brought, so that the base-level remains virtually stationary, or in some cases there may actually be a falling base-level, as the river cuts into and rejuvenates the peripediment; in such instances the pediment itself will be left upstanding as a terrace-like feature.

THEORIES OF PEDIMENT FORMATION

The theories which have been proposed to account for rock pediments may for convenience be divided into two categories. In the first are those that emphasise water action (and in particular lateral erosion by stream- and sheet-floods); in the second are those which, although taking due note of the contributory role performed by surface run-off, place greater emphasis on other factors and processes.

Theories of water erosion

The earliest and most comprehensible theory is that of McGee (1897), who attributed the gently inclined and planed rock-surfaces of deserts to sheet-flood erosion. Few authorities would dispute that this process is responsible for fashioning the details of many pediments, but for the reasons given on p. 297 sheet-floods cannot be regarded as the main formative agent, or even as the only modifying agent. Indeed, stream-floods may be more active on many pediment slopes, for Balchin and Pye (1955) noted that the pediments of the Sonoran Desert of Arizona often had 'a very hummocky surface, and were by no means as smooth as might have been expected from the literature'. These pediments were in fact usually 'runnelled', and dissections of up to 15–20 ft in depth were not uncommon.

A more recent and closely argued theory is that of Johnson (1932). Johnson laid great emphasis on the transitional location of pediments between (i) dissected uplands, where streams are engaged predominantly in vertical corrasion, thus giving steep-sided canyons, and (ii) low-lying areas (peripediments) where deposition of alluvium takes place on a very large scale. Thus pediments lie in that zone where desert streams undergo a change from an 'underloaded' (degrading) to an 'overloaded' (aggrading) condition. In other words, the pediments coincide with the stream sections where the running water is neither underloaded nor overloaded but is fully loaded; that is, the stream's energy is totally consumed in the movement of the water and its load, and there is none available for vertical corrasion. Johnson accepted, like Gilbert (p. 56), that such graded streams are capable only of lateral corrasion, and was thus able to explain the 'planed-off' appearance of many pediments.

Johnson argued that his theory could be successfully tested by observation in the field. In the embayments where streams debouch on to the piedmont (and where the change from the underloaded to the fully loaded state is supposed to take place), lateral corrasion ought to produce pediments that are slightly convex in profile parallel to the general line of the mountain-front, though still concave in profile at right angles to it. Johnson claimed that 'rock fans', as he termed these 'relatively flat semi-cones', do in fact exist in the desert areas studied by him in the field (Fig. 122). However, he admitted that their recognition is not always easy, owing to a covering layer of detritus which causes them to resemble alluvial fans. Johnson also stated that, at either edge of the rock fan, channels ('piedmont depressions') often occur, and mark the courses of present-day stream-floods. Such channels are, in his view, associated with the trimming back of the mountain-front, in the manner described on p. 296.

The main objections to Johnson's theory may be briefly summarised as follows:

(i) It is by no means fully accepted that graded streams are capable only of lateral erosion—indeed, in the light of modern studies of fluvial action the whole concept of grade, at least as envisaged by Gilbert, must be regarded with suspicion (pp. 57–9). There is, of course, the possibility that Johnson's detailed arguments may have been unsound, but that his central thesis—the existence of a piedmont zone where lateral corrasion by running water is the dominant process—was correct. In fact, there are some reasons to doubt even this.

(ii) The existence of rock fans, central to the lateral corrasion hypothesis, has been disputed by many geomorphologists, who have claimed that such fans are in reality no more than the alluvial fans that they are supposed to resemble so closely.

(iii) If Johnson's theory were correct, the typical mountain-front should be indented by numerous pediment embayments, separated by projecting spurs in the process of being eroded back by laterally wandering streams. In fact, many mountain-fronts are comparatively straight in plan, with the maximum pediment slope at right angles to the line of the knick. The presence of water running along the knick cannot be countenanced in such circumstances.

(iv) Residual masses of rock (sometimes called 'nubbins') disrupt the smooth surfaces of some pediments. It is difficult to see how such features could survive if the pediment were the product of a laterally migrating stream. Admittedly this argument is countered to some extent by Johnson's statement that stream-floods alone do not cause pedimentation. When arriving at the head of the pediment, mountain streams often become extensively braided; 'they divide and subdivide indefinitely, the minor channels reuniting and interlacing and frequently shifting both their positions and their connections'. This migration of small channels was regarded by Johnson as a fundamental process in the regrading of pediments. 'Since there are a multitude of interlacing channels, no one of them need shift very far in order that lateral planation should take place over the entire surface of a fan or pediment.' The suggested mechanism is not unlike that postulated by Davis (p. 298).

Howard (1942) is yet another American geomorphologist who regards lateral planation as the major process involved in pedimentation. Howard has successfully demonstrated in some places the reality of basal stream undercutting, resulting in well-defined knicks surmounted by mountain slopes that are steeper than the average for the areas concerned. Like Johnson, Howard argues that lateral planation by both stream-floods and sheet-floods, as well as by streamlets and sheet wash (a less voluminous and vigorous form of run-off than the sheet-flood proper), leads to the continual extension and lowering of the pediment surface. Howard's main contribution is probably the notion that the processes leading to the formation of the pediment (running water in various forms) and retreat of the mountain-front (weathering and transportation) are quite different, but that none the less they work in close harmony with each other (Fig. 128). Thus he considers that the rate of recession of the mountain slope is normally equal to the rate of headward extension of the pediment by laterally eroding surface run-off. As a result, it is rare to find a very marked oversteepening of the lower part of the mountain-front owing to exceptionally rapid planation on the pediment. Nor is it common to encounter large permanent accumulations of detritus at the base of the mountain slope, where they would tend to protect the underlying rock whilst recession of the upper slope proceeded unhindered, thus causing an overall decline of slope angle.

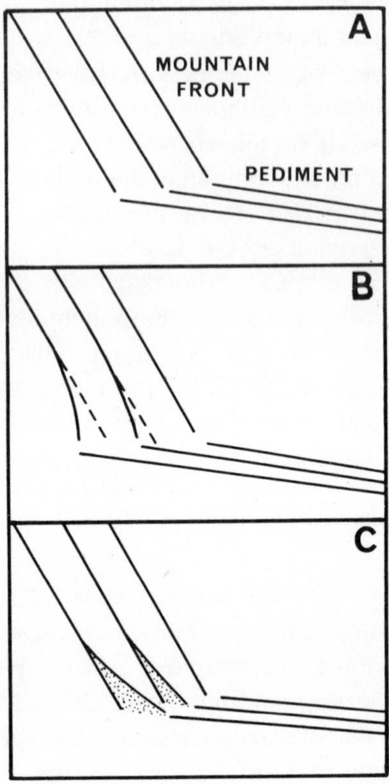

128 The relationship between the rate of mountain-front recession and the rate of pediment extension

Composite theories

One of the earliest and most convincing theories of pediment formation was proposed by Lawson in 1915. Lawson considered hypothetically the recession of a fault-scarp subjected to attack under arid conditions. The scarp would be driven back by weathering from its initial position, and the disintegrated rock would accumulate at the foot of the scarp unless removed by streams. With the passage of time, this detritus would grow upwards and steadily encroach on the receding scarp-face, the lower part of which would be progressively buried and insulated from further weathering. Only the upper exposed part of the scarp would continue to retreat, furnishing at each stage a smaller and smaller quantity of weathered rock to be spread ever more thinly over the surface of the growing detrital slope. The process as a whole is shown in Fig. 129, from which it will be seen that the inevitable outcome is the formation, beneath the accumulating debris, of a solid rock bench of convex profile. The lowest (oldest) part of the bench will be steepest in angle, approaching that of the initial fault-scarp, and the highest (youngest) part will be the most gentle, in fact closely resembling the slope of the overlying detrital cover, which at this point will be extremely attenuated.

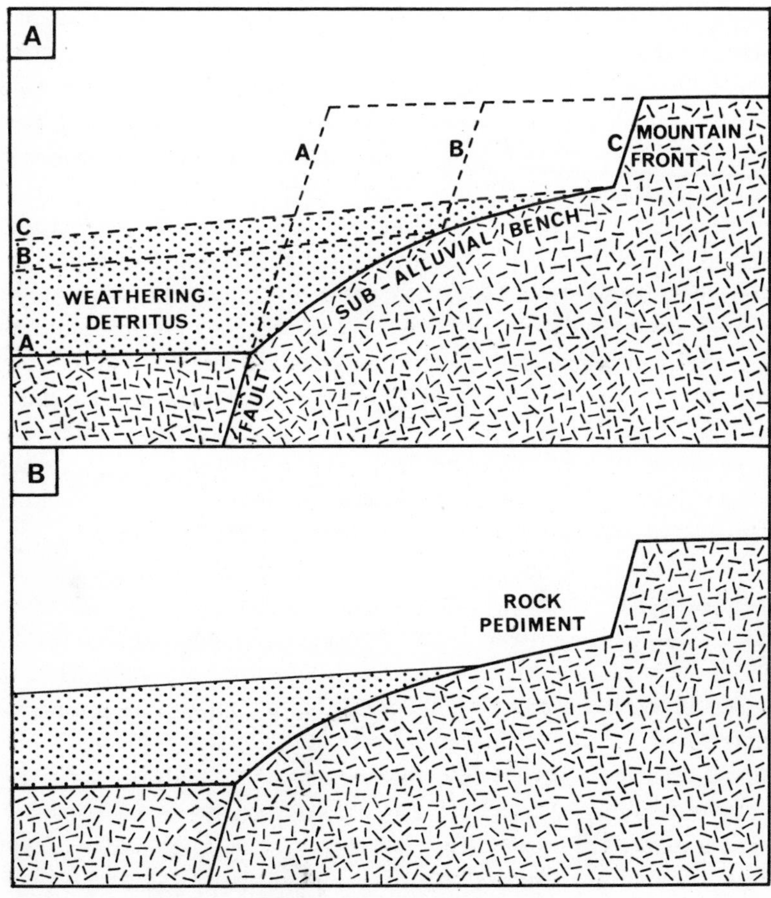

129 The development of pediments (after A. C. Lawson)

It will be appreciated that removal of the uppermost layer of the detritus, or alternatively the regrading of the weathered material to give a slope of lower angle, must lead to the exposure of the rock bench immediately adjacent to the foot of the scarp.

Lawson argued that these hypothetical conditions are actually fulfilled in the deserts of the U.S.A. Many of the mountain-fronts originated as fault-scarps which have been driven back by weathering over a long period. The vast quantities of detrital material so released have been transported, largely as alluvium by stream- and sheet-floods, into structural basins with no outlet. Here they have accumulated and steadily encroached on to the retreating scarps. 'Sub-alluvial convex benches', the reality of which is attested to by well borings (p. 308), have been widely developed. However, removal or regrading of the alluvial cover has in many places led to the exposure of the upper parts of the benches, forming the familiar rock pediments. Detrital material from the mountains has continued to cross (hence the alluvial cover of most pediments), and comparatively minor modification by water erosion has given the pediments their characteristic concave profiles.

Lawson's theory seems to be readily applicable to certain types of geological structure (fault-block country) and in areas where there is a rising local base-level caused by protracted alluviation. However, some writers have not accepted Lawson's inference that rock pediments are exposed only by accident, and have argued that they are so ubiquitous under arid conditions as to constitute a 'normal' landform. Even more important is the fact that pediments are not confined to the margins of enclosed basins, or similar morphological conditions which favour a continuous build-up of alluvium, but are equally well developed where the local base-level is stationary or even falling as a result of stream incision. As L. C. King has shown, this is the usual situation over much of Africa, where pediments form not at the bases of fault-scarps but beneath the retreating walls of stream-cut valleys. Even in the south-west U.S.A., Bryan and McCann (1936) have demonstrated, in the Rio Puerco valley of New Mexico, a sequence of pediplains and pediments, developed in relation to an intermittently falling base-level since Eocene times.

This, together with other evidence, leads us finally to consider a theory of pedimentation which has been revived by King (1948) but which originated in the writings of the German geomorphologist Penck. These authors regard the pediment as essentially a basal slope, developed at the foot of a steeper slope element undergoing parallel retreat (pp.

212–14). It is, therefore, a slope of transportation over which debris weathered from the steep slope (mountain-front or 'scarp') is moved either into a stream (as in the cycle of pediplanation) or to an area of aggradation (such as a bolson). Just as the steepness of the mountain-front is determined by the angle of repose of the weathered material lying on it (p. 305), so the gradient of the pediment is controlled by the calibre of the material which must be transported across it. It has been pointed out above that the angles of slope of alluvial fans and rock pediments are normally similar (up to 7°). The former clearly develop at the maximum angle of repose of alluvial material, whereas the latter are at the minimum angle needed for the transportation of alluvium by running water; these two angles are, of course, virtually identical.

It will be apparent that, if this theory is correct, pediments should be particularly evident in rocks which readily undergo block and granular disintegration (for instance, intrusive igneous rocks). The production of blocks permits the development of a steep parallel retreating mountain slope, and the further breakdown of these blocks *in situ* into grains allows the basal slope of transportation to assume a very gentle angle, hence the contrast between the two main slope elements and the sharp break of gradient between them. It is relevant to note that, in their description of the Sonoran and Mohave Deserts of Arizona and California, Balchin and Pye refer to the close association of steep mountain-fronts and pediments with 'granite, gneiss and other crystalline rocks'.

One of the many attractions of the Penckian theory is that, whilst differing from Howard's hypothesis in its acceptance of weathering and transportational processes as responsible for both slope recession and pedimentation, it embraces the possibility of some fluvial erosion. The transport of fine detritus across the pediment is attributed solely to the action of running water, and it is impossible to believe that such water, at times well armed with gravel and sand, can never have any corrasive effects. Clearly it can, as the typical concave profile, the occasional trenching, and the undercutting of slope bases in pediment embayments all show. In brief, although primarily a slope of transportation, the rock pediment is in some measure also a slope of fluvial erosion, both vertical and lateral.

Another merit of this theory of pedimentation is its applicability to a wide variety of structural, base-level and climatic conditions. Thus basal slopes are as likely to develop at the foot of structurally determined scarps as beneath slopes initiated by fluvial erosion. A stationary base-level offers perhaps the most favourable conditions for pedimentation, but the process may only be modified by a rising base-level, leading to pediment interment, perhaps later to be followed by exhumation. Falls of base-level will cause a renewal of pediment formation at a lower elevation, and the rejuvenation of existing pediments. Finally the occurrence of pediments not only in arid lands, but in savanna and possibly other climates, underlines the fact that pedimentation may be a slope process that operates in a variety of environments, producing the basal concavities regarded as typical of fully developed slope profiles. However, because of the particular nature of the weathering and transportational processes operating in deserts, the contrast between the gentle basal slope and the mountain-front rising above it is unduly emphasised; and it is accordingly in arid climates that the rock pediment has attracted so much attention and excited so much controversy.

10

PERIGLACIAL LANDFORMS

INTRODUCTION

At the present time one-fifth of the world's land-area is estimated to be affected by a climate that cannot be termed 'glacial', in that it does not nourish extensive ice-fields and glaciers, but that is none the less characterised by long periods of frost during the winter months and, in some instances, by large winter snowfall. These 'Arctic' conditions are especially widespread in the northern Hemisphere, where they are associated with the tundra of Alaska and northern Canada, parts of northern Europe (for instance, Lapland) and large tracts in Siberia—indeed, something like 50% of the total area of the U.S.S.R. may be underlain by permanently frozen ground.

One of the most important results of the growth of the continental ice-sheets during the glacial periods of the Quaternary (chapter 11) was the shifting of the climatic belts towards the Equator. Thus the climatic conditions now typical of the high-latitudes which have been referred to were displaced southwards, and affected parts of North America, the British Isles, Europe and Asia which today enjoy temperate climates. These areas were then subjected to phases of very active weathering and transportation, such as are operating now in 'Arctic' lands. Far-reaching modifications of the landscapes of, for example, southern Britain and northern France (both of which escaped direct glaciation) were made. The imprint left on such temperate areas by the severe climatic conditions, particularly those associated with the last (Wurm) glacial period, has not been erased, for the post-glacial period has been of short duration (approximately 10,000 years) and has encompassed little weathering and erosional activity.

The term which has been increasingly used to describe the denudational processes at work in such a cold non-glacial climate is 'periglacial' (meaning literally 'around the ice'). In a sense the term is misleading, in that some areas which experienced severe cold during the Quaternary (such as high upland masses in comparatively low latitudes) were far removed from the zones marginal to the continental ice-sheets. The same may be said of many areas of cold climate at the present day, for example in Siberia. The term 'periglacial' has, not surprisingly, also been used in a climatic rather than a geomorphological sense. The difficulty here lies in the fact that there can be much variation in the climatic régimes that favour the operation of typical periglacial processes. Thus at one extreme lies the so-called 'Maritime Arctic' (Icelandic-type) climate, in which precipitation is heavy and takes the form mainly of winter snow, and in which temperatures are comparatively moderate (the annual average temperature may in fact exceed 0 °C). At the other extreme is the 'Continental Arctic' (Siberian-type) climate, in which precipitation is light, and paradoxically falls as rain in summer. Temperatures are severe, particularly in winter when figures of −60 °C may be recorded. In summer, on the other hand, relatively mild conditions are experienced, with July mean temperatures always exceeding freezing-point.

There is much evidence to show that, during the Quaternary, marked variations of climate both in time and space occurred in periglacial areas. For instance, over much of the last glacial period south-east England was affected by extremely low temperatures, allied to comparatively little precipitation. However, at the very end of the glacial a short period of milder temperatures (with the July mean exceeding 5 °C) and heavy winter snowfall seems to have intervened. Again, in southern England as a whole some contrast may be detected between the south-west peninsula, where only the land above 1000 ft was severely affected by the cold, and south-east England, where typical periglacial features are found in abundance down to present sea-level.

In spite of these variations, it is still possible to make certain broad generalisations about periglacial climates. The first and most obvious is that the winters are always cold, and often bitterly cold, and encompass long periods when the temperature lies well below 0 °C. During such times both the ground and any surface water becomes frozen, and weathering and erosional processes become virtually inactive. The second is that the summer period, long or short according to the precise latitude and whether or not the climate is maritime or continental, is relatively mild, so that even at night frosts will not occur and the ground surface is in a state of thaw. Streams may flow and perform some erosional work, and processes such as soil creep may operate, though rather slowly in late summer where the warmth succeeds in drying out the ground. The third and most important generalisation is that the intermediate seasons between winter and summer are always, in the geomorphological sense, the most active. The reasons for this are threefold. In the first place, during the transitional seasons the ground is neither continuously frozen nor continuously thawed, but tends to freeze at night and thaw during the daytime. Thus a diurnal freeze-thaw cycle is set up, and this is of fundamental importance in the mechanical disintegration of rocks. Such frost weathering (sometimes referred to as 'congelifraction') is one of the dominant processes in periglacial areas. In the second place, at the end of the winter, as the ground becomes thawed semi-permanently it is highly charged with water, derived from the melting of ice within the soil ('ground ice') or from winter snowfall. Because of this abundance of water, which acts rather in the manner of a lubricant, soil and weathered material is able to 'flow' downhill, even on comparatively gentle slopes. This process, solifluxion, is the second dominant process under a periglacial climate. Often its action is short-lived—for instance, it may cease as soon as all the winter snow is melted—but the results are striking because of the relative rapidity of soil movement. In the third place, in areas of marked winter snow accumulation the 'thaw season' may see not only the saturation of the upper layer of the ground, but also the development of many surface streams, perhaps even meltwater torrents, which may perform both a transportational role (removing the products of frost weathering) and also effect in a short time considerable vertical and lateral corrasion.

It should be apparent that, in general, more active periglacial weathering and erosion will take place in a maritime climate, if only because here the transitional seasons are likely to be more prolonged than in continental interiors, and because winter snow accumulation, favouring both solifluxion and sudden meltwater activity, will usually be much greater.

PERMAFROST AND ITS EFFECTS

Probably the most important single phenomenon resulting from periglaciation is the deep and permanent freezing of the subsoil and underlying rock. This feature has been given many names (for example, 'pergelisol', a term derived from the Latin *per* = throughout, *gelare* = to freeze, and *solum* = soil or ground; or 'perennial tjaele', a Swedish-Norwegian word meaning 'frozen ground'), but the most convenient and widely used term is 'permafrost'. The permafrost layer varies considerably in nature and thickness, depending on existing climatic, morphological and hydrological conditions. In well-jointed rocks, forming interfluves or ridges, the amount of ground water may, because of the free subsurface drainage, be comparatively small and the permafrost will be 'dry' (that is, associated with little ice formation). Where the water content of the rock is higher, as beneath valley bottoms or in rocks retentive of water, the permafrost will be 'wet'. A great deal of ground ice will develop, as films, grains, wedges, lenses and other irregular masses. The depth of the permafrost will depend on a variety of factors. In general it reaches maximum development under continental conditions, owing to the extreme severity of the winter temperatures and the lack of snow cover (which acts as an insulating layer protecting the ground from very low temperatures). Under more maritime conditions the permanently frozen ground may be discontinuous, occurring only on shaded slopes or in valley bottoms. Indeed, for up to three months of the year the ground may be wholly thawed out, so that permafrost is replaced by what has been termed 'annual tjaele'.

It has been calculated that the mean annual temperature of an area must be as low as −1 to −4 °C for permafrost to develop. Under the influence of temperatures lower than this, the frozen layer may attain great thicknesses. At the present day, the permafrost of northern Siberia and Alaska may approach a depth of 1000 ft; indeed there is a report of a thickness of nearly 3000 ft from the Indigirka valley, Siberia. The depth of the permafrost in areas subjected to periglaciation during the Quaternary, but which now experience a temperate climate, cannot be judged as no evidence remains.

In terms of the direct weathering of rocks permafrost is not important, since as the ground remains always frozen there are none of the alternate expansions and contractions associated with diurnal freeze-thaw cycles. Indirectly, however, permafrost plays a very important role in the geomorphological development of periglacial areas. In permeable rocks it disrupts the underground circulation of water, for percolation is impeded and springs dry up. Surface run-off during times of melt is correspondingly enhanced. In valley bottoms and areas of subdued relief, soil drainage becomes impossible, and waterlogging and the development of extensive lakes, swamps and bogs is a common phenomenon. Even on slopes the saturation of overlying soil is promoted and the process of solifluxion thereby aided.

The active layer

It will be readily apparent that, even in a severe periglacial climate where permafrost is developed to an advanced degree, the ground cannot remain wholly frozen throughout the year. With the onset of spring, when daytime temperatures rise above 0 °C, the upper layers of the soil are thawed. To begin with, this soil will be refrozen by nocturnal frost, but as the warm season progresses these frosts will peter out and the thawed layer will gradually increase in depth until the beginning of the next winter. At its maximum the thawed layer may attain a depth of between 5 and 20 ft, depending on the warmth of the summer and the composition of the subsoil (for example, a coarse gravel because of its high conductivity favours deep thawing, whereas peaty material has the opposite effect). There is evidence in the form of disturbed gravels to show that, during the last glacial period, thawing of the permafrost in southern England proceeded annually to a depth of from 5 to 10 ft (Fig. 130).

130 Plateau-gravel near Fawley, Hampshire. The lower part of the deposit comprises well-bedded sands and flints; the upper 3–4 ft have been greatly disturbed by cryoturbation processes. [R. J. Small]

The thawed layer has been variously named. The American geomorphologist Kirk Bryan referred to it as the 'mollisol' (derived from the Latin *mollere* = to make soft, *solum* = soil or ground), to differentiate it from the underlying pergelisol. Others have used the term 'active permafrost'. However, the most simple, self-explanatory and acceptable term is 'active layer'. It is within this layer that frost weathering ('congelifraction') occurs, as a result both of diurnal freeze-thaw cycles early and late in the warm season and of the annual freeze-thaw cycle to which the layer is subjected. Furthermore, it is within the active layer that cryoturbation (*cryo* is derived from the Greek word for frost) or congeliturbation (p. 180) is concentrated. This includes the transportational process of solifluxion, which is often assisted by the expan-

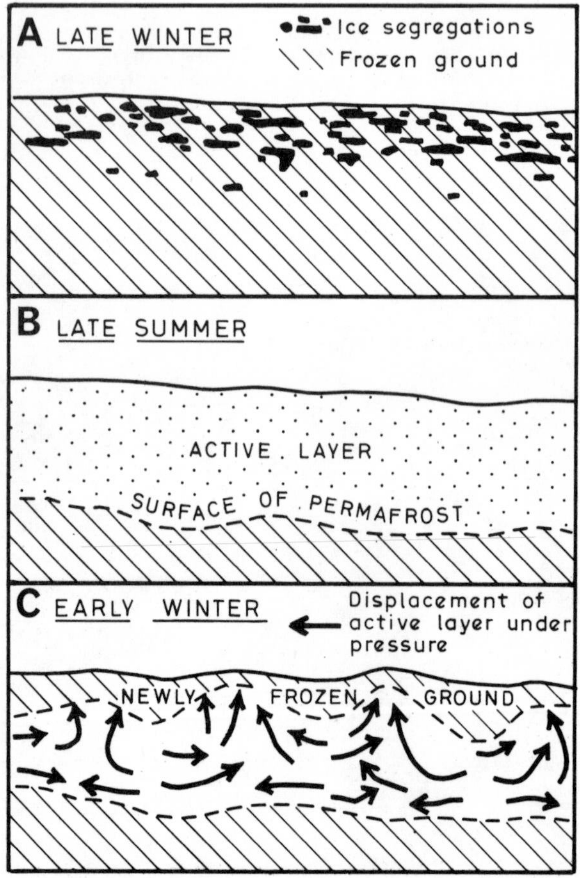

131 Permafrost and the active layer

sions associated with the formation of ground ice, together with two allied processes not previously mentioned. These are 'frost heave', involving the upward movement of rock detritus, and 'frost thrust', a direct lateral pushing of material, similar in effect to solifluxion but much more localised and occurring frequently on level ground. The mechanism of these two processes is complicated, and a number of controlling factors are probably involved.

Firstly, it is necessary to restate that, with the onset of winter, the active layer again becomes frozen. Naturally, the ground surface, radiating heat to the atmosphere, cools and freezes first, and subsequently the freezing penetrates downwards towards the permafrost. For a time, the intervening zone remains unfrozen and mobile, and as it is confined between two layers of ice one of which is undergoing growth and expansion is subjected to considerable pressure. Lateral thrust of the constituent material may be induced, and upheaval may take place at points where the upper ice layer is thin or weak (Fig. 131). This upheaval may be on a small scale, producing contortions (otherwise referred to as 'cryoturbations', 'convolutions' or 'involutions')—features commonly visible today in once horizontally bedded gravels in southern England (Fig. 132)—or it may result in the formation of dome-like structures ('hydro-laccoliths') with a core of mud or even water, which often becomes frozen to give a large ice-lens. The eventual collapse of such domes (which may exceed 100 ft in height) gives rise to circular hollows, surrounded by a ring of ridged-up material and sometimes containing ponds (Fig. 133). In Arctic America such subsurface domes of ice are referred to as 'pingos' (an Eskimo word). The previous existence of pingos in parts of Wales and the Belgian Ardennes has recently been demonstrated. Another process which leads to the upheaval of material in the active layer is the development of small localised ice-masses and ice-needles ('pipkraker'). Once ice begins to form underground it continues to develop by attracting water from its surroundings by some form of capillary action. This process, known as 'ice segregation', necessarily leads to some displacement of unfrozen material. It has been observed that ice segregations are common beneath large stones, probably because these are good conductors of heat and allow frost to penetrate more deeply into the ground. The growth of the ice leads to the upheaval of the stones, which may not return to their original positions during a period of thaw because finer soil flows in from either side to fill the cavity occupied by the ice. Because of the continual operation of this process, there is a tendency for the coarser material of the active layer to be pushed to the surface, whereas the finer material remains at some depth. Furthermore, the individual stones as they are raised become rearranged in such a way that their major axes are vertical or near-vertical. The sorting of detrital material in this way is not inevitable, however, for in some circumstances the finer material below may come under pressure (for example, because of the freezing of the surface layer) and be almost literally pumped out through the overlying stones.

Patterned ground

In addition to this vertical sorting, a marked feature of periglacial areas is the lateral sorting of fine and coarse particles in the active layer, to give highly distinctive ground patterns, particularly where there is no continuous vegetation cover. Very broadly, two types of patterns may be recognised.

Firstly, on near-level ground polygonal or circular markings are

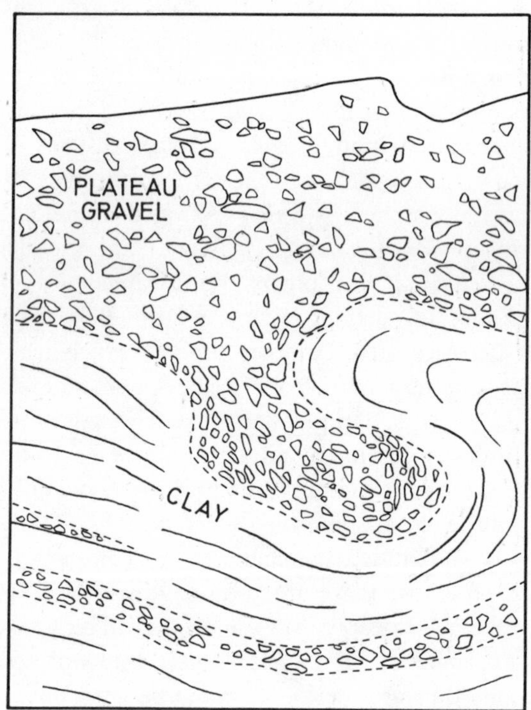

132 Evidence of cryoturbation in plateau-gravel, New Forest (after J. Lewin)

developed. The best known of these are the so-called 'stone polygons', which comprise borders of frost-shattered and up-ended stones and a central area of moist mud. Usually stone polygons occur in large numbers, the stone borders of individual polygons interlacing to give what may be termed a 'stone net'. Sometimes the stone borders are circular rather than polygonal, giving 'stone rings'. Other similar types of patterned ground include 'mud-flat polygons', which are developed in homogeneous materials and probably result merely from the drying out and contraction of the wet surface of clay and alluvium. 'Fissure polygons' are large-scale markings which, like mud-flat polygons, exhibit no sorting of material; instead they are bounded by large cracks which are occupied by wedges of ice, the expansion of which has led to some ridging up of the earth at the margin of the polygon. 'Fossil ice-wedges' may be observed in the gravel deposits of south-eastern England, the ice having been replaced by stones and other material which has fallen in from above (Fig. 134).

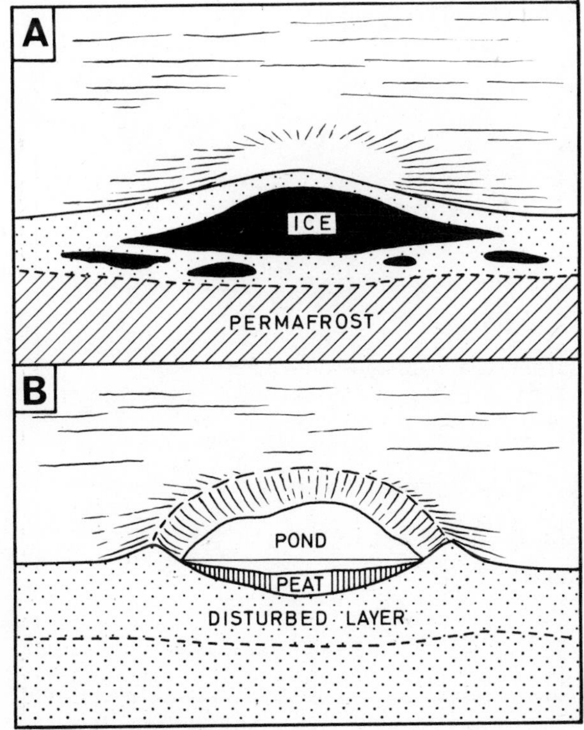

133 The development of a pingo

Secondly, on slopes exceeding an angle of 5°, 'stone stripes' are common features of a periglacial landscape. Like stone polygons and rings these clearly result from the lateral sorting of detrital material, for they comprise alternate lines of coarse stones and 'fines'. The latter, like the centres of stone polygons, are often slightly domed; furthermore, unlike the lines of stones, they hold moisture well, and because of their mobility are clearly affected by solifluxion. It is, in fact, widely believed that stone polygons and stone stripes may be closely related in terms of origin. Thus on a very gentle slope polygons may become rather elongated, with their major axes aligned parallel to the maximum gradient. If the slope becomes a little steeper, the polygons give way to 'stone garlands' (Fig. 135). These are lobate forms, containing a domed centre of sticky mud which is held up by a retaining 'wall' of large stones; this wall is highest and best developed on the downslope

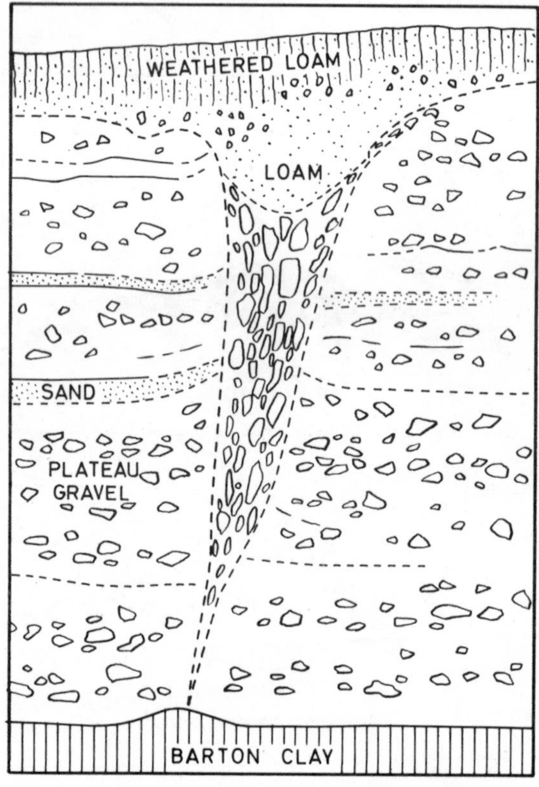

134 A fossil ice-wedge, Highcliffe, Hampshire (after J. Lewin)

side, so that in cross-section the slope begins to take on a terraced appearance. On even steeper slopes the mud may burst through this stone barrier, to form a long tongue of material bounded on either side by the remains of the original stony margins. Thus rudimentary striped ground will be formed, and as solifluxion proceeds further lateral sorting by frost will enhance the striped pattern. Finally, on very steep slopes, exceeding 30°, stone stripes become rare, perhaps because downhill slumping and sliding is so rapid that lateral movements are relatively insignificant and patterns derived from the gentler slope above are deranged.

Although it is perfectly clear that the processes of cryoturbation are responsible for the development of patterned ground, certain problems remain to be solved. For example, the regularity of the stone net in many areas is very striking, and requires some explanation. Furthermore, the reason why stone polygons as a whole vary so much in size (their diameters range from 0·5 to 15 metres) is not understood. Yet again, polygons are very irregular in their distribution, and are not found at

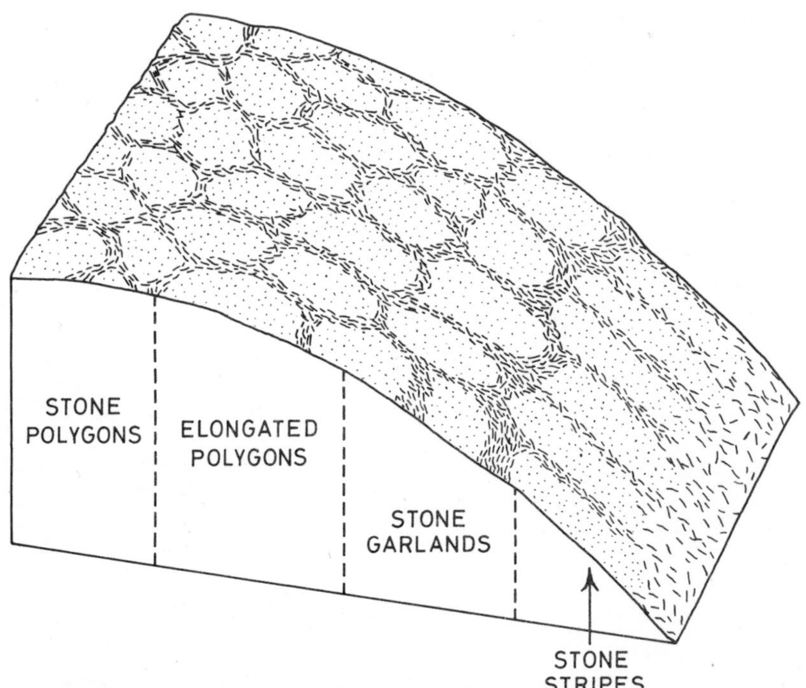

135 Idealised relationship between angle of slope and stone polygons, stone garlands and stone stripes

all in some places which experience periglacial conditions. It is hardly surprising that, in detail, the theories which have been proposed to explain patterned ground vary so much. Some authors have suggested that polygons and stripes develop only where the bedrock is covered by a heterogeneous layer of weathered material. In this the coarser fragments will at certain points be heaved to the surface and thrust laterally (Fig. 136). Those at the surface will slowly slip sideways if the underlying fines are domed or 'pumped' upwards, and thus patches of muddy material bordered by stones occupying gentle troughs will be exposed. As a result of these processes rudimentary polygons or circles may be established. Subsequently, the contrasts between the centres of fines and the stony margins will become more accentuated for two reasons. In the first place, the stony areas will be characterised by free drainage of meltwater. Any fine material remaining here will be quickly washed out, and little will be added since the stones, which are able to dry out comparatively rapidly, will not be greatly affected by further frost weathering. Secondly, the centres of the incipient polygons will be more

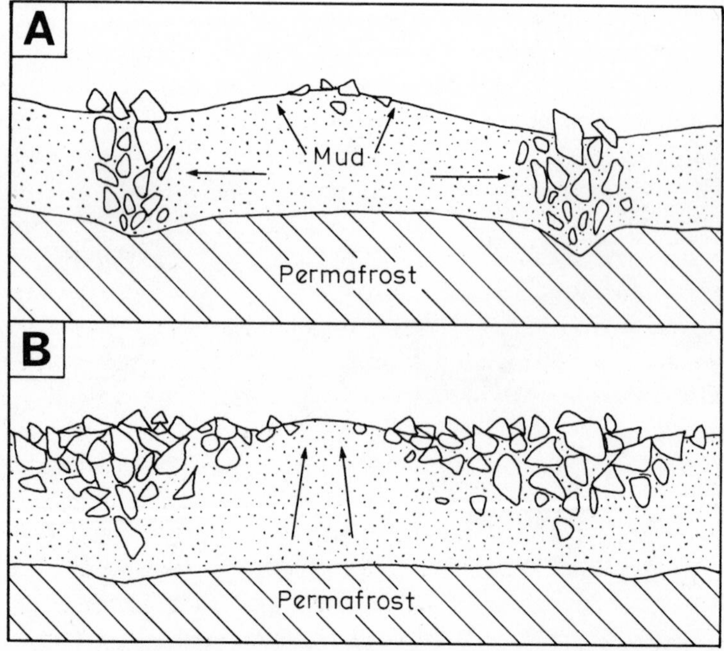

136 The development of stone polygons by frost thrust (A) and the heaving of fines through a stone layer (B)

retentive of moisture. Frost action will continue to comminute the material further, and the development of ground ice will, through the processes of heave and thrust, accelerate the expulsion to the surface and the margins of any remaining large fragments.

Other authors have noted the development of polygonal features as the weathering of 'live' rock proceeds. It has been suggested that providing the initial weathering is 'differential' patterned ground can result. Thus in an area where the rock surface is uneven damp patches will occur in hollows, either because of surface drainage or because these hollows are occupied in winter by small snow accumulations. This dampness will favour advanced comminution by frost weathering, and a good deal of fine detritus will be produced. Intervening higher ground will dry out more quickly, and frost action may be able to effect only an initial breakdown of the rock into large splinters and blocks. The processes described above may then enhance the contrasts between the initial depressions and ridges, and in time true polygons may result. Obviously, the main problem in this hypothesis is that of explaining the regularity of so many polygonal patterns, for initial hollows will tend to occur randomly.

Some geomorphologists have postulated that convection currents are able to develop in the active layer, and that these can play some part in the sorting of detritus to give patterned ground. It is well known that water decreases in density as it is cooled from 4 °C to 0 °C. Near the surface of the active layer temperatures are more likely to approach the former temperature, whereas just above the permafrost the soil water will be only just above freezing-point. The resultant differences of density might conceivably promote a pattern of very slowly moving currents of mud and water. Presumably the centres of stone polygons will be points of upwelling, and the lateral surface movements from these centres may transport coarse fragments to the stony margins. Another suggestion is that as the moist centres of polygons freeze, the development of ground ice leads to the capillary attraction of water from surrounding areas. This movement of water on a minute scale could, so it is argued, lead to the transportation of very fine particles into the polygon and so aid slightly the process of sorting.

It is apparent that much observational and experimental work is needed before the formation of patterned ground can be fully understood. Stone polygons and stripes are in themselves admittedly minor features of the landscape, but are deserving of detailed attention for the light they shed on the operation of cryoturbation processes as a whole.

SOLIFLUXION AND ITS EFFECTS

The process of solifluxion

Permafrost may be regarded with justice as the most striking phenomenon of periglacial areas, but solifluxion is probably the most important and characteristic geomorphological process. As will be shown, its effects on the development of valley-side slopes are far reaching, and indirectly it can affect the rate of river incision and thus the development of valleys as a whole. As stated above, solifluxion occurs only in the active layer, particularly during the thaw period, and is aided by the presence of permafrost, which impedes drainage of the soil and so maintains the fluidity of the thawed material. Other factors which promote solifluxion are as follows.

During the winter period of freezing, the extensive formation of ground ice leads to an appreciable expansion of the active layer. Pore spaces and other cavities are opened up by ice-grains and lenses, and a state of 'minimum density packing' is achieved. Because of this loosening of soil texture, when the spring thaw arrives each particle can be cushioned by its own film of water, acting as a lubricant. Friction between the particles is thereby diminished, and downhill movement is comparatively easy. Experiments have shown that active solifluxion tends to be concentrated during the early part of the thaw season, and if the soil layer dries out during the summer even subsequent heavy rain and saturation of the ground may not cause renewed solifluxion, simply because the soil has again been compacted.

The formation of ground ice can itself directly aid the downhill movement of slope detritus. The formation at or near the ground surface of small masses of ice, such as pipkraker, commonly has the effect of heaving up small stones at right angles to the slope. When thawing takes place the stones drop vertically, so that in the course of a single freeze-thaw cycle they are displaced slightly downhill. On an altogether larger scale, the development of lenses and wedges of ground ice can lead to downslope movements even in spreads of angular stones and boulders and in the stony margins of stone garlands and stripes. Such movement may not be solifluxion in the strict sense of the term, for comparatively fine material ('soil') is normally involved in that process. Even so, true solifluxion can affect heterogeneous debris, and the flow of the matrix of fines can lead to the downslope transportation of large rocks, either contained within the moving mass or 'floating' on its surface. When this takes place, an interesting result is the orientation of the individual stones with their long axes pointing in the direction of solifluxion flow.

The actual rate of solifluxion movement varies somewhat, according to the nature of the debris affected, the steepness of the slope (attention is often drawn to the fact that solifluxion can operate on gentle slopes as low as 2–3° in angle, but clearly the process is more effective on steeper slopes than this), the presence or absence of a binding vegetation mat, the water content of the soil, and the type of soil movement. In this last connection it must be noted that solifluxion can affect more or less uniformly the waste layer over the whole of a slope, or it can be concentrated along certain restricted lines (as in the case of the fine mud separating stone stripes). Often it takes the form of sudden localised slumps or rapid mud flows, especially where the ground water content is abnormally high. Measurements of solifluxion have revealed rates of movement in the order of 2–5 cm per year on slopes of medium steepness (10–15°). These figures may not appear very high, but they should be contrasted with those quoted for soil creep on p. 224 (up to 2 mm a year). Solifluxion can, in fact, act about twenty times as quickly as the most common form of mass movement in present-day humid-temperate areas. This fact must be borne in mind when the effects of recent periglaciation on landscape evolution in these parts is considered. Moreover, local solifluxion movements can be even more rapid, and lobes of thawed material may move at a rate of 10–15 or even more centimetres per year. Finally, it must be stated that solifluxion does not always affect the whole of the active layer, but may be concentrated in the topmost 50 cm or so. The reason is that during the winter freeze the formation of ground ice, leading to minimum density packing and maximum lubrication, is greatest in this upper zone.

The effects of solifluxion

The role of solifluxion in a periglacial region is to transport downslope the products of frost weathering. Since the two processes are comparatively rapid in their operation, especially in fissured and porous rocks which are subjected to both block and granular disintegration by expanding ice-wedges and crystals, the amounts of rock debris released and moved are considerable. As a result of this 'mass wasting' there will be a general tendency for hill-tops and interfluves to be both lowered and rounded, with bare rock outcrops being reduced by frost attack and 'submerged' by moving solifluxion debris. In this way smoothly graded slope profiles may be eventually produced. Towards the base of such slopes, where gradients are gentler, large-scale deposition of soliflucted waste may take place, giving detrital fans or aprons with rectilinear or concave profiles. So great is the amount of weathered material arriving at the valley bottoms that the existing streams, which throughout much of the year are either frozen or depleted in discharge, may lack the power to transport it downvalley. In these circumstances aggradation may be marked. Infillings of 'Coombe Rock', comprising several feet of unstratified chalk fragments, chalky paste and shattered flints, can be observed in many dry Chalk valleys of southern England. These accumulations are evidently the product of the periglacial conditions that affected the Chalk during the Quaternary glaciations.

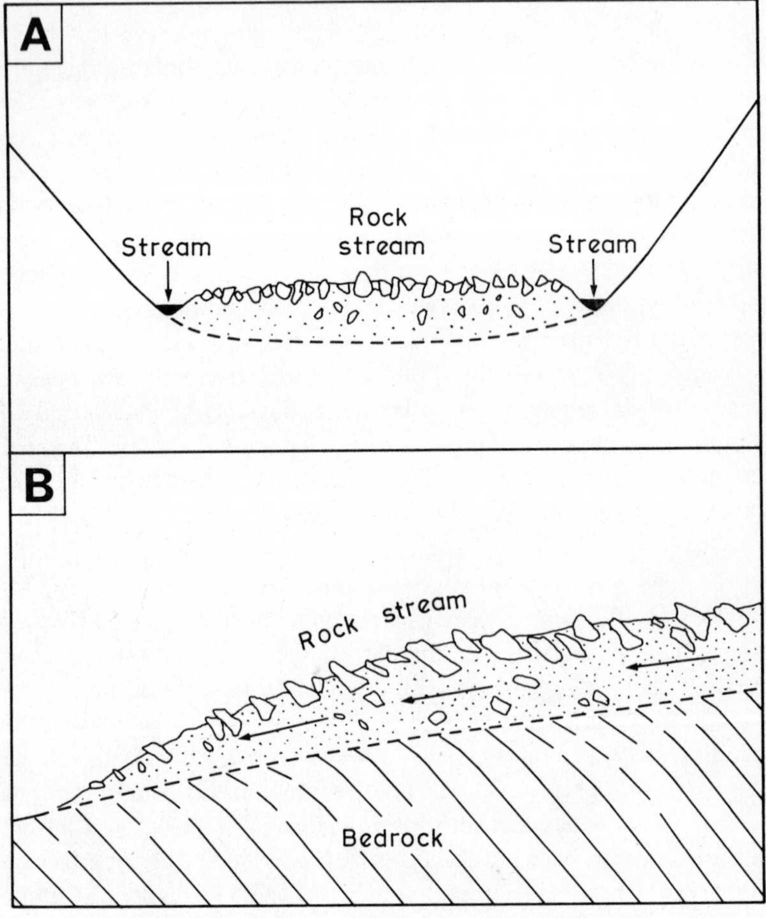

137 **Cross-profile (A) and long-profile (B) of a periglacial rock stream**

On occasion, valley-floor infillings are themselves affected by down-valley solifluxion. In the process coarse rock fragments are heaved to the surface by frost action and backtilted by the flowing subsoil; and the moving mass takes on a convex cross-profile, with small streams flowing at the foot of each valley wall (Fig. 137). This phenomenon is known as a 'rock stream'. Fossil rock streams exist in many temperate lands today, and have been studied in certain valleys of the High Ardennes of Belgium. Similar features may also be produced from unsorted solifluxion tongues by the subsequent washing away of the fines, leaving the immovable larger rocks lying on the surface. The remarkable accumulations of sarsen-stones in certain valleys of the Marlborough Downs of Wiltshire (notably Clatford Bottom) may have been formed in this way (Fig. 138).

In general periglacial processes have the effect of 'toning down' the relief, both by degradation of interfluves and by aggradation of valley

138 An accumulation of sarsen-stones in Clatford Bottom, Marlborough Downs, Wiltshire. The stones rest on a considerable thickness of coombe rock, and have been transported downslope to their present positions by solifluxion in the Quaternary. [R. J. Small]

floors (Fig. 139A). The gently undulating landscapes of Salisbury Plain, Picardy and Artois, characterised by shallow valleys, smoothly curving slope profiles and extensive deposits of frost-shattered and transported chalk, are attributed by many authorities to the action of mass wasting in the Quaternary and are therefore to be classed as 'relict' or 'palaeo-Arctic'. However, the smoothing effects of periglaciation must not be

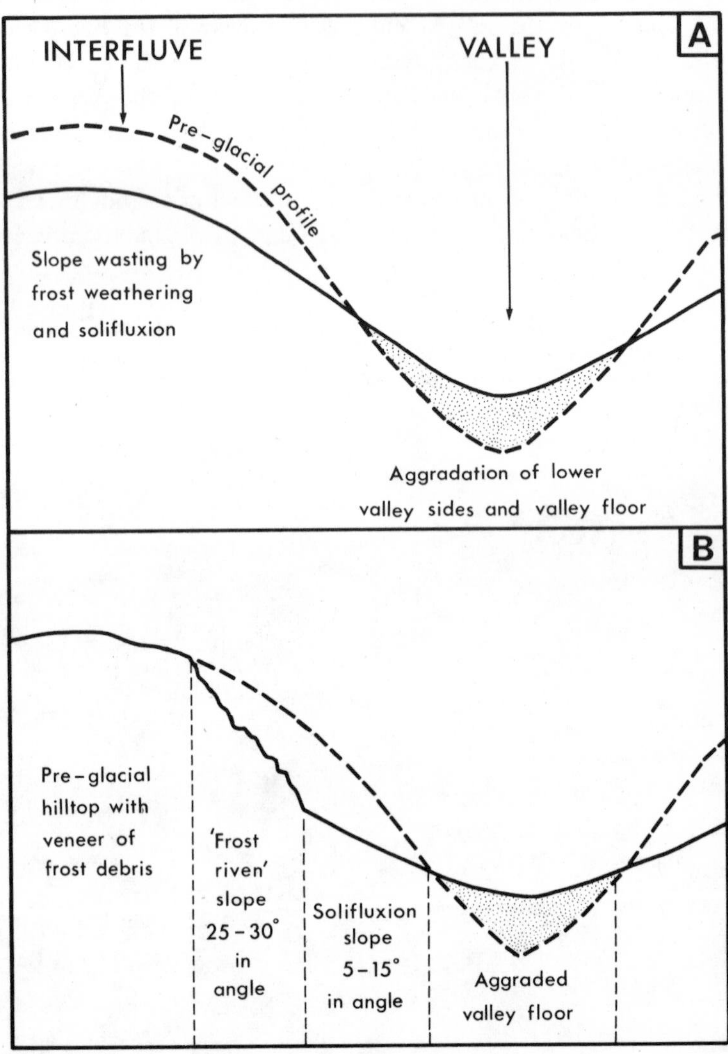

139 The development of slopes in a periglacial area: by divide wasting and valley aggradation (A); by the formation of frost-riven slopes (B)

overstressed. In detail solifluxion can produce irregularities of slope profile, through the building up of slope detritus at certain points or along well-defined lines. Important aggradational features, with steep frontal slopes several feet high, can form on gentle slopes. These 'solifluxion terraces' are gradually extended downslope as material flows over the tread and accumulates on the riser. Again, the development of the slope profile as a whole does not always proceed in quite such a straightforward manner as the general comments above would imply.

PERIGLACIAL SLOPE DEVELOPMENT

The forms and angles of slopes in periglacial regions, like those of other climatic régimes, pose questions which have as yet not been finally answered. Important controlling factors here appear to include the initial form of the slope, the nature of the underlying rock and the aspect of the slope.

The initial form of the slope

In his hypothetical cycle of periglacial erosion (pp. 180–2), Peltier (1950) has envisaged as an initial form, or 'pre-periglacial surface', a maturely dissected landscape (Fig. 78). He considers that the onset of periglacial mass wasting would not lead simply to a reduction of relief and a decline of slope angle, with the basic convexo-concave form being maintained, but that new slope elements, some steeper than any previously in existence, would be formed (Fig. 139B). It is quite possible that some rocky faces in the present landscape of Britain have been initiated by periglacial action (Fig. 140). For example, the late-Tertiary surface of the southern Pennines may well have been smoothly rounded, with the solid rock masked by a thick cover of weathered material. The arrival of cold conditions in the Quaternary may have led to the stripping away of this debris layer by solifluxion, and subsequently to differential frost weathering of the exposed rock. The impressive gritstone scarps of the Millstone Grit, separated by gentler slopes in the intervening shales and overlooking aprons of soliflucted angular blocks released by frost attack on the jointed grits, may thus have been formed. Locally, where the joint pattern of the grit is closer and the rock more susceptible to ice wedging, more advanced recession of the scarp has led to its fragmentation into individual tor-like masses or even to its complete obliteration and replacement by a gentle slope graded across the resistant stratum (Fig. 141). This suggested sequence of landform development (which in

140 St Albans Head, Dorset. The upper cliff comprises a series of free faces developed in the Portland Limestone; the lower cliff is a debris slope made up of blocks of limestone, weathered from the free face above, which obscure the underlying Portland Sand and Kimmeridge Clay. It is probable that both the free faces and the debris slope were formed by frost action during the Quaternary, though some small active fans can be seen. [Aerofilms Ltd]

many respects is similar to that proposed for Dartmoor by Palmer and Neilson (1962) (pp. 137–8)) is in general accord with Peltier's contention that it is only in the more advanced stages of periglaciation that the landscape is characterised by low relief and/or gentle graded slopes.

However, it is improbable that Peltier's hypothesis can apply to all situations. In areas where the pre-periglacial landscape was at the old age stage, the cold climate must have led simply to an increase in the rates of weathering and slope transportation, but (in the absence of rejuvenation, which seems to have been an additional factor in the southern Pennines and Dartmoor) no major changes of form. On the other hand, in many youthful periglacial landscapes, such as those of parts of Spitzbergen at the present time, steep and towering cliffs bearing the imprint of intense frost action stand above massive scree fans. It is difficult to believe that these slopes ever formed part of an original maturely dissected landscape. In the Chalk country of southern England, where Peltier's initial conditions may have obtained, it is rare to find what might once have been a frost-riven slope (p. 181) with a low-angled solifluxion slope beneath it. Instead the various elements of the slope (upper convexity, central rectilinearity and lower concavity) tend to run imperceptibly into each other. It is tempting to suppose that frost weathering and solifluxion have been active over the whole of the slope. This is supported by the fact that maximum angles, usually in the order of 32–34° in deeply dissected areas, evidently represent a past condition

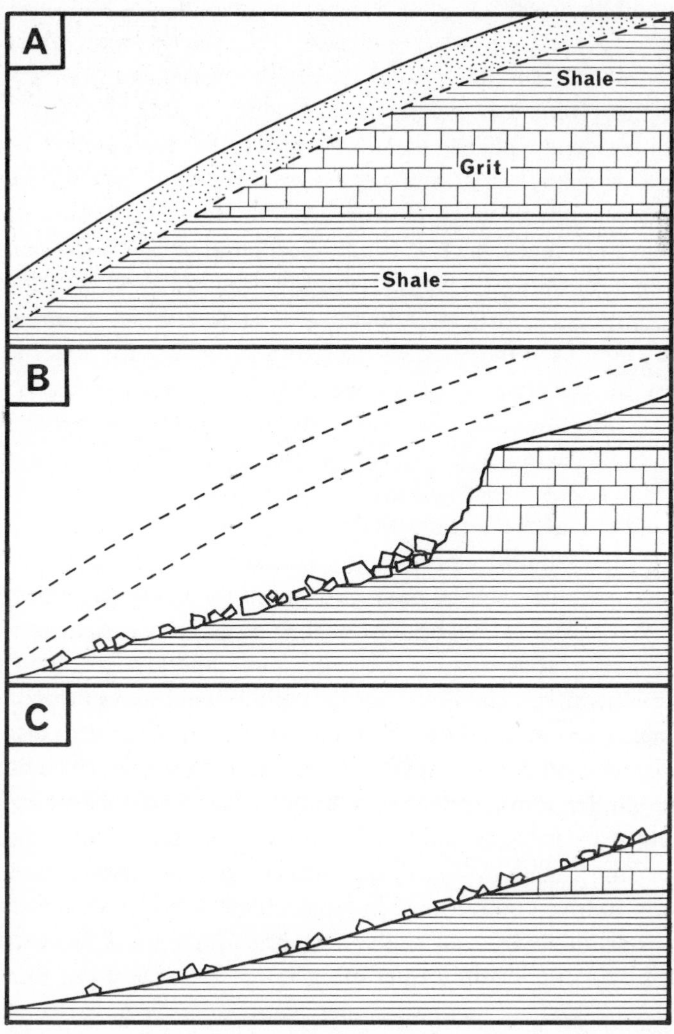

141 The development of gritstone edges and tors in the southern Pennines

of delicate balance between the rate of weathering of the rock by freeze-thaw and the rate at which the detritus could be transported downslope by solifluxion. Only in this way can the absence of free faces and the lack of thick solifluxion deposits from the steeper chalk slopes both be satisfactorily explained.

The influence of rock-type

The great importance of lithology in periglacial slope development can be illustrated by reference to both Dartmoor (pp. 126–39) and the Grands Causses of southern France (pp. 139–56). In the latter area a typical slope profile comprises an alternation between steep, even vertical or slightly overhanging, free faces and rectilinear slope sections at 30–35°. There are good reasons to believe that these slopes, like their gentler counterparts in Dartmoor, were fashioned essentially during the Quaternary by frost weathering and solifluxion rather than by the solution processes which have produced the more individual landforms of the Causses. In the first place, the lower slopes here are often mantled by deposits of angular limestone fragments, apparently the result of frost disintegration. Secondly, the existing relationship between slope form and rock-type can best be explained if development under periglacial conditions is admitted. The most positive elements, the free faces, are associated closely with the massive dolomites (pp. 143–4), which possess few joints or bedding-planes and are very resistant to frost attack—though not to solution, as the fantastic etching of some dolomite outcrops clearly shows. The rectilinear slopes at 30–35°, on the other hand, are mainly developed on the well-bedded and closely jointed sub-lithographic limestones (p. 143), which because of their structure and permeability are very prone to freeze-thaw weathering and thus give rise to the less positive elements in the landscape. An additional point is that the sub-lithographic limestones contain numerous seams of marl, which when wet help to lubricate the surface weathered layer and accelerate solifluxion. Under non-periglacial conditions, the two rock-types would probably react in quite different ways. In a warm humid climate, for instance, the dolomite would not have such a positive influence, for although massive and not readily permeable it would succumb easily to attack by acidulated surface water and the imposing free faces would be replaced by rochers ruiniformes or even debris-covered slopes.

The aspect of the slope

It is now widely recognised that, under periglacial conditions, the aspect of a slope may have a considerable influence on its form and development. This view is based on the common occurrence, in areas which have been subjected to periglaciation, of asymmetrical valleys. It is clear that these are not simply the result of uniclinal shifting by streams in the direction of rock dip, nor are they caused by the accidental lateral swinging of a stream in such a way as to undercut continually one of its valley sides. In fact, valley asymmetry in any one locality may display a marked orientation in one direction. For example, in the Chilterns, where in general the valleys run from north-west to south-east, the valley slopes facing south-westwards are often appreciably steeper than those which face north-eastwards.

To account for the development of an asymmetrical valley from a symmetrical valley, it must be assumed either that one side of the valley has undergone steepening or that the other has experienced decline. It is, however, very difficult to discover, from the evidence available, which of these alternatives is correct. Not surprisingly, several theories have been formulated to account for asymmetrical valleys (Fig. 142). These theories fall into two broad categories: those which suggest that the slope most affected by periglacial processes undergoes a steepening of angle; and those which assume that frost action and solifluxion lead to a decline in slope angle. There is little agreement over the precise way in which climatic factors can influence the periglacial processes at work on the two sides of a valley. Many authorities argue that the different amounts of insolation received by the two slopes is the crucial factor, whereas others consider that the prevailing wind direction plays a major role.

Among the more reasonable theories of asymmetrical valley formation are the following.

(i) It is suggested that slopes which face south and south-west are more prone to frost weathering and solifluxion simply because they receive greater amounts of insolation than the slopes facing north and north-east. The latter, as a result of their more shaded nature, may remain frozen throughout the day and are therefore 'inactive'. The slopes facing in the opposite direction, on the other hand, may well thaw out during the afternoon, though at night they will again become frozen. These slopes will therefore be affected by diurnal freeze-thaw cycles, causing disintegration of the rock, and during the daytime the weathered material will be transported downslope by solifluxion. The south- and south-west-facing slopes are in other words 'active', and may undergo both retreat and steepening of angle (Fig. 143B). This is essentially the

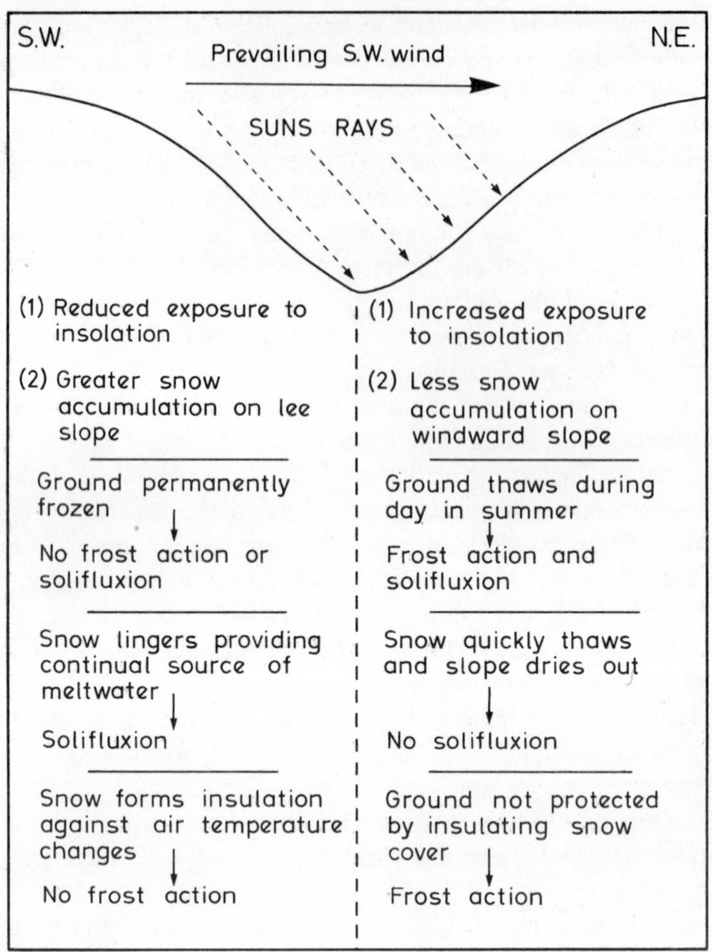

142 Possible relationships between aspect and process in a periglacial area

theory which has been proposed to explain the steep south-west-facing slopes of the Chiltern valleys.

(ii) It can be argued equally well that, under the conditions just described, more intense frost weathering and solifluxion should lead to the decline of south- and south-west-facing slopes. In brief, the gentle slopes may be regarded as the more active. According to this theory, the steeper slopes—on which periglacial processes are retarded—should face to the north and north-east. It has further been postulated that the tongues of solifluxion debris moving down the gentler slope will actually push any stream flowing along the valley bottom into the base of the opposing slope, which may thus be steepened by undercutting and the valley asymmetry accentuated (Fig. 143A). Clearly this theory could not be applied to the Chiltern valleys referred to above, but it has possible application farther to the west, in the Marlborough area, where some valleys have gentle slopes which face south-westwards and are mantled by solifluxion debris (Coombe Rock).

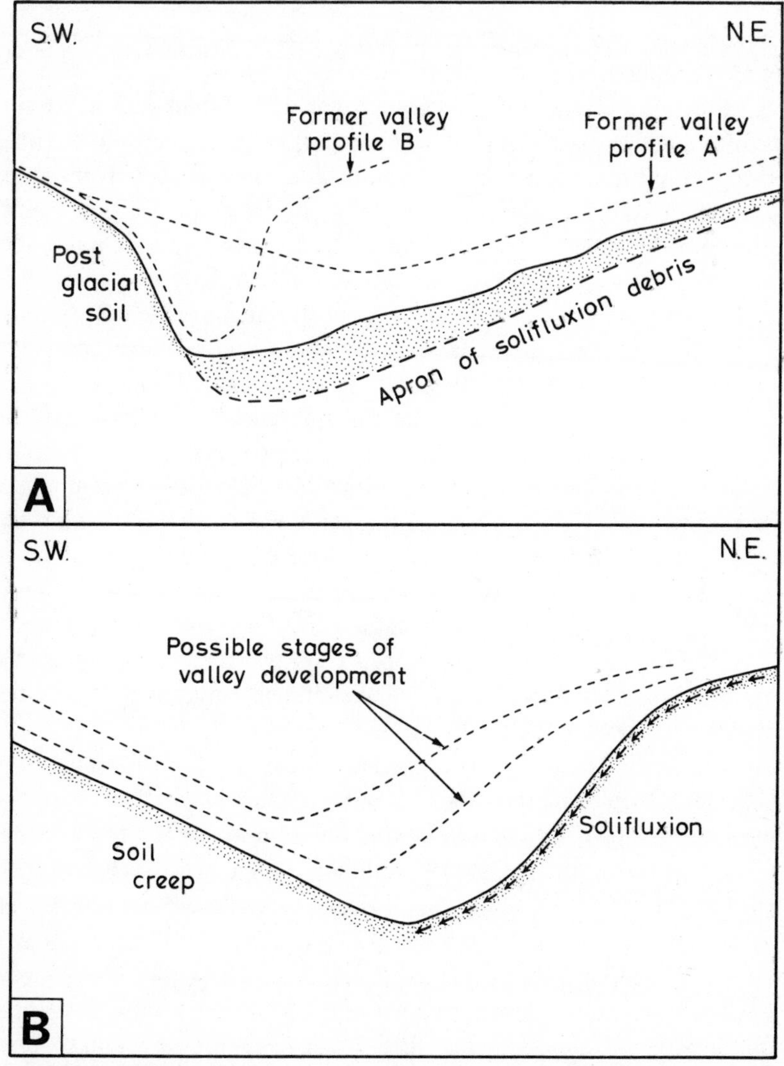

143 Possible mechanisms of asymmetrical valley development: maximum solifluxion on the gentle slope (A) and on the steeper slope (B)

(iii) Some authorities have argued that slopes facing north and north-east had, under periglacial conditions, a greater thickness of snow cover, and that this endured longer, than the slopes facing south and south-west. On the latter snow would be melted quickly as a result of the greater insolation, and the ground would be dried out, thus retarding or halting solifluxion. On the north-east-facing slopes, however, the snow would melt slowly, providing a continual source of moisture to aid solifluxion. These slopes would therefore experience more active development—though whether this would lead to a decline or a steepening of the slope is again a moot point.

(iv) Conversely, it has been suggested that slopes which face north and north-east are less active than their counterparts because they are protected by their snow cover. It is argued that the snow effectively insulates the ground from fluctuations of air temperature about freezing-point, with the result that frost weathering of the underlying rock cannot take place. The south-west-facing slopes, on the contrary, lose their protective snow more quickly, and as a result can be attacked by frost action; they are thus more active, and may undergo decline (according to one theory) or steepening (according to another).

(v) Finally, it has been argued that the differential development of the snow cover on either side of a valley is not the result only of differential insolation effects, but may be determined more by the direction of the prevailing winds. For instance, in an area where these are westerly the slopes which face west (that is, to windward) may be swept clear of snow, whereas those facing eastwards (to leeward), because of their sheltering effect, may become occupied by thick and extensive snow drifts and banks. Differential exposure of the rocks underlying the opposing slopes, or differential rates of solifluxion, may then lead to the development of asymmetrical valleys.

It will be obvious that the problem of asymmetrical valleys is a particularly perplexing one, and that the mechanisms which are involved are varied and complex. It may be that different causes are to be sought in different areas, for it is a very significant fact that the orientation of valley asymmetry is by no means the same in all periglacial regions.

OTHER PERIGLACIAL PROCESSES

Nivation

In present-day periglacial areas it is common to find deep hollows, especially on sheltered north-facing slopes, which are occupied permanently or for the greater part of the year by patches of snow. There are good reasons for believing that these are not just pre-existing hollows which preserve, because of their lack of exposure, snow throughout the thaw season, but that the snow is in fact 'eroding' back into the hillside. It may be that the process is initiated by the chance occurrence of a gentle depression, but this is clearly deepened and extended by snow-patch erosion or 'nivation'. In areas which experienced periglacial conditions in the recent past, rounded or elongated hollows—evidently not the result of surface run-off since they do not resemble normal valley forms—may also have been produced in the same way. For example, Bull (1940) has attributed the formation of corrie-like coombes on the scarp-face of the South Downs near Eastbourne to nivation during the Quaternary. Yet again, many valleys in southern England that are mainly the result of fluvial action possess well-rounded heads, which could be accounted for in terms of modification by snow-patches.

Although nivation is a widely recognised process, its precise mechanism is not wholly clear. It is generally agreed that little or no actual downslope movement of the snow occurs, so that the possibility of corrasion as such can definitely be ruled out; in other words the term 'snow-patch erosion' is something of a misnomer. Lewis (1939) has suggested, after study of snow-patch hollows in Iceland, that in summer melting occurs at the base of the snow, whereas in winter the whole patch becomes frozen. Thus it is feasible to envisage freeze-thaw activity affecting the uppermost layer of the ground immediately beneath the snow. Any finely comminuted detritus so produced can be washed out by 'sub-nival' meltwater rivulets during the summer season. In this way the snow-patches are slowly let down into the ground, the process being one of weathering removal (Fig. 144).

The main difficulty confronting this explanation concerns the possible insulating effects of snow. Williams (1949), who has studied snow-patches in the San Gabriel Mountains of California, has discovered that underneath deep snow accumulations the minimum temperature of the ground is never less than 32·5–33 °F, and that sub-nival freeze-thaw weathering is therefore impossible. However, it is dangerous to generalise from particular examples, for beneath the snow-patches studied by Lewis the subsoil was frozen even during July.

Williams has gone on to suggest that chemical weathering may materially aid nivation. In general, chemical action is considered to be of relatively little importance in periglacial areas. Although water is abundant in summer in flat and low-lying parts, and the resultant boggy vegetation is associated with the formation of organic acids, rates of chemical reactions are much reduced by the low temperatures. None the less, meltwater emerging from beneath snow-patches has been found to contain dissolved calcium bicarbonate, the product of carbonation (p. 20). Williams has attributed this to the high solubility of carbon dioxide in water at low temperatures (water at just above 0 °C can hold twice as much dissolved carbon dioxide as water at 30 °C); and the high concentrations of carbon dioxide observed to exist within snow-patches appear to support this view.

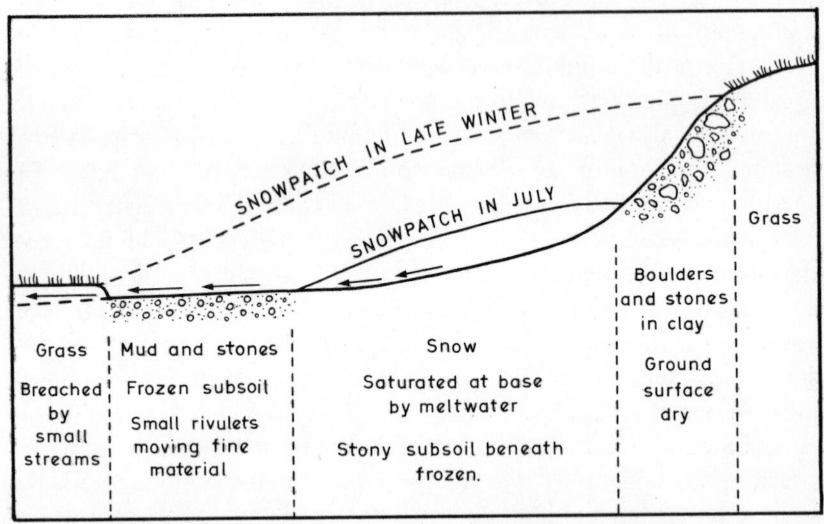

144 **The development of small nivation hollows (after W. V. Lewis)**

These facts are incontrovertible, but it is still difficult to believe that chemical processes are in any way dominant in nivation. Around most snow-patches are areas of moist bare ground, left as a result of summer shrinkage of the patch. Here both frost action and solifluxion can operate freely and to much greater effect than carbonation and other chemical processes beneath the snow. Furthermore, providing the snow-patch is underlain by incoherent material and summer thawing of the ground can occur, sub-nival solifluxion may be of importance, especially with the help of meltwater percolating through the snow from above.

Stream action

It is sometimes stated that the action of running water, at least as an erosive agent, is by comparison with frost action and solifluxion of negligible importance under periglacial conditions. During the more advanced stages of the periglacial cycle it may be subsidiary even to wind action. This view is based largely on the argument that effective stream corrasion is prevented by the very large amounts of detritus fed into valley bottoms by the processes of mass wasting. As the latter is concentrated on the slopes and so leads to a widening of the valley, so the valley floor becomes encumbered with solifluxion debris. The contemporary streams, which may cease to flow during part of the year, are therefore grossly overloaded and aggradation results.

In reality, the situation seems to be more complex than this, for periglacial streams can in some instances achieve striking erosional effects. These have been described by Jenness (1952) in parts of Arctic Canada, where the annual precipitation may be less than 10 inches—though this comes largely in the form of winter snowfall, which may accumulate over a period of seven months or more, giving a snow cover up to 3 ft in thickness over wide areas. This store of water is suddenly released at the onset of the warm season, and rapid and violent run-off takes place. Jenness has described the formation of gullies 3 ft in depth during one such thaw. More important, he draws attention to the existence of ravines, sometimes as much as 200–300 ft in depth, which cannot be 'pre-glacial' and which therefore testify to the powers of vertical corrasion by meltwater streams.

Furthermore, many authorities have invoked stream action to explain important erosional forms in areas which have been affected by periglacial conditions in the recent past. A much quoted example is that of the dry valleys of the English Chalklands (but see also pp. 123–5). One attractive theory to explain these is that of Bull (1940), who postulates incision of the valleys by meltwater streams fed by snowcaps occupying the higher parts of the Downs. Recent research in east Kent has revealed that small but striking valleys near the village of Brook were eroded in a period of only 500 years at the end of the last glacial period, when a comparatively mild and humid periglacial climate existed in this area (Fig. 145). The material eroded from the valleys has been built into fans lying at the foot of the Chalk scarp. Beneath these fans lie marsh deposits which have been dated with precision by the radiocarbon method. The processes involved in the formation of the valleys seem to have been frost shattering, solifluxion and erosion by meltwater torrents coursing down the steep scarp-face; a composite term to describe these processes is 'niveo-fluvial'. It must not be thought, however, that all Chalk dry valleys were formed in this way and at this particular time.

Many are far too large to have been eroded in the short space of 500 years. Others (for example, that at Birling Gap, near Eastbourne) contain quite thick deposits of Coombe Rock, indicating a phase of aggradation under periglacial conditions rather than rapid incision.

The French geomorphologist Corbel (1961) has argued that the precise nature of the periglacial climate may determine whether or not stream corrasion can occur. He suggests that, in a Maritime Arctic climate, mass wasting is an important process, but that the comparatively heavy precipitation supports powerful streams. These are capable both of removing all the solifluxion detritus derived from the weathering back of the valley-side slopes, and of using this heavy load in effecting considerable vertical corrasion. In such areas, valley cross-profiles are therefore normally V-shaped. In a Continental Arctic climate, on the other hand, frost shattering produces vast accumulations of scree which smother the bases of slopes. Solifluxion is comparatively restricted and takes the form of 'streams' of rocks and stones. Running water is very

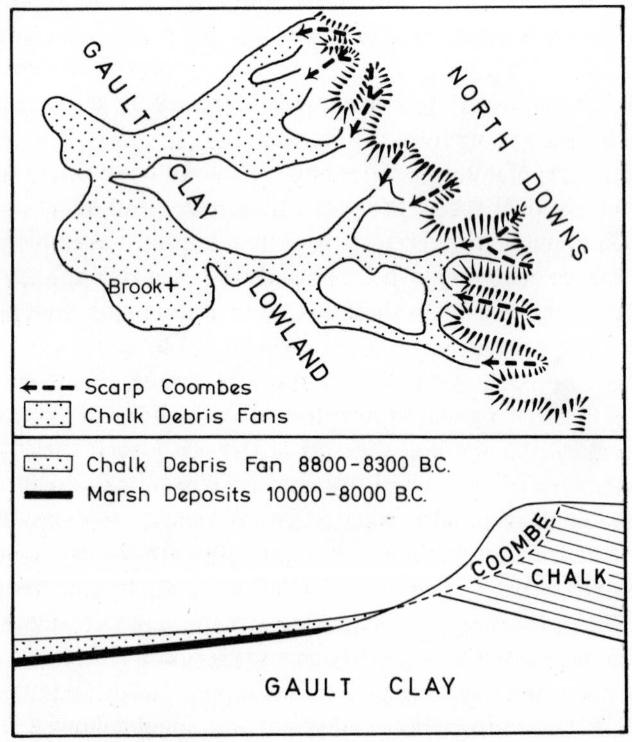

145 Late-glacial escarpment valleys near Brook, Kent

limited in volume, owing to the aridity of the climate, and streams are both feeble and intermittent in flow. They are quite unable to cope with the amounts of coarse debris shed by the slopes. Valley bottoms are therefore choked by waste and are steadily built upwards. U-shaped valleys, in some ways resembling desert wadis, are commonly formed in this way.

Wind action

Peltier has suggested that, during the ultimate stages of the cycle of periglaciation (pp. 180–2), the processes of frost weathering and solifluxion will reduce all slopes to a gentle angle of 5° or less. By this time the slope detritus will have been greatly comminuted, and will be vulnerable to wind transportation. Thus a late-periglacial landscape will become characterised by loess and sand deposits, and by 'wind-swept pebble pavements'.

The significance of wind action at certain periods of the Quaternary is attested to by the vast spreads of loess, referred to by J. K. Charlesworth as 'by far the most important periglacial accumulation', which exist in parts of Europe, Asia, and North and South America. Loess is usually a brownish-yellow sandy loam, rich in lime, homogeneous in structure, friable and giving rise to highly fertile soils. In Europe, it occurs as a belt running from west to east from northern France (where it is referred to as 'limon'), through Belgium, along the northern margins of the Hercynian uplands of southern Germany, and into Poland. Usually it is from 10 to 20 ft in thickness, but locally, especially in large river valleys such as that of the Rhine, it attains a much greater depth. The northern edge of the loess belt is clearly defined, and lies some way south of the line reached by the Wurm ice-sheet at its greatest extent. The deposit tends to blanket the landscape and gives rise to few interesting landforms as such, though it has caused the diversion of some small streams whose former valleys have been choked by loess. In parts of Asia, the loess reaches much greater thicknesses—over wide areas as much as 500 ft, and locally 1000 ft or more—and the sub-loessic landscape, which may have been much dissected by valleys, has been completely altered.

The origin of loess has excited much controversy. However, most authorities accept that it is a Quaternary deposit, and that it is clearly related to the extension of the great continental ice-sheets. In part the loess may be water deposited (where it shows signs of stratification, or contains stones too large to have been moved by the wind), but the bulk of the material has undoubtedly been transported and deposited by aeolian action. This is shown by the 'non-sorting' of the constituent grains (a characteristic of wind deposits), by the common occurrence of 'terrestrial' fossils, and by the fact that the loess has accumulated on hill-tops, valley sides and valley floors. The most attractive theory is that, during cold continental conditions, winds blew over the dried outwash deposits lying south of the ice-margins. Much of the finer material was removed by deflation and deposited as sheets of loess elsewhere. In some areas, reworking of the deposits by the action of running water has probably occurred. This would explain why, in southern Britain, loess commonly overlies fluviatile gravels, forming layers of 'brickearth'.

CONCLUSION

It will be apparent from this account of frost weathering, solifluxion and allied processes that periglacial climates can promote rapid geomorphological activity. However, it is pertinent to conclude this chapter by emphasising two important points. Firstly, many of the phenomena which have been described are quite minor features (for instance, stone polygons and stripes, solifluxion terraces and pingos), and often the emphasis has been on the modification rather than the initiation of landforms (for example, the grading of slope profiles, and the development of asymmetry in valleys). Secondly, it must be remembered that the duration of periglacial conditions has in many areas been remarkably brief. Thus southern Britain experienced in the Quaternary a total of approximately 500,000 years of seriously reduced temperatures, when the processes which have been described were able to operate. Yet much of the area had previously been subjected for many millions of years to continuous or near-continuous erosion during the Tertiary era.

The question must inevitably be asked: was the Quaternary era, despite the modifications to the existing landscape that it obviously wrought, none the less a comparatively minor episode in the geomorphological history of regions experiencing periglacial rather than true glacial conditions? The answer to this question is not easy. Some authorities would point to the widespread preservation of old planation surfaces and pre-glacial drainage patterns in localities that, to judge from their superficial deposits, were much affected by the Quaternary cold climate. (See, for example, Dartmoor, where Tertiary planation surfaces are still a very obvious element in the landscape (pp. 128–30).) Others have argued that the role of periglaciation has been primarily one of transportation, or denudation in the strictest sense of that term. There is much evidence to show that in late-Tertiary times much of western Europe was affected by climatic conditions a good deal moister and warmer than those of today. Deep chemical rotting of the rocks occurred, producing a thick and uneven regolith, some of which remains today in areas such as Brittany (where the episode has been graphically referred to as the 'maladie Tertiaire'). For the most part, however, this weathered material has been stripped away by periglacial mass wasting. Still other geomorphologists take the view that, at the end of the Tertiary era, the landscape of western Europe was largely made up of undissected planation surfaces, across which streams ran without effecting much corrasion—simply because their loads, of dissolved salts and fine clay particles derived from chemical weathering, were inadequate for the task. However, with the onset of the Quaternary cold conditions, leading to rapid mechanical weathering, these streams were armed with much coarser detritus, forming large bed loads that they were able to use in corrading their beds and deepening their valleys. The whole process of valley incision would have been aided too by the intermittent falls of base-level which have taken place during the Quaternary (pp. 254–6).

Clearly this issue, as is so often the case in geomorphology, must remain for the moment an open one. What may be said categorically is this: that no geomorphologist who is attempting a detailed study of the landforms of an extra-glacial region can afford to be unaware of the many influences of past periglaciation.

11

GLACIAL LANDFORMS

INTRODUCTION

Ice-sheets, ice-caps and glaciers are considered to form 10% of the world's land-area at present. Most of this ice is concentrated in the great Antarctic ice-sheet (which occupies some 12,500,000 sq. km), in the smaller Greenland ice-sheet (1,750,000 sq. km), and in the numerous ice-caps and valley glaciers occupying mountains and uplands such as the Himalayas, the northern Rockies, the Alps, the Norwegian plateau and central Iceland. During the Quaternary era, which lasted for approximately 1½ million years, the extent of the ice was greatly increased; indeed at their stage of maximum growth the Quaternary ice-sheets spread over some 30% of the world's land area, including wide tracts of the temperate lowlands of the northern Hemisphere. When one considers the vast erosional, transportational and depositional capabilities of these ice-sheets, plus their indirect effects in terms of the displacement of climatic belts (p. 291), the extension of periglacial conditions (chapter 10), eustatic changes of sea-level (p. 423), and isostatic depression and recovery of parts of the earth's crust (p. 424), Flint's assertion that 'the latest epoch of geologic history has witnessed changes in the physical aspect of the Earth . . . such as are not recorded in any earlier span of time of comparable length' is seen to be fully justified.

It is outside the scope of this book to consider in any detail the causes and history of the Quaternary glaciations. Suffice it to say that, during the Quaternary era, four or five (and possibly more) major advances and recessions of the ice, separated by interglacial periods in which climatic conditions were similar to or warmer than those of the present day, have been confidently identified. Within each of these glacial periods, important fluctuations in the position of the ice-margins also occurred. For example, during the last glacial (usually known in Europe as the Wurm or Weichsel) at least two distinct ice advances, separated by an 'interstadial' involving a considerable measure of deglaciation, have been recognised. Again, a more limited re-advance towards the close of the

Wurm, in Zone III of the so-called 'late glacial', was associated with the development of small cirque glaciers in upland Britain. There can be no doubt that the earlier glacial periods (originally referred to as the Gunz, Mindel and Riss, but now known by various local names) were also characterised by successive advances and retreats of the ice.

Geomorphologically, these glacial fluctuations are of the utmost significance. Firstly, it cannot be sufficiently stressed that most glacial landforms are 'polygenetic', in the sense that they are the product of several episodes of glacial moulding, and have also been modified under periglacial and interglacial conditions. Thus, in studying the origin of cirques (pp. 377–81), it must be remembered (i) that at different times these have been occupied by glaciers of greatly varying sizes, (ii) that in milder periods the glaciers have been replaced by snow-banks (so that the cirques have become, in effect, temporary nivation hollows), and (iii) that in interglacials the cirque headwalls have been exposed, with the total disappearance of the ice and snow, to normal frost weathering by atmospheric temperature changes. Again, glacial drainage channels (pp. 393–402), which are often explained as if they were the result of the last deglaciation only, must almost certainly have undergone some development by meltwater during earlier episodes of glaciation. Secondly, in each glacial period the ice-sheets advanced to different limits. In some instances, they overrode and profoundly modified earlier tills and outwash deposits; in others previously formed glacial drifts were left exposed, and were deeply eroded by meltwater streams issuing from the ice-margins or affected by the various processes associated with a periglacial climate. In eastern England, for example, the Wurm ice-sheet appears to have extended only as far south as Hunstanton, in north Norfolk. The extensive East Anglian boulder-clays of earlier date were therefore left open to considerable modification. It is, in fact, possible to make a broad division of the glacial drifts of this country into the so-called 'Newer Drift' (deposited during the ultimate glaciation, and giving rise to well-preserved features such as end-moraines, drumlins, kames, kame-terraces and eskers) and, lying generally farther to the south, the 'Older Drift' (the product of earlier glaciations, and so affected by subsequent weathering and erosion that it is rare to find typical forms of glacial deposition remaining intact).

Finally, it must be said that, although in detail the history of each glacial period was different from that preceeding it, the *general* pattern of glaciation must have been repeated with every glacial advance. Within the British Isles all the major upland masses, except those of the South-West Peninsula of England, acted on each occasion as centres of ice accumulation, from which large valley glaciers flowed outwards, amalgamating into large ice-sheets over the lowlands of eastern and central Scotland, eastern England, the English Midlands, the Irish Sea basin and the central lowland of Ireland. In eastern Britain, the indigenous ice was also joined by the great Scandinavian ice-sheet, which had crossed the North Sea Basin from east to west. In the upland areas glacial scouring was dominant (though even here extensive morainic deposits are found, as is shown by any Geological Survey One-inch Drift Geology map), whereas in the lowlands, particularly near the limits of the ice-sheets where forward movement was offset by ablation or where actual ice stagnation occurred, great quantities of glacial till and outwash deposits were laid down, often burying completely the 'pre-glacial' surface. It has been estimated by G. W. Lamplugh that one-eleventh of the country would be submerged by the sea if all the glacial drift were removed, and that in addition a further one-sixth is so thickly mantled with drift that the sub-glacial surface has no influence on the form of the present land-surface.

GLACIAL EROSION

It is arguable that no other form of erosion has so great an impact on the physical landscape as glaciation. One has only to compare, say, the landforms of a glaciated upland, where the sculpturing action of glaciers is at its most potent, with an area of comparable height and relief that has escaped glaciation to realise the fundamental validity of this statement. It is true that early in this century certain authorities, notably Garwood (1910), held that glaciers performed essentially a protective role, and that the highly distinctive landforms of glaciated uplands, including those of the Alps, were in fact largely the product of fluvial erosion. However, this view is, for reasons that will be shown below, hardly tenable today. Under certain circumstances, where the ice is abnormally thick and immobile (as at the centre of a large ice-sheet) glacial erosion must be minimal, but for the most part—and especially in dissected mountain areas where the glaciers are *guided* by pre-existing river valleys—glaciers can be shown to have had considerable and sometimes quite staggering effects. By way of introduction to the theme of glacial erosion it is appropriate to consider, with reference to selected areas, the forms of both glaciated uplands and glaciated lowlands.

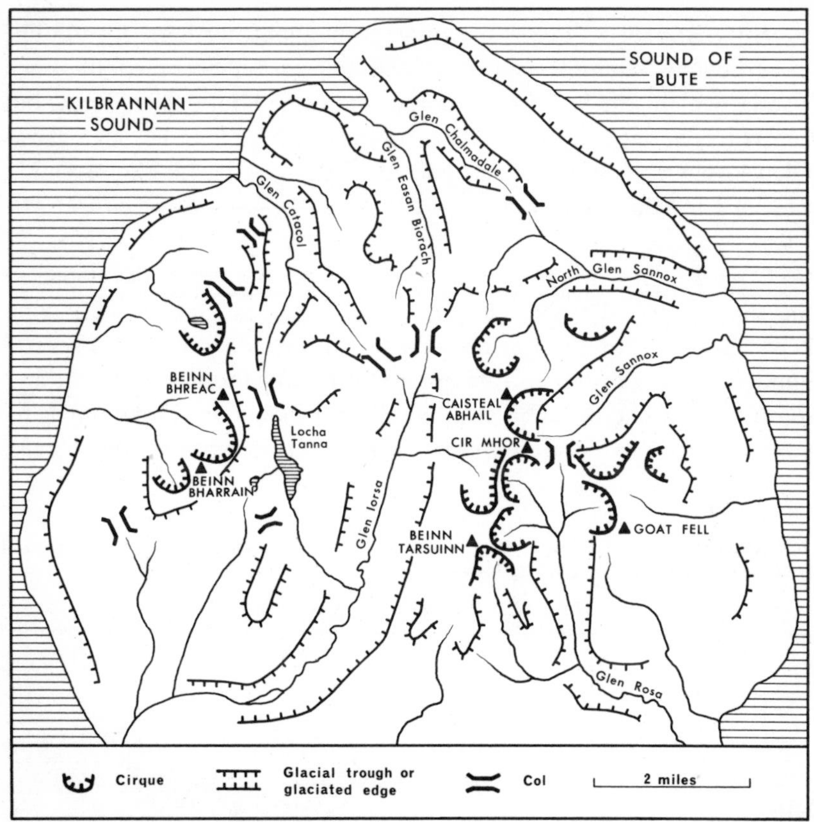

146 **The glacial features of northern Arran**

Glaciated uplands

Many of the most important landforms resulting from glacial erosion may be found in the northern granite uplands of the Isle of Arran (Fig. 146).

A *Deep glacial troughs.* Valleys having a more or less U-shaped cross-section are well developed in the eastern 'coarse' granite (where Glen Sannox and Glen Rosa (Fig. 147) are especially fine examples, trending north-eastwards and southwards respectively from the sharp peak of Cir Mhor) and also in the 'fine' granite to the west (notably the great trench of Glen Iorsa). These valleys cannot, however, be regarded as wholly typical, in the sense that their long-profiles are smooth and breaks of gradient and rock steps absent.

B *Hanging valleys*. These are found throughout northern Arran. The most obvious example is the curious depression occupied by Loch Tanna, which stands some thousand feet or so above Glen Iorsa, but hardly less noteworthy are Coire a Bhradain, whose floor lies high above Glen Rosa on the western side, and the valley of Gharbh-choire-Dubh, west of Cir Mhor (Fig. 148).

C *Cirques*. These are most numerous and generally best developed in the deeply fretted eastern granite. The finest example is that lying between Cir Mhor and Caisteal Abhail, with its steep headwall several hundred feet in height and its fine rock sill, but those to the west and south (Garbh-choire Dubh and Fionn Choire) are also important. The growth of these three amphitheatres has almost isolated the granite mass

147 Glen Rosa, Isle of Arran. View northwards towards Cir Mhor (left) and the col leading into the head of Glen Sannox (see p. 356). [Eric Kay]

148 Coire a Bhradain, a glaciated valley hanging approximately 1000 ft above Glen Rosa (out of picture to the right), Isle of Arran. [Eric Kay]

of Cir Mhor, which seems to represent an early stage in the formation of a 'pyramidal peak'. Cirques are also found in the western granite, where the north-west-facing Correin Lochan, containing a small lake and hanging high above the coast, is an almost perfect case.

D *Aretes*. Typical sharp-edged ridges are not a prominent feature of the Arran granite. The ridge to the north of Goat Fell, with its upstanding tor-like masses, may be broadly classified as an arete, but the nearest approach to the sharp and precipitous forms found in other glaciated areas (for example, Crib Goch on Snowdon) is afforded by A'Chir, trending north-eastwards from Beinn Tarsuinn.

E *Cols*. These are a very important feature of northern Arran. They fall generally into two groups: (i) the high-level cols developed between cirques and resulting from the headward extension of the latter into the 'pre-glacial' upland, and (ii) the deep and well-defined gaps between adjacent U-shaped valleys. Excellent examples of the latter are the pass

149 Val d'Hérens, Switzerland. View southwards towards the Pigne d'Arolla. Note the broad aggraded floor of the main valley in the foreground, contrasting with the almost V-shaped section of the Arolla tributary valley (partly in shade in the background). [Eric Kay]

between the north-trending Gleann Easan Biorach and the south-trending Glen Iorsa (which unite in effect to give a through valley bisecting the Arran granite), and the col linking the heads of Glen Sannox and Glen Rosa (Fig. 147). In order to understand the first of these features it is necessary to realise that during the Quaternary the Arran hills were eroded not only by local ice, feeding into the valleys from the cirques referred to above, but were at times overridden by non-indigenous ice from the north. This was naturally guided by pre-existing valleys, and where these occurred exactly opposite to each other the watershed was broken and the valleys united (pp. 373–4).

It is interesting to compare the glacial landforms of northern Arran with those of a Swiss valley, the Val d'Hérens, where erosion has been on an altogether larger scale and has affected a far more complex geological structure (Fig. 149). The most obvious contrast lies in the sheer size of the Val d'Hérens, which in its middle section near Evolène is, from divide to divide, some 4–5 miles in width (compared with about 1 mile in the case of Glen Rosa) and approximately 5000 ft in depth (compared with the 1500–2000 ft of Glen Rosa) (Fig. 150).

In cross-profile too the Val d'Hérens shows some notable differences (Fig. 151). The characteristic U-profile, where developed at all, is confined to the lower 2000 ft of the valley; and above the shoulders of the trough the 'alp benches' slope upwards for a further 3000 ft at a gentler angle. The immediate inference is that glaciation has transformed only the bottom part of a very large 'pre-glacial' valley cut by a predecessor of the present-day Borgne. However, some glacial modification of the alps cannot be ruled out. An important feature of the Val d'Hérens, also to be observed in other Alpine valleys, is the total absence at some points of the so-called 'typical' U-section, and its replacement by the V-profiles indicative of fluvial erosion (see profiles 5, 7 and 8 in Fig. 151). Such profiles are associated with 'deepened' sections of the

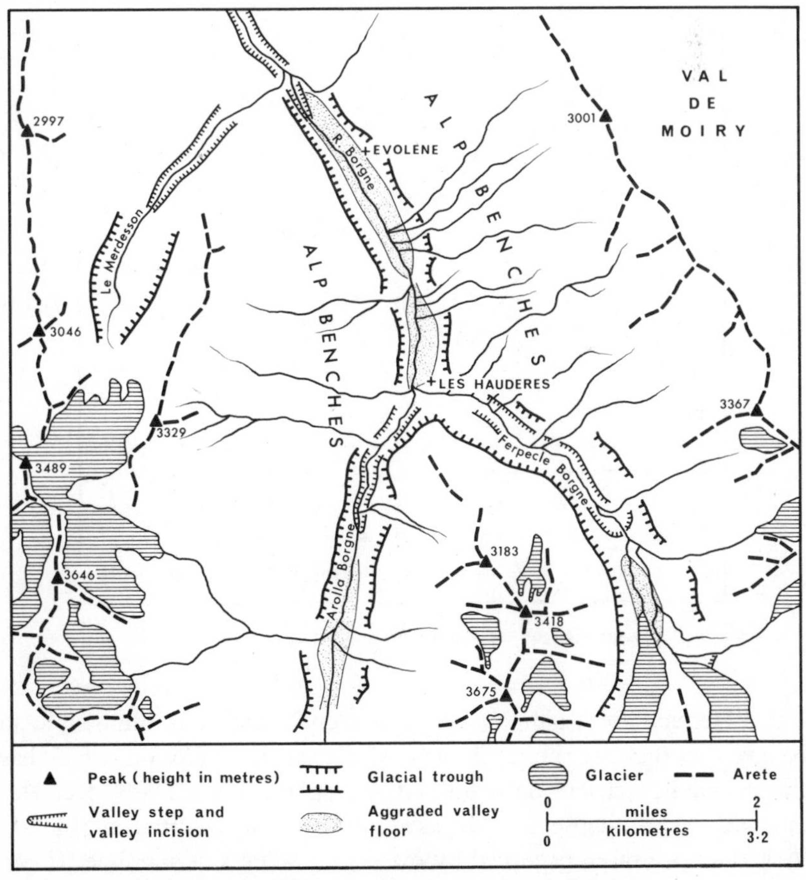

150 The glacial features of the Val d'Hérens, Valais, Switzerland

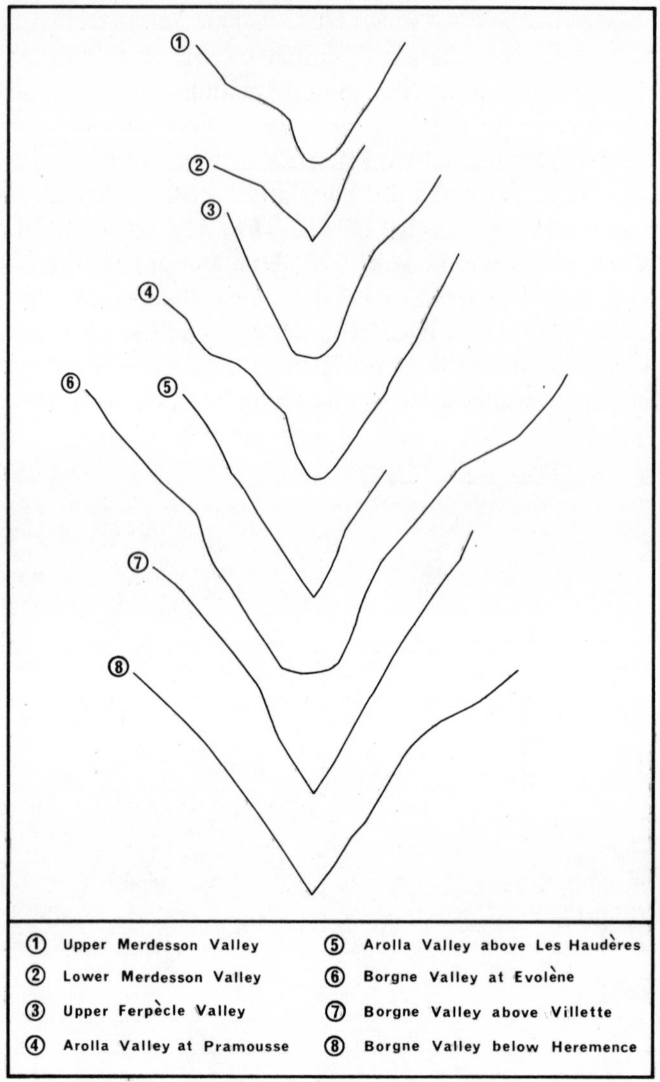

151 Cross-profiles of the Val d'Hérens

Val d'Hérens (for example, below Evolène), but it is impossible to believe that they are the result of 'post-glacial' river erosion, which has transformed glacial troughs formed during the last ice advance. Not only has the time-lapse since deglaciation been quite insufficient, but the V-sections are often occupied by masses of valley-side moraine. It can only be inferred that, for some reason, ice passed through these V-profiles in large quantities without altering their form, whereas in other places (as between Les Haudères and Evolène) the more normal U-profile was developed with ease. It is not enough to adopt Garwood's hypothesis, for in the Val d'Hérens as a whole there is abundant evidence of the potency of glacial erosion.

Another notable feature of the Val d'Hérens and its tributary valleys is the nature of the valley long-profiles. As Fig. 152 shows, these are very irregular, with near-horizontal sections alternating with high and steep valley steps. In some instances the latter are allied with V-profiles (as in the lower part of the Merdesson valley), whereas the gentler parts of the long-profiles coincide with U-sections. One such step has recently been uncovered at the head of the Ferpècle valley by recession of the Ferpècle glacier. There is little doubt that most of the Val d'Hérens valley steps are of structural origin, for they are closely related to resistant gneissic and granitic rocks which contrast greatly with the easily fractured 'schistes lustrés' lying downvalley of them.

Lastly, the hanging nature of the small tributaries of the Borgne is

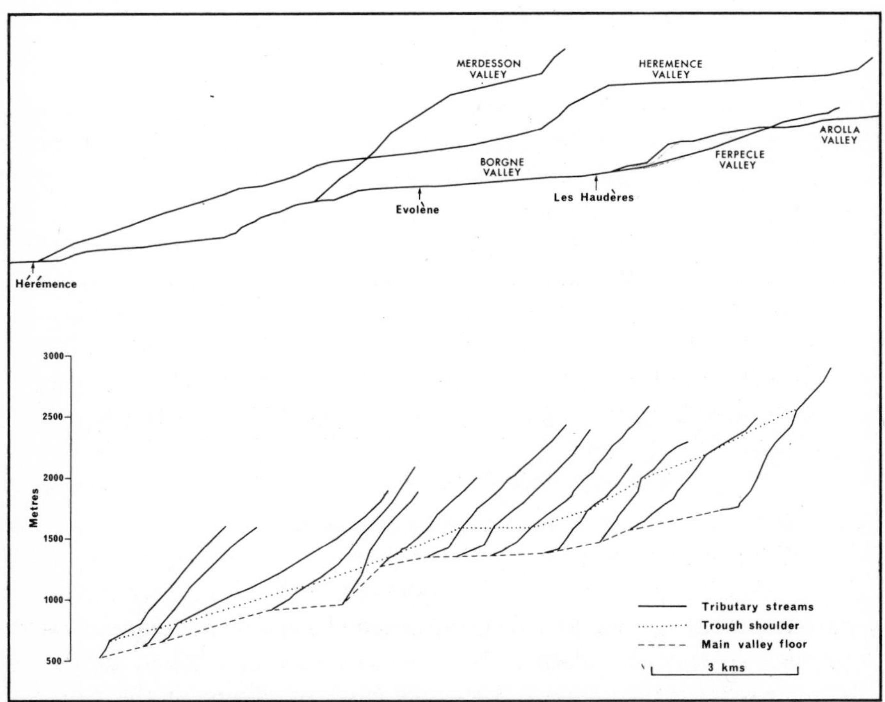

152 **Long-profiles and tributary valley profiles of the Val d'Hérens**

153 View towards Suilven, Sutherland, Scotland, over 'knock and lochan' country developed in the Lewisian Gneiss (see p. 361). [Eric Kay]

worth commenting on. Profiles of those entering the Val d'Hérens from the east are shown in Fig. 152. Most show breaks of slope coincident with the trough shoulders, and from these it is possible to make a tentative estimate of the amount of valley deepening which has been effected by the former Val d'Hérens glacier. It is true that this estimate is based on the assumption that the alp benches drained by the tributary streams have not been significantly lowered by glacial erosion, and may not therefore be justified. However, it is interesting that the amount of deepening indicated shows a progressive increase from the lower part of the valley (where it is in the order of 300 feet) to the upper (where it may be as high as 1500 feet). This is comparable with the trends noted by some authorities in other glaciated valleys, and indicates that the estimates made here may not be totally unrealistic.

Glaciated lowlands

It is generally assumed that glacial erosion is of less consequence in lowland areas, where glaciers often seem to perform a largely depositional role. However, there is some divergence of opinion on this point, and in any case it must be emphasised that many low-lying areas show *some* evidence of erosion by ice. Well-known examples on a large scale are the Laurentian Shield of North America and the Fenno-Scandian Shield of Europe. In these differential scouring, taking advantage of fault-lines, joints, shatter-belts, and other lines of geological weakness, has led to diversification of the pre-existing relief, so that many ice-moulded and 'streamlined' hills and knolls rise above rocky basins containing lakes and bogs. The result is a landscape of confused aspect, with a badly disorganised drainage network. However, one must be careful not to overestimate the role of glacial erosion in such landscapes. R. F. Flint, for example, has suggested that in the Laurentian Shield the ice has merely removed the 'pre-glacial' weathered mantle and effected moderate erosion of the basal weathering surface. This view is based largely on the fact that the sub-Ordovician surface, developed across the pre-Cambrian rocks of the Shield and widely exposed by sub-aerial denudation prior to the Quaternary glaciations, is morphologically similar to its present-day buried counterparts. Furthermore, the drainage of the Laurentian Shield continues to show a high degree of adjustment to structure; indeed the 'pre-glacial' drainage 'remains unaltered by the glaciation save for the excavation of shallow rock basins in some of the larger valleys'. Flint's conclusion is that the average thickness of rock removed by glacial erosion from the Shield does not amount to more than about 30 ft. In fact, it is safe to generalise that 'in mountains with abundant snowfall glacial erosion has been deep, whereas in regions of slight relief overspread by ice-sheets glacial erosion has been comparatively slight.'

Flint's views are not, however, accepted by all geomorphologists. Linton (1963), for example, has suggested that the remarkable surface of the Lewisian gneiss in north-western Scotland and the Hebrides owes much to glacial erosion (Figs 153 and 154). The landscape here is diversified by numerous rocky hills ('knocks'), often abraded by the ice to form elongated 'rock drumlins', and hundreds of hollows containing small lakes ('lochans'). In late-Tertiary times this area may have been mantled by a thick detrital layer, resulting from prolonged chemical decay of the gneiss, but none of this remains, proving that the ice has cut below the level of the basal weathering surface. Furthermore, the relief, which locally exceeds 600 ft, is much greater than would be expected of such a surface, so that considerable glacial over-deepening, perhaps in the order of 200–400 ft, may be inferred. Linton has also argued that elsewhere in Britain major relief features have been profoundly modified by massive ice-sheet erosion. A striking case is the

Chalk escarpment of East Anglia. This stands well over 500 ft in elevation to the west of the Hitchin gap, but changes in an astonishing way to the east, where the main escarpment of the Upper Chalk is replaced by a gentle slope leading up to the boulder-clay plateau of Suffolk (generally at 300–400 ft O.D.). Linton argues that this 'complete failure' of a major scarp for no lithological or structural reasons can only be attributed to massive erosion by an ice-sheet deploying from the Fens basin. Certainly the Chalk here has been overridden by the ice, as the abundant glacial drift testifies, but it is not clear whether this was made possible by the 'pre-glacial' form of the scarp. Sparks (1957) has demonstrated the existence of well-preserved Quaternary marine benches to the south-west and east of Cambridge in positions where the scarp should be. The 'stepped' profile of the Chalk may well have

154 The relief and drainage pattern of part of Lewis

facilitated ice movement to the south and south-east, and thus the part played by ice in the obliteration of the scarp may have been overestimated by Linton. Other examples quoted by Linton are perhaps more convincing. The important Carboniferous Limestone escarpment overlooking the lowland of Gort, in western Ireland, has evidently been fragmented by the ice into 'isolated little hills, some of which have been streamlined parallel to the drumlins of the plain', whilst the Namurian 'escarpment' adjacent to the lowland of the Fergus and pierced by the Shannon at Foynes has lost its scarp form entirely, being replaced by streamlined or fish-shaped low ridges some 2–4 miles in length.

The processes of glacial erosion

Glaciers and ice-sheets have formed their distinctive erosional features through the operation of four main processes, only two of which can be classified as erosional in the narrow sense of the term.

A *Weathering-removal.* As indicated already, many areas affected by the Quaternary glaciations were mantled by thick weathering deposits, produced under the warmer and wetter conditions prevailing during the Tertiary era in present-day temperate lands (pp. 135–6). Similar weathering phases may have also occurred during interglacial periods. One of the first tasks of the glaciers and ice-sheets was to remove this overburden before attack on the underlying sound rock could begin. Generally they were able to achieve this task with ease, though it has been argued, from the occurrence of tors and pockets of chemically decomposed rock on rounded 'pre-glacial' divides high in some glaciated mountains (for example, the Cairngorms of Scotland), that vestiges of this former era, and the landscape associated with it, remain. The weathered material, which may have reached a thickness of 20–30 ft or even more over wide areas, must have made a substantial contribution to the great spreads of boulder-clay and other drift laid down at or near the glacial margins.

B *Abrasion.* It is usually accepted that one of the most important of glacial processes is the grinding away of bedrock by ice armed with rock fragments of various sizes. The reality of ice abrasion is indicated by polished rock surfaces (produced by the rubbing of fine grains of silt trapped between englacial boulders and the bedrock), deep grooves or furrows (usually found in relatively soft rocks such as limestone or shale, where they may attain a depth of some 6 ft and a length of 50–100 yards), and finer striations (which are fine-cut lines on the surface of the exposed rock inscribed by stones and boulders contained within the overriding ice) (Fig. 155). Other evidence is afforded by the finely ground rock material ('rock flour') washed by meltwater streams from the snouts of present-day glaciers.

Many authorities, however, have cast doubt on the efficacy of the abrasion process. For instance, it has been argued that, because of important changes that occur in the physical state of ice under great pressure, the bed load of a glacier cannot be moved with sufficient force to have much effect on the solid rock beneath. The observations of Carol (1947), who penetrated nearly 50 metres below the surface of the Ober Grindelwald glacier in Switzerland, are interesting in this context. Carol observed that where the basal ice passed over roches moutonnées,

155 Exposure of glacially abraded surfaces of gneiss by recent retreat of the Ferpècle glacier, Val d'Hérens, Switzerland. [M. J. Clark]

the forward velocity of the ice was approximately doubled (71·8 cm per day, as compared with 36·8 higher in the glacier). The ice seemed, in fact, to take on a plastic quality, and Carol noted that it did not hold stones and gravel sufficiently firmly to striate and grind away the smooth upstream slopes of the roches moutonnées.

In considering the effects of abrasion it must always be remembered that polishing, grooving and striating can be minor processes modifying forms that had been formed essentially in other ways. Many of the polished surfaces observable in glaciated valleys represent pseudo-bedding planes which have been revealed by removal of overlying sheets of rock by a quite different mechanism. It seems likely, in fact, that abrasion is a major erosive process only in rocks that are physically weak or are not prone to other forms of glacial attack, and that it is comparatively unimportant in hard, massive and widely jointed rocks.

C *Quarrying or plucking*. One of the most distinctive features of areas affected by glacial erosion is the abundance of rocky cliffs and slopes which appear to have been produced by the wedging or even plucking away of large joint-bounded blocks. On a large scale such cliffs and slopes are found at the heads of cirques or are associated with major steps in the valley floor; on a smaller scale they characterise the downstream sides of roches moutonnées both in the bottoms and on the sidewalls of glaciated valleys. In the view of many authorities quarrying is the most important erosional process effected by glaciers, and is responsible more than any other for the development of 'typical' glacial landforms (cirques, rock steps, rock basins and roches moutonnées). This conclusion is based both on theory and on actual field evidence. As Flint has pointed out, it takes much less work per unit volume to quarry out most kinds of rock than to wear down that rock by abrasion. Furthermore, he has been able to demonstrate from a study of the joint patterns of roches moutonnées that far more rock material has been removed from the quarried lee slopes than from the abraded 'stoss' slopes. Yet again, glaciated valleys are often most deeply incised in zones where the underlying rock is most closely jointed and thus most amenable to the quarrying process. Flint in fact takes the view that 'generally, and with local exceptions, quarrying has removed far more rock than abrasion'.

Even so, there are difficult problems posed by the quarrying process. These will be explored more fully in the discussions of rock steps and cirques later in this chapter, but for the moment certain relevant points may be briefly made. The process of quarrying may be effected in one of two ways. Firstly, the glacier ice may freeze on to the solid rock and, since it is moving, literally wrench away blocks defined by joints or other lines of weakness. Secondly, the percolation of water into joints, followed by freezing and expansion of that water, might lead to the wedging away of blocks, which can then become incorporated in the flowing ice. It is, however, very doubtful whether ice, particularly at depth where 'extrusion flow' may operate, has the strength to achieve the former. Moreover, it is not known whether (and if so how) temperature fluctuations about freezing-point can occur at the base of a thick glacier, where atmospheric influences cannot easily penetrate. That some frost wedging does take place beneath glaciers has been shown by Lewis (1947), who tunnelled beneath the Storjuvbreein glacier in the Jotunheim ice-field of Norway and found that 'all around was evidence of freeze-thaw, and all the evidence pointed to this as the effective agent in disintegrating the rock bed'. However, precisely what is involved remains something of a mystery. To suggest, as has Boyé (1949), that the selective block disintegration of the rock exposed in the valley floor is achieved by frost weathering in a periglacial period immediately prior to glaciation seems hardly realistic, especially when it is remembered that many large glaciers have deepened their valleys by hundreds or even thousands of feet.

D *Pressure release* (*dilatation*). The great potential importance of pressure release as an aid to glacial erosion (and in particular to the quarrying process) has only recently been realised. In 1954 Lewis described the remarkable 'bursting up' of layers of gneissic rock uncovered by the retreat of the Svellnosbreein glacier in the Jotunheim, and suggested that some of the rock fragments broken free by the process are so sharp that the action must have occurred *after* the retreat of the glacier. In Lewis's words, 'the gneiss had been consolidated under the pressure of several thousand metres of overlying strata. The gradual removal of these confining strata had been completed by the glacier in lowering its bed. The strength of the gneiss presumably enabled it to resist bursting up until the final release of all superincumbent load on the last retreat of the glacier.' Lewis went on to suggest that if the glacier advanced again it could remove the loosened layers of rock. However, he did not consider that pressure release effects were only important at or beyond the snout of the glacier, but argued that dilatation could take place wherever the weight of the ice on the valley floor was locally reduced. For example, caves are often found on the downstream side of roches moutonnées and rock steps, and pressure release here may weaken the rock by favouring joint development, which in turn can assist the process of glacial quarrying. It is even possible that in such a situation the tensile strength of the jointed rock will be less than that of the ice, so that true plucking without the aid of frost wedging and by simple frictional drag can be achieved by the moving glacier. Lewis's ideas have been extended by Linton (1963), who has postulated that deep glacial troughs are initiated by pure abrasion, but that later the role of mass quarrying becomes dominant, as evidenced by the formation of rock steps and quarried basins. The problem is to explain the efficacy of the process at depths where the rock is sound and temperature fluctuations presumably minimal. Linton argues that, as the glacial valley is deepened, the substitution of a certain volume of rock by a similar or smaller volume of ice one-third of its density must lead to extensive dilatation and parting of the rock along sheet joints. In other words, pressure release is an active aid to erosion beneath all glaciers, whatever their thickness. The suggestion is particularly interesting in the light of Flint's hypothesis that downward corrasion by ice may virtually cease at the level where joints are closed up. He had gone on to infer, rather inconsistently after his contention elsewhere that quarrying is the most important glacial process, that abrasion can be more effective than quarrying in *deep* valley formation. If Linton's thesis is correct, joints are never closed up beneath glaciers, and quarrying can proceed on the floors of the very deepest troughs—a fact which makes the development of those troughs less problematical.

The origin of glacial erosional landforms

A *Glacial troughs.* Virtually all U-shaped valleys in glaciated areas have been produced by the glacial modification of pre-existing river valleys. In plan these become simplified, for the truncation of spurs by massive ice erosion leads to a more 'straight-ahead' course. The conversion of the river valley entails both vertical downcutting (evidenced by the presence of overdeepened basins, with a 'closure' sometimes exceeding hundreds of feet, and intervening rock steps on the floors of the trough, together with the considerable bed loads revealed by upthrusting and ablation near the glacier snout) and lateral extension (revealed by roches moutonnées on the trough walls and large masses of lateral moraine, some of which is produced by frost shattering of the exposed slopes above the glacier surface). In explaining the characteristic U-shape of the glacial trough, some writers have likened the latter to the channel of a river, which will normally have a sub-rectangular or approximately semicircular profile.

The most obvious feature of the long-profile of glacial troughs is the very pronounced headward steepening, which contrasts vividly with the more gentle curve of most river valleys. An outstanding example of this, described by Linton, is afforded by the Unteraar glacier of the Swiss Alps. Above the glacier the headwall rises 2300 ft in half a mile, and plunges a further 1300 ft beneath the upper part of the ice at about the same angle. The lower part of the trough, however, flattens suddenly to an average gradient of only 1:150, with the result that the greatest thickness of the glacier (some 1450 ft) is found within 1 mile of its upper limit. One can only infer that the Unteraar glacier must have deepened its bed far below what would be possible for any mountain stream—indeed 'it is clear that the floors of glacial troughs and the profiles of the pre-glacial stream valleys on the same sites diverge more and more as they are traced upstream'. It is possible, by comparing valleys of approximately similar size in the same area, some of which possess profiles evidently little affected by glacial erosion and others which have been severely scoured, to gain some idea of the amount of glacial incision. By contrasting Glen Prosen and Glen Clova in the eastern Grampians, Linton was able to compute that the latter had been lowered in its upper part by some 600–1000 ft. An even greater amount of glacial erosion would be necessary to explain the Norwegian and Alaskan fiords. In Alaska the Lynn Canal reaches a maximum depth below sea-level of 2800 ft. Only a small proportion of this, perhaps 300 ft, can be accounted for by the post-glacial submergence, so that it is apparent that the Canal was produced by glacial erosion below sea-level. In this context it should be remembered that (i) rivers cannot erode below base-level, so well over 2000 ft of glacial erosion in the Lynn Canal is implied, and (ii) that as the density of glacier ice is about 0·9, a valley glacier 1000 metres thick would continue to erode its floor even when submerged to a depth of 900 metres. Since the steep walls of the Lynn Canal trough rise a further 2000 ft or more above sea-level, the total glacial incision here may well have been in the order of 4000 ft.

The other important feature of many glacial troughs is the occurrence of breaks of gradient and rock steps (already referred to in the brief description of the Val d'Hérens), together with rock basins which sometimes contain lakes and sometimes have been infilled by glacial and fluvio-glacial deposits. The rock steps, which show a great variety of form, are of particular interest (Fig. 156). Some are incised to a greater

156 The Mont Miné glacier, Val d'Hérens. Note (i) the ice fall of the upper glacier, marking a valley step which has been partially uncovered, (ii) the zone of ablation on the northern (right-hand) side and at the snout of the glacier, and (iii) the complex pattern of small terminal moraines beyond the glacier snout. [M. J. Clark]

or lesser extent by stream action, whilst others are near-vertical cliffs, apparently the result of glacial quarrying. A by no means untypical example is the abrupt step, some 500 ft in height, at the exit from the Lota corrie, cut into the gabbro of Sgurr nan Gillian, Skye. Lewis has written that the step is so steep that it resembles the headwall of a corrie, with its face chiselled and plucked away along joints and other planes of weakness; nowhere is there significant evidence of smoothing or grinding. Another particularly fine step is found at the head of the Nant Ffrancon valley, immediately below Llyn Ogwen, in Snowdonia.

Various theories have been proposed to explain glacial valley steps, but few have been totally convincing. Davis likened them in north Wales to the falls and rapids of an ungraded stream profile. However, this is of little help in any attempt to understand the processes responsible for step formation, and moreover it contains the unwarranted implication that with time glacial erosion will smooth the steps away. It is more likely that, owing to differential glacial deepening of the trough floor, the steps may become gradually more accentuated. Garwood (1910) attempted to explain the steps of Alpine valleys from a protectionist standpoint. He argued that each step marked the point where down-cutting by powerful rivers emerging from the glacier snouts was halted during retreat stages (in other words, during interglacials) of the Quaternary glaciations (Fig. 158A). However, this theory, though it explains the V-profiles described on p. 359, is countered by a great deal of evidence. Meltwater streams do not suddenly form at the glacier snout, but result from the amalgamation of many sub-glacial streams which themselves should, if Garwood's emphasis on the great powers of stream incision is correct, be capable of eroding channels beneath the ice, and not merely beyond its margins. Again, one of the most striking features of actual glaciers is the vast quantity of morainic debris, much

157 Leirdalen valley, Jotunheim, Norway. The valley floor has been heavily aggraded by outwash deposits, and the stream pattern is typically braided. [Eric Kay]

in the form of large boulders, which is released by ablation at the snout. This material may be spread over the valley floor immediately downstream, since it cannot be transported far by the manifestly overloaded —and therefore non-eroding—meltwater streams (Fig. 157). Finally, it is clear that valley steps do not coincide with interglacial 'positions' of the ice-margins, but are in most cases associated with resistant bands of rock (p. 359). This latter fact also militates against Solch's hypothesis that the Alpine steps are merely modified pre-Quaternary rejuvenation heads initiated by the late-Tertiary uplift of the mountain mass.

The fact that most valley steps represent inequalities of rock resistance is the basis of more modern theories of step formation. Thus Matthes (1930), writing of the evolution of the Yosemite valley, suggested that the treads or basins of the valley profile were produced by glacial quarrying of more closely jointed and therefore weaker rock, whereas the steps (which on their upper surfaces showed smoothing by glacial abrasion) coincided with the outcrops of more resistant rock (Fig. 158B). A contributory factor would be the nature of glacier flow. The ice would naturally thicken over the treads, with the result that velocities would be greatest at the base of the glacier, thus accentuating the plucking process. The reality of this 'pressure flow' (or 'extrusion flow') was indicated by R. Streiff-Becker's observations on Clariden Firn in the Alps over a period of 21 years. Careful measurements of the surface velocities of the glacier revealed that these were insufficient to account for the removal of all the snow and ice accumulating in the névé region. Thus a more rapid movement at depth was inferred by Streiff-Becker. Over rock steps, on the other hand, normal gravity flow of ice will occur. In this the maximum velocities are located at or near the surface of the glacier, and large tensional crevasses penetrating to a depth of some hundred or so feet are opened up.

In 1947, Lewis, who regarded rock steps and roches moutonnées as genetically similar forms, argued that plucking, or as he preferred to term it 'sapping', of the downstream face was the most important process in step formation (Fig. 158C). From his observations in Norway he concluded that

> roches moutonnées and steps of various sizes which abound under and alongside shallow glaciers are shattered and sapped on their downstream sides, and that in this way large quantities of rock are removed from valley sides and floors. Grinding and smoothing may shape projecting knobs and hummocks, but the amount of material removed in this way must be small compared with that removed by freeze-thaw and plucking.

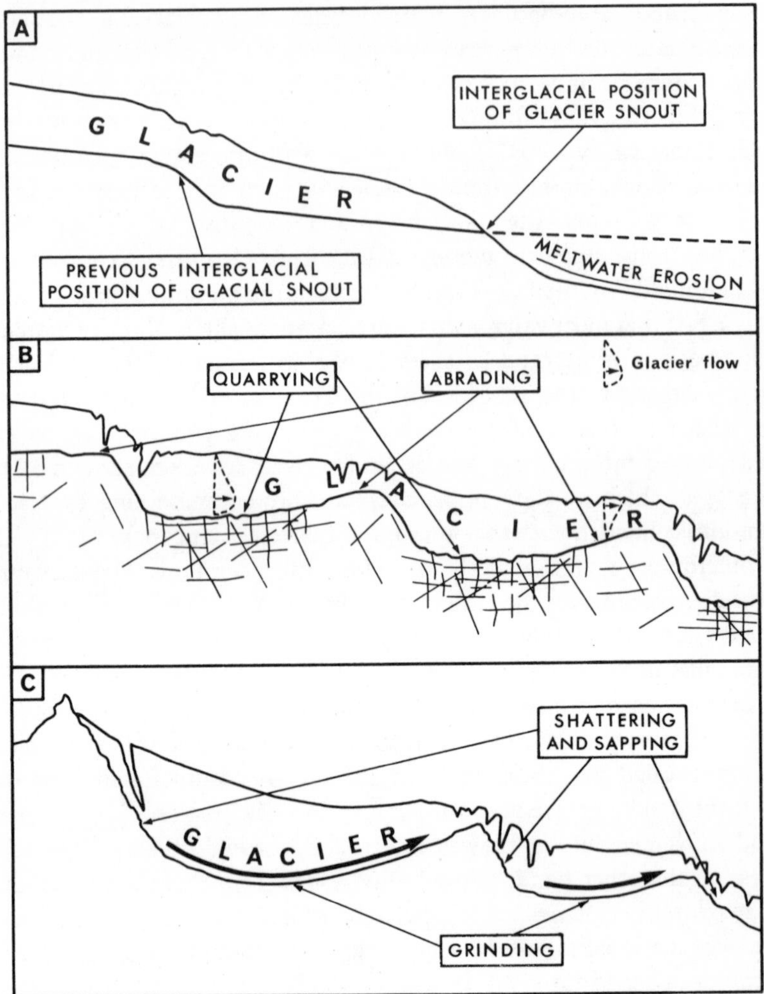

158 Possible methods of valley step formation: glacial protection (A); quarrying associated with extrusion flow (B); grinding and plucking associated with rotational slip (C)

The problem was to explain how such freeze-thaw shattering of the face of the rock step could occur. Lewis considered that beneath thin glaciers, where lithologically determined breaks could appear through the surface of the ice or where crevasses could penetrate to the base of the glacier, such frost action was feasible (Fig. 156). However, the operation of frost shattering beneath thick glaciers was far more difficult to envisage. Yet all the evidence suggests that the rate of erosion is greatest where thick glaciers are concerned. A very tentative solution offered by Lewis was that, owing to the lowered freezing-point of water at great pressure, meltwater can percolate beneath deep ice and enter rock crevices there without freezing, despite having a temperature of slightly less than 0 °C. However, as glaciers pass over rock steps there is a relief of pressure, and any supercooled water will then freeze. Unfortunately, the effects in terms of frost weathering will only be very small, for the temperature of the frozen water will be just below freezing-point and minimal ice expansion in rock crevices will occur. On the other hand, the pressure-release mechanism (pp. 366–7) could well achieve even at considerable depths beneath the ice what normal freeze-thaw weathering cannot. Thus Lewis's contention that steps are produced by fracturing of the rocks forming their faces may remain valid. The rock basins often separating steps were explained by Lewis not in terms of quarrying but, by analogy with cirques which also contain erosional hollows, as the outcome of abrasion or grinding by rotationally slipping ice (p. 381).

It is appropriate to conclude this discussion of glacial troughs and associated features by referring briefly to the classification of troughs proposed by Linton. Four main categories are recognised.

A *Alpine* troughs are those fed either at present or in the past by areas of snow and ice accumulation surrounding the immediate valley head. The former tributaries here are modified into a series of 'convergent' cirques which hang above the main trough. The Ullswater valley in the Lake District is a good type-example.

B *Icelandic* troughs occur where the ice accumulates on extensive plateaus and is discharged by steep ice-falls into the heads of valleys dissecting the plateau margins. The valleys leading down from the Grampian plateau (such as Glen Callater, Glen Muick and Glen Esk) fall into this category, as do many of the glacial troughs fed by distributaries of the Iceland ice-cap.

C *Composite* troughs are those in which pre-existing river valleys are only partially used and some wholly new courses are added by glacial erosion. The essential process involved is that of 'watershed breaching' by 'diffluent' or 'transfluent' ice, and results from the fact that in many upland masses the ice accumulated in the Quaternary was so great that it could not be adequately discharged through existing valley routes.

Glacial diffluence occurs where a valley glacier finds its normal outlet blocked, perhaps by the junction with a larger and more powerful glacier that forms in effect an impenetrable wall, and consequently is forced to build up its level to escape across a neighbouring watershed as a 'distributary ice tongue'. In some cases high-level cols are eroded, but quite commonly the diffluent ice carves troughs down to the level of the main valley. The effects of glacial diffluence on the trough pattern of the Cairngorms are illustrated in Fig. 159. Glacial transfluence is found where the impeded ice cuts out of a valley system not by a lateral distributary but at the very head of the valley. The process will be materially aided where the headward extension of adjacent cirques has given rise to a col well below the level of the former watershed. The breach between Glen Rosa and Glen Sannox, in Arran (pp. 355–6), may well be of this type. Glen Rosa itself was clearly initiated as an Alpine trough, with at least three cirques (Coire Daingean, Fionn Choire and Dearg Choirein) feeding ice into its upper part. The lower section of Glen Rosa is, however, severely constricted, and the penetration of the ice northwards between Cir Mhor and Goat Fell into the head of Glen Sannox seems to have occurred.

D *Intrusive* troughs are formed where the glaciers have cut against the prevailing pre-glacial gradients, in particular where ice movement has been from a lowland into an upland. A good example is Glen Eagles, in Perthshire, where ice of west Highland origin has passed through Strath Allain and eroded deeply into the northern margins of the Ochil Hills.

It might be argued that a fifth category, that of fiords, should be added. Although often displaying features of one or more of the above types, fiords are characterised by (i) possible development along lines of weakness such as major faults and shatter-zones (hence the rectangular pattern of certain Norwegian fiords), and (ii) exceptional glacial scouring which has penetrated well below the contemporary sea-level (p. 368).

Hanging valleys

These features have perhaps been over-emphasised in accounts of glacially eroded landforms. Certainly hanging valleys are not confined to glaciated mountains, but are found wherever a main valley is over-deepened (perhaps because it follows a line of exceptional geological weakness) by comparison with its tributaries. Hanging valleys in the Alps were explained by Garwood as the result of glacial protection of

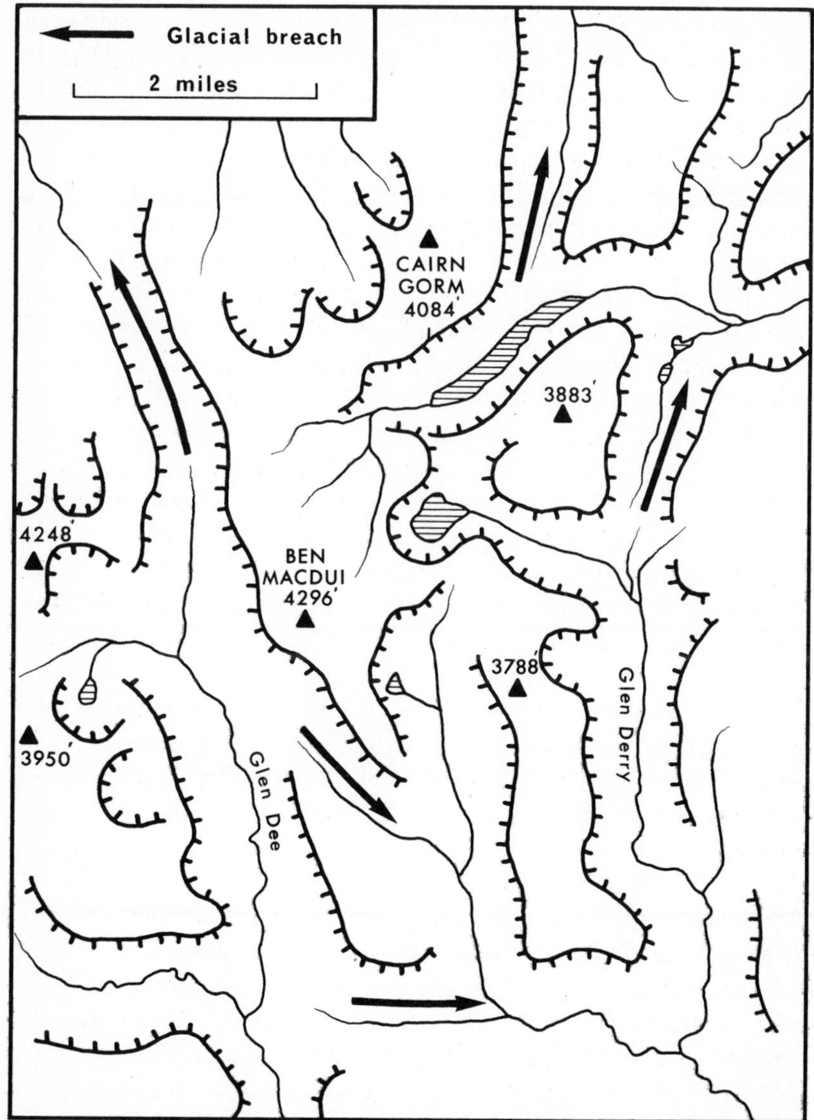

159 **Glacial diffluence in the Cairngorms (after D. L. Linton)**

small tributary valleys at a time when the main valley, free of glacier ice, was being incised by fluvial action (Fig. 160A). However, as in the case of glacial valley steps, the protectionist interpretation of hanging valleys has little to commend it today.

Once the reality of glacial erosion is accepted, as it surely must be,

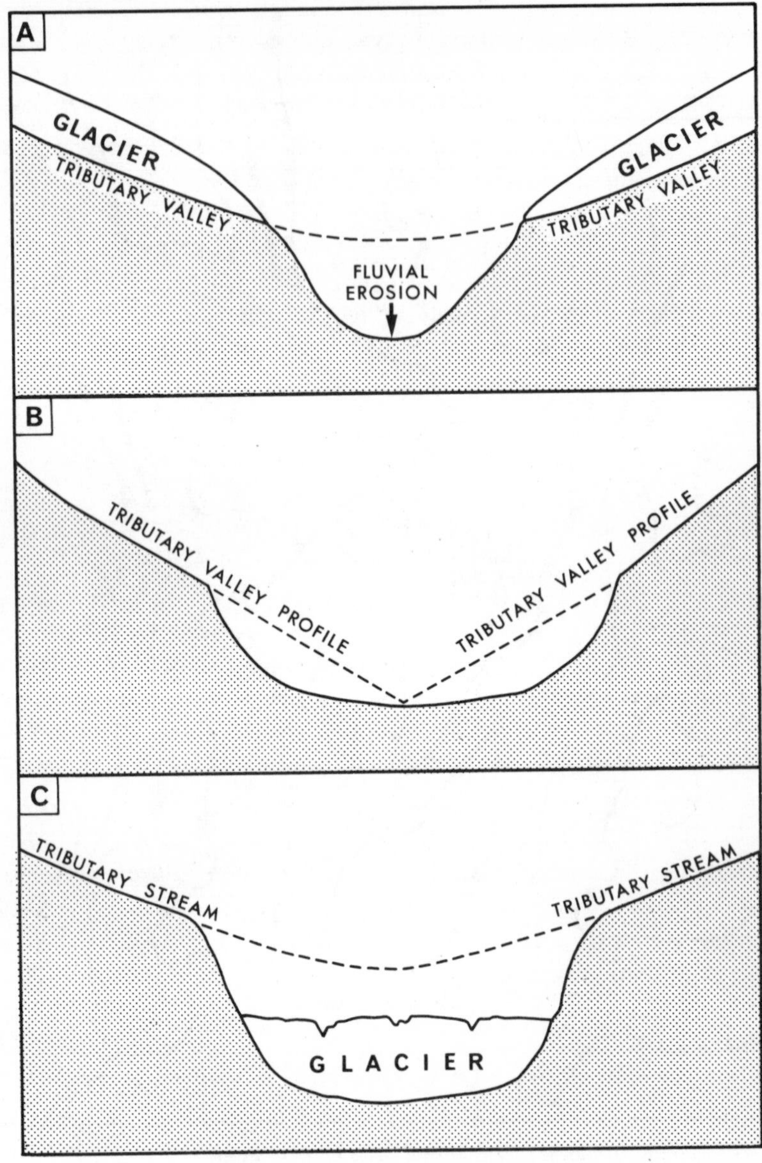

160 The development of hanging valleys: glacial protection (A); valley widening (B); glacial erosion of the main valley (C)

hanging valleys are seen to be quite normal and expectable features. The conversion of a V-shaped valley into a U-shaped trough, by glacial deepening, widening, or a combination of the two, will entail truncation of the lower courses of tributary valleys (Fig. 160B). It is true, of course, that these valleys themselves may experience some glacial erosion, but this will be insufficient to maintain accordance of junction with the floor of the main valley. Two factors will, in fact, cause the degree of discordance of level to be increased. Firstly, there is little doubt that larger glaciers, exerting a greater frictional drag on the underlying bedrock and carrying massive loads of morainic material that can be used in abrasion, can deepen their valleys at a far more rapid rate than small tributary glaciers. It is also possible that some valley steps may be caused by a sudden increase in the corrasional powers of a large glacier formed by the uniting of two smaller ice flows. Again, the steep 'trough-ends' of many glaciated valleys may result from the vastly greater efficiency of a single glacier resulting from the amalgamation of three or four cirque glaciers originating high up on the mountain slopes. Secondly, there is the fact that at the junctions of glaciers the ice-surfaces will tend to be accordant, so that if one is substantially smaller than the other its base (or in other words the level of the valley floor over which it moves) has to be at a higher level. Certainly, the surface of the smaller glacier cannot be below that of the larger glacier. Not only would its movement cease altogether, but the large glacier in the main valley would spill laterally into the tributary. In either case, glacial deepening of the tributary valley would tend to cease.

A final cause of hanging valleys which is often overlooked lies in the fact that large glaciers, fed by copious snow and moving at a high rate, can move so far down a main valley that they enter regions where the tributary valleys are unoccupied by ice and are being only slowly eroded by running water (Fig. 160C). This is, in fact, the very converse of the Garwood protectionist theory. These conditions are actually met with, on a small scale, at the head of the Val d'Hérens, where the Ferpècle glacier occupies a deep trough whose eastern shoulder is free of ice; instead, the bare slopes are crossed by small streams, in ill-defined valleys, that are fed by small cirque glaciers high on the western side of the Dent Blanche.

Cirques

These amphitheatre-like hollows are perhaps the most characteristic of all glacial erosional landforms, in that (unlike, say, U-shaped valleys and hanging valleys) they have no real counterpart outside glaciated uplands. At the same time, they are probably the most problematical of glacial landforms, for the processes that produce their precipitous headwalls and their rock basins (giving what Willard D. Johnson called their 'down-at-the-heel' appearance) are by no means fully understood.

Most writers are agreed that cirques are initiated by the process of nivation or snow-patch erosion (pp. 343–5). In sheltered localities on north- and north-east-facing slopes snow may accumulate in gentle hollows, small gullies or even on the open slope. Associated freeze-thaw weathering and transportation of the comminuted debris by meltwater will lead to the formation of a quite deep and rounded depression, in which snow can gather to a greater and greater thickness. Such snow, when more than one year old and remaining because of ineffective summer melting, will be recrystallised into 'granular' snow known as 'firn' or 'névé'. With further build-up and compaction, so that enclosed air is forced out, the firn will be converted into ice. With still more build-up, plastic deformation of the ice can occur, and thus downward flow begins.

It seems likely that nivation hollows may need to reach a considerable size and depth before the final stages in the formation of the cirque glacier can occur. In some instances, the hollows may themselves resemble true cirques, with steep headwalls riven by frost and reaching a height of 100–200 ft. The disintegrated material may slide across the bank of firn below, and accumulate to give moraine-like piles of debris ('pro-talus') running approximately parallel to the headwall (Fig. 161A). Excellent examples of such late-stage nivation hollows may be seen indenting the edge of the Pennant Grit plateau west of Cwmparc, in the Rhondda valley, south Wales.

There is no doubt that continued frost shattering of the cirque headwall, wherever it is exposed, continues after the appearance of the cirque glacier. However, it is also perfectly clear that disintegration of the rock face must continue *below* the level of the glacier, for the frost-riven walls of abandoned cirques may be several hundreds of feet high, and extend at their base almost to the floor of the hollow. A possible mechanism was suggested by Willard D. Johnson as long ago as 1899. At the head of a cirque glacier is normally found a major crevasse, the bergschrund, which may penetrate to a depth of some 150 ft in favourable circumstances. The bergschrund, the result of tension as the cirque glacier draws away from the headwall (to which some ice remains attached), can in some instances reach the rock at the base of the glacier (Fig. 161B).

Johnson argued that, at such a point, freeze-thaw weathering could take place, leading to the release of blocks which would become incorporated within the ice and be used to abrade the cirque floor.

Unfortunately, Johnson's hypothesis has certain limitations. The problem of weathering remains unsolved where the bergschrund does not expose solid rock. Moreover, in the 'bergschrund hypothesis' only a very localised attack can be envisaged, whereas in fact rock shattering has evidently operated in most cirques over the whole of the backwall. I. Bowman argued that 'sapping' must occur regardless of the existence of the bergschrund, and this led Lewis (1940) to suggest that meltwater and rainwater may percolate down behind the névé and the glacier, sometimes far below the limit of the bergschrund. His well-known description of the resultant process is as follows:

In the summer months, melt-water, from direct precipitation and from the winter snowdrifts on the headwalls and upper névé, can be seen pouring here

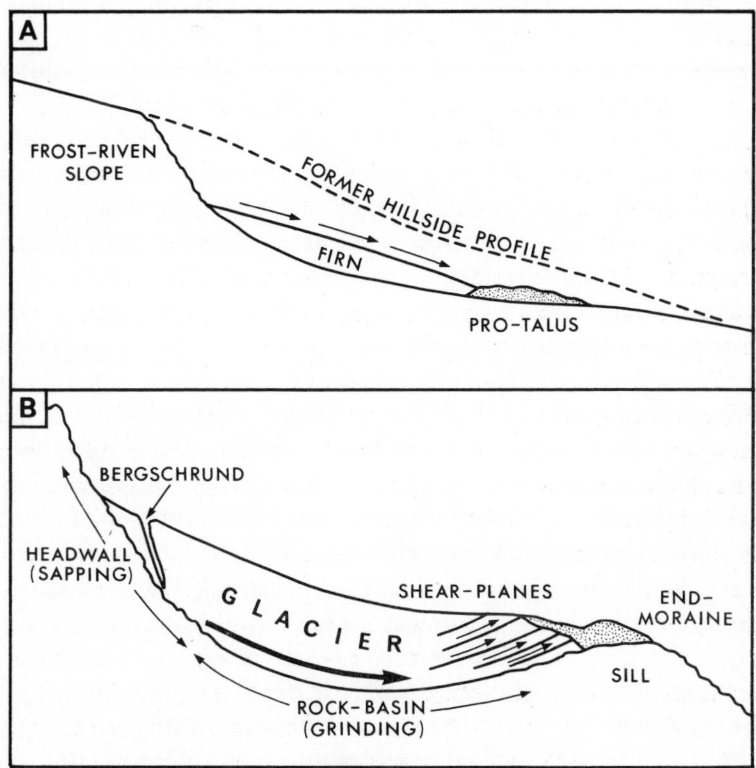

161 **Large nivation hollows (A) and cirques (B)**

and trickling there down behind the glaciers. At night and during cold spells some of the water freezes, and if it occupies crevices in the rock the expansion on freezing exerts great pressure, tending to split the rock apart. The pressure exerted depends on the degree of the cold and at very low temperatures it is capable of bursting open a steel bomb. After repeated thawing and freezing, the loosened rock fragments break off and are finally removed by the glacier, which may also render some aid by pulling away blocks frozen into the ice. This latter would seem to be but a subsidiary factor in removal, because the tensile strength is low compared with that of the hard rock composing the cirque walls, though the almost universal presence of joints must be allowed for.

Convincing though Lewis's explanation is in a general way, there are specific problems that remain unsolved. Once again the chief of these is: how can significant temperature fluctuations, capable of causing freeze-thaw, operate beneath hundreds of feet of ice at the head of the cirque glacier? Observations made by Battle (in Battle and Lewis 1951) reveal the seriousness of this difficulty. Battle found that, during August-September 1951, temperatures in a bergschrund on the Jungfraujoch ranged only from 0 °C to −1 °C, despite the fact that outside temperatures varied between 11·5 °C and −12° C. Even under winter conditions, the temperatures in the bergschrund ranged only from −2 °C and −4·4 °C. Temperature changes such as these seem insufficient to cause powerful shattering of the rock, and in any case they are presumably still less pronounced in the crack ('randkluft') between the névé and the headwall. Nevertheless, Lewis observed what appeared to be 'prizing and sapping' of the rock in sub-glacial caverns behind the ice by the repeated freezing of meltwater. In 1954 he argued that the latter process might be materially aided if the rock of the headwall were already fissured by the development of pressure-release joints. More recent work in the Jotunheim by Battey (1960) seems to offer some confirmation of this hypothesis. Battey found that in a small cirque the strike of the major vertical joints followed round the headwall in a remarkable manner, as if they had arisen by spontaneous dilatation of the rock in a direction normal to the cirque wall.

The second major problem posed by cirques is that of their rock basins and bounding sills. Obviously a localised downwearing is implied, though the causes and mechanism of this are unclear. It is conceivable that extrusion flow could lead to either quarrying of well-jointed rock underlying the floor or to concentrated abrasion, with the help of a bed load comprising morainic blocks sapped from the cirque headwall. The non-erosion of the sill remains unexplained, unless one assumes that it is composed invariably of massive and ill-jointed rock (which seems too much of a coincidence to be readily acceptable). Alternatively, in the case of a cirque glacier *sensu stricto* (that is, one that has never escaped the cirque hollow), limited erosion at the glacier snout might be explained by the so-called 'shear-planes' which are commonly observable in such a situation. These seem often to have the effect of bringing up the englacial detritus to the surface, with the result that the bed load available for abrasion is much reduced.

Lastly, the theory of 'rotational slip of glaciers' proposed by W. V. Lewis is worth considering. This was based on an analogy with landslips, which frequently follow curvilinear (in section) shear-planes, and are associated with the upthrusting of material at the toe of the slip. The curvilinear cross-section of cirques, and the apparent shear-planes (at 30–35° upwards) observed at the snouts of cirque glaciers on Vatnajokull, Iceland, suggested to Lewis a similar mechanism of glacier movement. Any tendency for the glacier to attain equilibrium, and so for the major shear-plane between the base of the ice and the underlying rock to become inactive, would normally be offset by (i) the additional weight of the snow accumulating each winter on the upper part of the glacier, and (ii) melting of the glacier snout during the summer season (Fig. 161B). As described already, Lewis envisaged that the headwall of the cirque is attacked by shattering and sapping, whilst the rock basin results from grinding by the rotationally slipping ice. The hypothesis is clearly attractive, even though more recent work has cast some doubt on the detailed mechanism of rotational slip. For example, McCall (1952) has demonstrated that 'thrust-planes' in glaciers are actually of sedimentary origin, and are associated with no shearing at all. Nevertheless, it is difficult to escape the conclusion that a small glacier, confined to a cirque containing a rock basin, must have some rotational element in its movement—though whether the cause or result of the cirque form must remain a moot point.

GLACIAL DEPOSITION

The results of glacial deposition *in toto* are no less striking than those of glacial erosion. In many lowland areas the Quaternary ice-sheets attained a maximum thickness of many hundreds or even thousands of feet, and in the zones of melting the quantities of detrital material released were truly phenomenal. Over thousands of square miles the drift cover may be anything from 50 to 200 ft or more in depth, and along 'pre-glacial' valley lines over 1000 ft has been recorded. None the less, the actual landforms associated with glacial deposition are less strongly characterised than the deep troughs, cirques, rock steps, aretes and sharp peaks of glaciated mountains. Indeed the average drift plain is extremely dull and, at first sight, relatively featureless. The reason lies mainly in the fact that the process of glacial deposition is not only complex, but is also 'self-destructive' in the sense that a feature built up by one type of action can be immediately modified or effaced by another. Many areas of glacial deposition comprise what can only be described as a 'melange' of deposits and forms, in varying states of preservation. Although attempts have been made to classify depositional landforms (into till-plains, end-moraines, outwash fans, kames, eskers and so on), these must be recognised as arbitrary, for in reality there exists an almost infinite variety of forms, which are often extremely difficult to interpret accurately. For these reasons it is not easy to exemplify the main features of glacial deposition by reference to one or two selected areas.

The processes of glacial deposition

In its broad sense glacial deposition is really an amalgam of processes which cannot properly be separated from each other. It is convenient to consider in turn (i) the direct depositional role of the ice itself, and (ii) the action of meltwater on, within, beneath, at the margins of, and beyond the glacier or ice-sheet. However, it must be remembered that since the ice cannot deposit without itself melting, these two actions are complementary.

A *Deposition by ice.* Any large body of ice will possess the capacity to move a correspondingly great load of detrital material, which has been derived either from the frost weathering of valley sides or isolated peaks ('nunataks') exposed above the surface of the ice to atmospheric temperature changes, or from abrasion and quarrying of the sub-glacial surface. This 'moraine' may be concentrated along certain lines (for example, the lateral and medial moraines of valley glaciers), or it may accumulate in greatest quantities at the bed of the glacier or ice-sheet (thus reflecting its most important source, as well as the subsidiary process of falling down crevasses). Often the lower part of the ice may become so charged with 'ground moraine' that, if slow pressure melting occurs, drift particles will be freed and become plastered or lodged on to the sub-glacial surface. In this way, a thick deposit of 'lodgement till' may be built up. The main characteristics of such a till will be (i) an absence of stratification or sorting of constituent materials, (ii) a 'fabric' in which most of the individual stones have their major axes parallel to the direction of ice flow, and (iii) the existence of low streamlined hills (drumlins).

Towards the margins of the ice, where it is affected by melting and evaporation and is thus gradually reduced in thickness, englacial debris will be exposed in large quantities on the surface, thus forming 'ablation moraine' (Fig. 156). With the ultimate disappearance of the ice, a layer of structureless detritus ('ablation till') will be left. During the final stages of ice-melt, the fabric of this till will be disturbed by collapse and slumping, and its finer constituents may be washed away by meltwater, so that its constitution and form will be different from that of underlying lodgement till.

Finally, at the actual margins of a glacier or ice-sheet which is moving forwards, melting may be accompanied by the strongly localised deposition of englacial rock debris to give an end-moraine.

B *Deposition by meltwater*. The streams of water released by the melting of ice perform a major task of transportation, sorting and deposition in glaciated areas. Supra-glacial streams near the ice-margins wash away material exposed by ablation; englacial and subglacial streams, formed by meltwater percolating down crevasses, transport rock debris either from one point to another beneath the ice or to and beyond the ice-edge; and in the 'pro-glacial zone' the meltwater streams running off or emerging from the base of the ice carry away and eventually deposit as 'outwash' as much of the glacial detritus as they are capable of handling (Fig. 157). In contrast to the non-stratified tills and moraines laid down by the ice itself, these fluvio-glacial deposits usually show clear evidence of sorting of constituents and stratification into beds. However, some may be later disturbed and severely contorted by a subsequent ice advance or by cryoturbation in a later periglacial episode, whilst those built up against the margins of the ice ('ice contact' features) will be affected by collapse when the buttressing ice melts.

The origin of glacial depositional landforms

The processes which have been described operate in both uplands and lowlands alike, and for that reason it seems undesirable to make a clear distinction between the depositional features typical of each. Even so, some landforms found commonly in lowlands and areas of moderate relief (such as kames, kame-terraces and eskers) are not characteristic of most mountain valleys. The reason is that in the latter the heavy snowfall and steep gradients promote constant glacier movement, and thus militate against the conditions of ice stagnation needed for the optimum development of these forms. Conversely, some depositional features normal to mountain valleys have no real counterpart in glaciated lowlands. For example, at the head of the Ferpècle tributary of the Val d'Hérens (p. 359), a substantial recent recession has divided the Ferpècle and Mont Miné glaciers (Fig. 162). There is no end-moraine here, but merely a large spread of rocky debris which is in the process of being reworked by anastomosing meltwater torrents. A bare tract of boulders, gravels and sand separates the two glacier snouts, and is pitted by numerous circular water-filled hollows ('kettles') marking small masses of ice that calved from the glaciers, became trapped in the debris, and subsequently melted. Much of the detritus on the valley

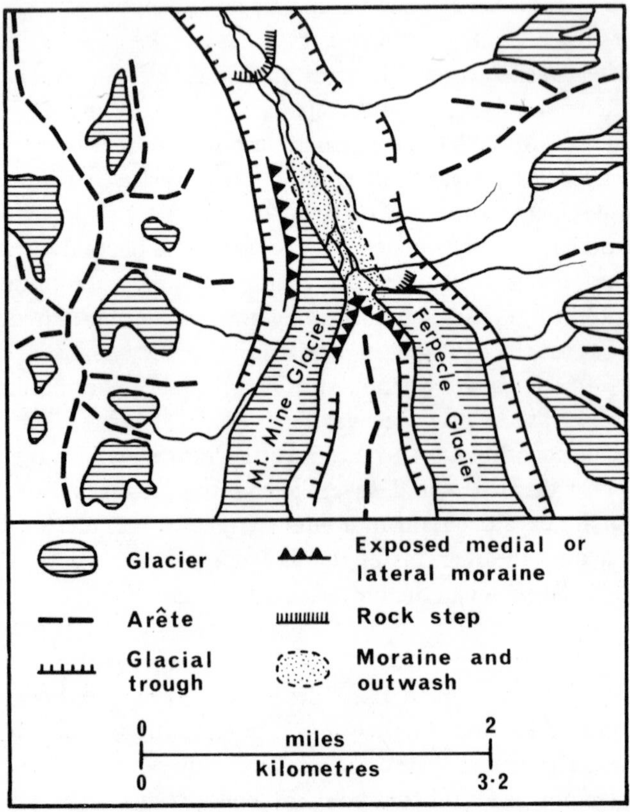

162 Glacial features of the upper Ferpècle valley, Val d'Hérens

floor has evidently been derived from a medial moraine, the very head of which is preserved in good condition, that once separated the individual 'components' of the Ferpècle-Mont Miné glacier. Since medial moraines occupy central positions in valleys it is normal for them to be partially or completely destroyed, by fluvio-glacial stream activity, as soon as they are bared by ice retreat. To the west of the Mont Miné glacier, an equally fine lateral moraine, with a flat and narrow top and a steep edge, rises above the surface of the shrunken ice (Fig. 163). This is in a less vulnerable position than the medial moraine, but has none the less been rapidly gullied by rainwater and meltwater rivulets except where the weaker constituents have been protected by large boulders within the moraine. At such points, early stages in the development of 'earth-pinnacles' ('desmoiselles') can be observed. Features such as medial and lateral moraines are not found in glaciated lowlands, which have been affected by ice-sheets not valley glaciers.

It is, of course, in extensive lowland areas that glacial deposition is at its greatest. A tentative classification of the main forms commonly found here is as follows:

A *Till-plains (boulder-clay plains)*. These result from direct glacial deposition and are, in terms of area covered and degree of modification of the 'pre-glacial' landscape, the most important of all glacial depositional features. They are developed where considerable masses of unstratified drift smother the underlying rocks, which may or may not protrude through the glacial deposits at high points of the sub-drift surface. The till itself may be greatly variable in composition, depending on the precise origin, movement and power of the responsible ice-sheet and on the resistance of its constituents to grinding and crushing. It usually consists of a heterogeneous mass of rocks, pebbles, gravel, sand and clay; hence the term 'boulder-clay', though widely used, is not always an accurate or useful one. The coherence, and thus resistance to denudation, of a till will also vary. Those which contain less than 10% clay particles are usually loose and friable (though this is to some extent offset by their permeability); those with a greater proportion of clay are more massive and compact.

One of the best examples of a till-plain in Britain is afforded by the low plateaus of East Anglia (excluding the areas in the west where Chalk is exposed). The most widespread deposit here is the so-called 'chalky boulder-clay', a composite till of tough and unstratified stony clay with many fragments of chalk and flint, together with erratics derived from

163 A dissected lateral moraine, Ferpècle valley, Val d'Hérens. [M. J. Clark]

rocks outside the region (for example, from the Jurassic outcrops to the west and north-west). It gives rise to a morphologically dull and featureless surface, into which gentle valleys have been incised at quite wide intervals. The till itself is variable in depth. It commonly reaches a thickness of about 100–150 ft on the watersheds between the main streams, and is up to 230 ft thick in the rather higher areas around Bury St Edmunds, Haverhill, Clare and Lavenham. At some points the chalky boulder-clay was scoured at an early stage by deep channels, which have been attributed by some to erosion by pro-glacial streams but are more likely to be the result of sub-glacial flows (pp. 398–402). These channels were subsequently filled by younger glacial deposits, and their presence has been detected only with the aid of borings. The most striking example in East Anglia, lying beneath the floor of the Stour valley, reaches a depth of −347 ft below sea-level at Glemsford.

It must be added that the history of the glacial landscape in East Anglia has been very complex. Most workers have recognised the existence of several individual tills, which appear to have been laid down in a number of distinct ice advances. The chalky boulder-clay itself overlies an earlier and deeply weathered deposit, the Norwich Brickearth, notable for its content of Scandinavian erratics. This older till had been deeply dissected, sometimes to the level of the Chalk, prior to the onset of the next glacial advance. In the valley of the river Wensum at Norwich chalky boulder-clay is found resting well below the base of the Brickearth. A detailed examination of the orientation of stones within the East Anglian tills ('till-fabric analysis') by West and Donner (1956) has shed light on the sequence of glaciations. An early glacial advance (the Cromer Advance), in which the dominant ice movement was from north-west to south-east, was responsible for the formation of the Norwich Brickearth and the so-called Cromer Till. The next glaciation (the Lowestoft Advance), from west to east, deposited the Lowestoft Till, equivalent to the lower chalky boulder-clay. Thirdly, the Gipping Advance, broadly from north to south, laid down the Gipping Till, equivalent to the upper chalky boulder-clay. There is also evidence of a fourth and final glaciation, the Hunstanton Advance, which was of more limited extent and was responsible for the distinctive reddish-brown boulder-clay found in the extreme north-west of Norfolk. It is therefore evident that the glacial landscape of East Anglia represents a 'palimpsest', comprising individual tills and associated outwash deposits, together with various 'erosional surfaces' developed in several glacial, periglacial and interglacial episodes.

The degree to which till deposition leads to the development of an entirely new physical landscape must be briefly considered. Where the drift is of exceptional thickness, the pre-glacial valleys may be very effectively plugged and the system of streams and valleys initiated on the drift-surface may be a 'discordant' one. However, this is certainly not true of all glaciated lowlands. Brown (1959), examining the drift landscape of Hertfordshire, found that many of the channels cut into the main drift deposit, the Springfield Till, closely follow the lines of valleys in the sub-drift surface (represented by the Chalk dip-slope). In other words, the underlying topography has been by no means obliterated, but sizeable depressions must have existed on the boulder-clay to guide the 'post-glacial' streams in their task of re-excavating the former valleys. An interesting exception is the Mimram valley, the pre-glacial course of which is partially blocked by drift plugs. The successor to the original stream follows a new valley some half a mile to the south-west of and parallel to the former course, parts of which have been revived by left-bank tributaries of the Mimram.

B *End-moraines*. It is important to realise that the formation of end-moraines depends on a precise balance between glacier advance and the rate of marginal wasting, so that the ice-front remains stationary. Unless this is so, deposition will be spread over a broad area, and no clearly defined ridge or line of detrital mounds will result. If the ice-margins are stable for a very long period, the end moraine may in exceptional circumstances build up to a height of some hundreds of feet, particularly in mountain valleys where the steep valley gradient, rapid ice flowage, and intense glacial erosion combine to deliver an immense load to the glacier snout. Alternatively, if the ice is affected by episodes of retreat, separated by stillstands, a number of smaller, sub-parallel ridges ('recessional' or 'stadial' moraines) will be formed. Sometimes the ice may advance over morainic material, bulldozing some of it into a low ridge. Such 'push moraines' rarely exceed 20–30 ft in height, simply because larger accumulations will cause the ice to ride up over and destroy them.

In studying the various types of end-moraine, it must be remembered that (i) they are liable to modification or destruction by meltwater torrents issuing from the ice-front, and (ii) they cannot be formed, by definition, at the edge of a stagnant ice-sheet. In fact, it is now known that, at their margins and/or where they have moved against a considerable gradient, ice-sheets commonly become nearly or actually stagnant.

The absence of end-moraines from many glaciated lowlands is thus readily explicable.

The form of end-moraines varies greatly, depending on the relative roles of ice and water in their deposition, and on the extent to which they have been dissected, either contemporaneously or subsequently. In an idealised form, a moraine due to deposition by ice alone will comprise a curvilinear ridge or line of low hills, representing the lobate margin of the glacier or ice-sheet. Some breaks might occur to allow the passage of meltwater streams from the base of the ice. In cross-profile the moraine should be roughly symmetrical, unless deposition of outwash material by streams has led to the formation of fans on the outer side. In this case, the moraine will be strongly asymmetrical, with a steep face on the ice-contact side and a long gentle slope away from the moraine crest on the opposite side. Some so-called 'end-moraines' are in fact composed entirely of stratified drift, and seem to have resulted from the confluence of a number of outwash fans or deltas formed in ponded water at the ice margins. The Cromer Ridge in Norfolk, which may be broadly classified as an end-moraine possibly related to a recessional stage of the Gipping glaciation, is composed at least in part of outwash materials. Where exposed at Briton's Lane, to the south of Sheringham, the gravels dip northwards towards the former ice-margins, and have been interpreted in terms either of build-up by meltwater against the ice-wall or as a delta formed in a lake ponded between the ice and the main ridge. Another interesting feature of part of the Cromer Ridge is its close dissection by networks of small valleys, resulting either from 'post-glacial' spring sapping or rapid erosion in an ensuing periglacial climate.

C *Ice-marginal features (excluding end-moraines).* A number of different fluvio-glacial features are found at or very close to the margins of an ice-sheet. The best known of these are kames, kame-terraces and eskers (Fig. 164).

Kames are mound-like hills of stratified drift (chiefly gravel and/or sand). They vary in dimensions, form and degree of stratification (for many have been disturbed by subsequent collapse), and indeed appear to have been formed in several different ways. Some are undoubtedly deltaic features, built up in small ice-marginal lakes by supra-glacial streams washing fine-grade material off the glacier surface. Others mark large crevasses, infilled by alluvium, which is left as a mound or ridge with the disappearance of the ice. A few are evidently developed where supra-glacial stream erosion has perforated the ice and concentrated the accumulation of gravel and sand on the underlying surface. It will be apparent that ice movement will modify or destroy such features, so that an abundance of kames (and indeed of kame-terraces and eskers too) will point to ice stagnation during deglaciation.

The variability of kame form and origin may be further illustrated by reference to Sissons's work on the deglaciation of part of east Lothian, Scotland (1958). The kames here, occurring mainly in groups in a belt at 550–750 ft O.D., following the northern edges of the Lammermuirs, fall into four main classes. These are:

(i) Groups of flattish-topped, roughly parallel mounds separated by glacial drainage channels. In some instances, these appear to have been produced by the dissection and fragmentation of kame-terraces.

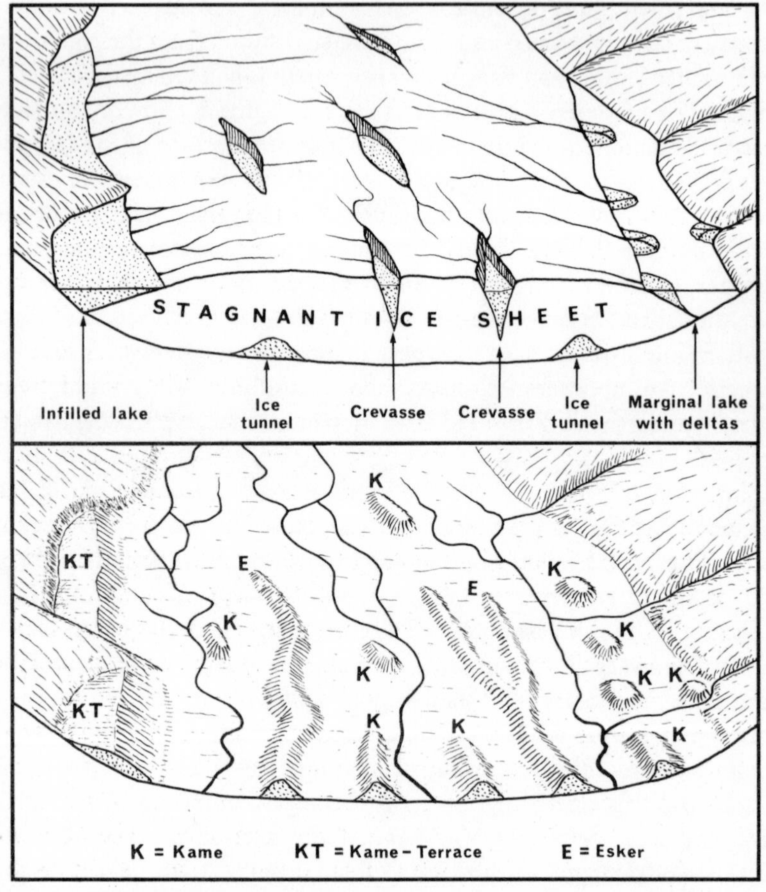

164 The formation of kames, kame-terraces and eskers (after R. F. Flint)

(ii) Groups of flattish-topped mounds occurring within a series of flat-floored, interconnected depressions up to 40 ft in depth and usually marshy. These are interpreted by Sissons as resulting from the decay of detached tongues of crevassed ice. The crevasse fillings have given rise to the humps, whereas the depressions mark the ice-blocks between the crevasses and are thus to be regarded as a species of kettle.
(iii) Chaotic assemblages of knolls, short ridges and depressions, which seem to represent an irregular accumulation of gravel and sand on thin ice, lowered to the ground with the melting out of that ice.
(iv) Kames which occur at the outer edges of kame-terraces. These rise a few feet above the greater level of the terrace, and evidently mark points at which meltwater streams emerged from the ice or flowed off its surface.

Kame-terraces are more continuous fluvio-glacial deposits laid down along the ice-margins. Sometimes they are narrow and discontinuous, for example where bedrock spurs projecting into the ice interfered with deposition or where subsequent dissection has occurred; in other instances, they are continuous over considerable distances. Sissons found that most of the kame-terraces of east Lothian were several hundreds of yards in length and a few tens of yards wide; the largest, however, is a mile long and 200 yards wide. The frontal slope of a kame-terrace (the 'ice-contact face') varies in angle, but is sometimes as high as 20° or more where 'post-glacial' slumping has been minimal. Some kame-terraces have been deposited by streams following the ice-margins, and receiving tributaries both from the 'ice' and 'ice-free' sides (Fig. 164). These terraces have an appreciable longitudinal gradient which can assist in the accurate reconstruction of past glacial drainage conditions. Other kame-terraces are horizontal, and have resulted from the coalescence of deltas formed in narrow ice-marginal lakes. The terraces of east Lothian seem largely to have originated in this way. In many glaciated areas there occurs a series of kame-terraces, one above the other. These mark successive margins of the decaying stagnant ice, and thus afford valuable evidence of main stages in the deglaciation of the area concerned.

Eskers are long and narrow ice-contact ridges, which are gently sinuous or even meandering in plan and sometimes divide into 'tributaries' and 'distributaries'. They are usually composed of gravel and sand which is horizontally bedded at the centre of the esker, but with pronounced dips at either margin of the cross-section. Eskers are best developed in areas of low relief and along valley courses, but occasionally run against the slope and surmount divides, usually by way of cols. Most authorities agree that eskers are the deposits of glacial streams, formed at the margins of or beneath the ice, and are usually associated—like kames and kame-terraces—with stagnant ice. Under conditions of deglaciation, the development of sub-glacial meltwater channels is at a maximum. The streams following these may transport a large load, particularly when under the influence of hydrostatic pressure, and at the exit from the ice, especially if this is to a lake which impedes free flow, the channel may be choked by debris. Subsequently, deposition of material which cannot escape will proceed headwards until most or all of the tunnel is infilled. Downwasting of the ice will then either expose the esker as a surface ridge or promote its destruction by additional meltwater flow. Another possibility is that some eskers represent a series of confluent deltas, developed in a similar manner to kames at the margins of a continually retreating ice-sheet. In this case, the protection for the growing ridge is afforded not by the overlying ice but by the submergence of each fan as soon as it is developed.

However, there is no doubt that esker-like ridges can be formed in many ways, and that it is not always possible to make a real distinction between eskers and kames. The so-called 'Blakeney Esker', running for some 2 miles north-eastwards from the Glaven valley in Norfolk, has been attributed by Sparks and West (1964) to the infilling by a westward-flowing stream of a major crevasse, opened up where stagnating ice was under tension on the brow of a sub-glacial ridge. Again, Sissons has explained an esker near Haddington, east Lothian, in terms of deposition by a sub-glacial stream running *into* the ice from its margins; at its southern end the esker is actually related to a kame-terrace, so that it is clear that the same stream formed both features.

Outwash plains and fans. Beyond the glacial margins meltwater streams transport and deposit a large proportion—and in some instances virtually all—of the rock debris released by ice-melt (Fig. 157). The efficacy of the process can be seen in Iceland, where many of the broad valleys leading seawards from the central ice-cap are choked by great fans and spreads of outwash debris. An actual example is the mile-wide gravel-strewn floor of Kalfafellsdalur, which declines by 450 ft in only 10 miles. This unusually steep gradient—which is three times that of the Spey, one of the swiftest rivers in Britain—is a measure of the intense aggradation of the coarse outwash. This has been necessary to provide the braided and inefficient meltwater channels with the energy needed to move their great loads. Not all outwash plains have this character. The well preserved fans at Kelling and Salthouse, in north Norfolk, are much gentler. Indeed the large Kelling plain, in which there is a typical gradation of deposits away from the former ice-front at Kelling Heath (where there are 'cannon-shot gravels', comprising perfectly rolled large flints) to Edgefield Heath (where sands predominate), has a slope of only 10 ft per mile.

Most outwash plains and fans related to past glaciation are not in a good state of preservation. For one thing, lowering of the ice surface by ablation causes the streams running off it to work to a continually lower level, with the result that earlier outwash deposits are incised. For another, the ending of glacial conditions, and a reduction in the amount of load fed into streams, will cause underloading and an excess of stream energy which will be used for downcutting. This is excellently illustrated on a small scale in the Val d'Hérens between Les Haudères and Evolène (Fig. 165). The valley floor here is occupied by a considerable thickness of moraine and outwash, supplemented by alluvium and scree from the valley walls. The 'post-glacial' Borgne has cut into these deposits to form a magnificent series of unpaired river terraces. On a larger scale are the fluvio-glacial terraces (the Low and High Terraces, and the Younger and Older Deckenschotter) of the Iller and Lech valleys in western Bavaria. Each of these terraces, when traced upstream, comprises an increasing proportion of unstratified drift and finally is composed of pure moraine. From this it has been inferred that the terraces represent important phases of glacial outwash and aggradation associated with the four main glacial advances (the Gunz, Mindel, Riss and Wurm) known to have occurred in the Alpine region. The preservation of the terraces well above the present rivers has been determined (i) by a prolonged uplift of the area throughout the Quaternary, and (ii) by effective stream incision during interglacial periods.

GLACIAL MODIFICATION OF DRAINAGE

In most areas which experience glaciation, the pre-existing pattern of rivers and valleys will not be totally obliterated or even radically altered. In glaciated uplands, glacier movement will in fact be largely guided by the available 'channels', and only where these are inadequate will the glaciers create wholly new courses through previous interfluves and watersheds (pp. 373–4). In lowland areas, on the other hand, the massive ice-sheets may be less influenced by the pre-existing relief; indeed, in many parts of Britain the main direction of ice advance has been

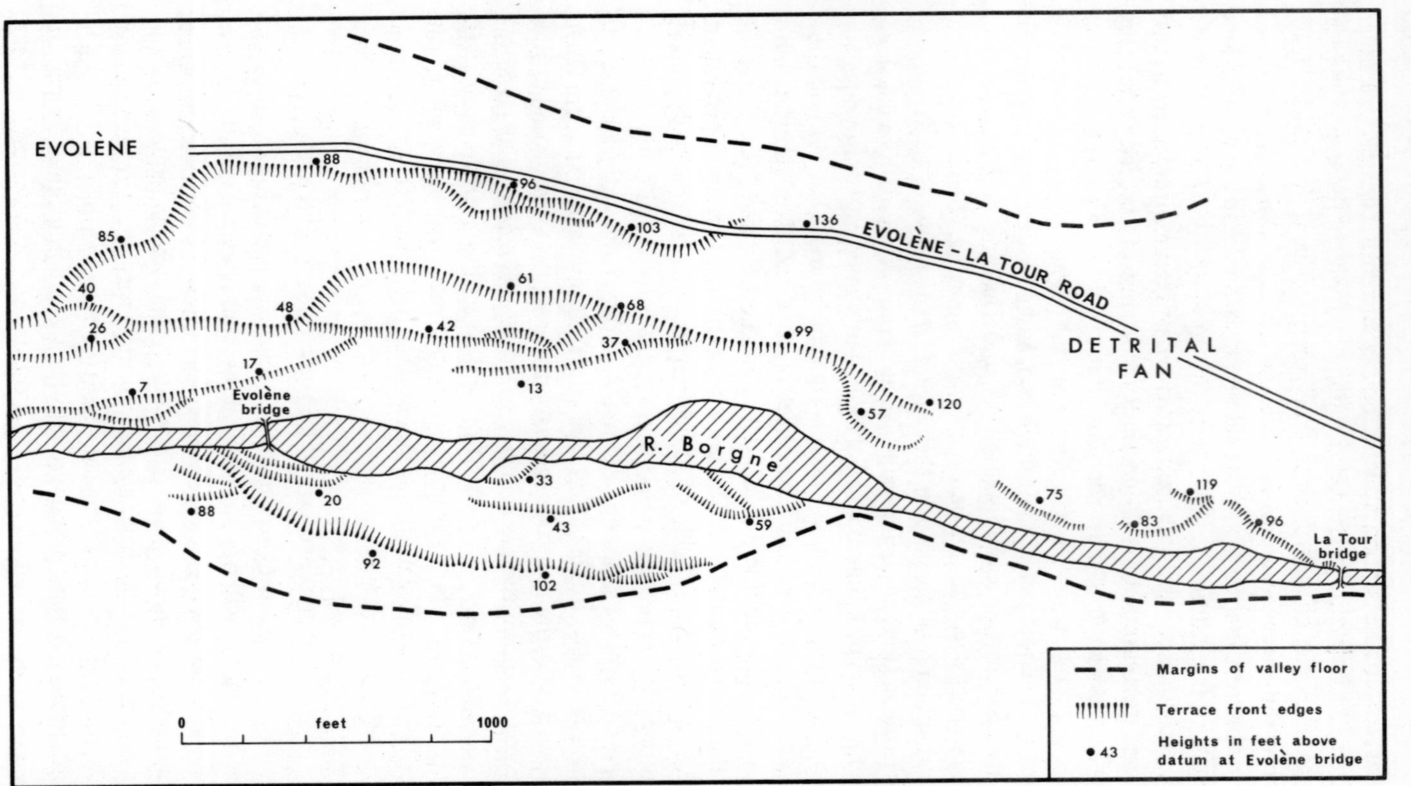

165 Glacial outwash terraces, Val d'Hérens

transverse to the stream valleys and divides. It is quite clear that, although Linton has argued otherwise, the old relief features have by no means been effaced. Rather they remained intact beneath the ice, and actually exerted a considerable influence on the configuration of the ice-sheet during the phase of deglaciation. Even so, there is a great deal of evidence to show that the 'pre-glacial' river pattern has been modified, both by the action of the vast quantities of meltwater present during retreat stages and also, in some instances, as a result of direct glacial erosion.

Pro-glacial drainage

The term 'pro-glacial', used above in connection with fluvio-glacial deposits, has been coined to describe the processes and features found at and close to the margins of a glacier or ice-sheet. It has been postulated that, providing conditions are favourable (and in particular where the movement of the ice is contrary to the slope of the land, so that 'natural' drainage from the ice-margins is impeded or blocked altogether), one of the most important phenomena developed here is the pro-glacial lake. This is a sometimes large body of fresh water fed both by streams draining the unglaciated slopes and by meltwater from the surface and edge of the ice. When melting is particularly intense, as during deglaciation, pro-glacial lakes will be at their largest and most numerous.

In this country, some of the earliest research into pro-glacial lakes and associated drainage phenomena was carried out by P. F. Kendall at the beginning of this century. His study of the Cleveland Hills of Yorkshire (Kendall, 1902) was for long regarded as a geomorphological classic, and the principles he derived were adopted and applied by many other writers (notably J. K. Charlesworth) in various parts of Britain and Ireland. As a basis for his reconstruction of pro-glacial drainage conditions Kendall used three main lines of evidence. These were: (i) the old strand-lines of lakes which have disappeared with the final melting of the ice-sheet, (ii) the deltaic and similar deposits laid down by contemporary streams entering the former lakes, and (iii) the steep-sided and often deeply entrenched 'glacial overflow channels' (or 'glacial spillways') marking points at which the waters of the lakes escaped, either into adjacent lakes or away from the pro-glacial zone altogether. The overflow channels are particularly important, in that as they were cut down the level of the lakes was lowered; at the same time, some of the smaller and less effective channels were abandoned, and their role assumed by the more successful and deeper channels, or even by new channels initiated in areas uncovered by the waning ice-sheet. In some instances, Kendall was also able to invoke as evidence lake-floor deposits. For example, the so-called 'warp' of the Vale of Pickering is composed of clear laminations, evidently corresponding to seasonal fluctuations in the volume and velocity of the meltwater streams feeding into the lake. This warp, together with other features, enabled Kendall to postulate the blocking of the Vale of Pickering at its eastern end by the North Sea ice and the formation of 'Lake Pickering', which drained south-westwards through the Howardian Hills by way of the great spillway, still followed by the river Derwent, of the Kirkham Abbey gorge.

In the Cleveland area as a whole, wave-cut notches as such are not well developed (in contrast with the famous case of Glen Roy, where the 'Parallel Roads' are strand-lines in a magnificent state of preservation), but accumulations of sand and gravel, interpreted by Kendall as deltaic deposits, exist at many points. However, the most reliable evidence was afforded by the overflow channels themselves, for these could be less easily modified or destroyed altogether by 'post-glacial' erosion. The principles on which Kendall interpreted the channels is shown in Fig. 166. This depicts a spur crossed by two spillways, A (with a floor at a maximum height of 475 ft O.D.) and B (with a floor height of 375 ft). The two channels can most satisfactorily be explained by assuming a lake with a level at 475 ft, draining through A at a time when B was not in existence—indeed, perhaps its site was occupied at the time by the ice. Later, owing to a recession of the ice-sheet and the lowering of its 'dam-like' front, the lake level clearly fell suddenly, channel A was abandoned, and a new exit at B, draining the lake at 375 ft, was opened up.

In detail, considerable differences were found by Kendall among the Cleveland channels: Firstly, there were the so-called 'direct overflows', formed where lakes impounded between the ice and a main watershed running broadly parallel to the ice-front spilled away over that watershed. The Kirkham Abbey gorge is a case in point, and also, in another area altogether, is the great spillway of the upper Severn at Ironbridge, created by water draining southwards from the former 'Lake Lapworth'. Secondly, there were the so-called 'lateral overflows', which carried water across minor interfluves between valleys running roughly at right angles to the ice-edge. In many instances such channels (which can be observed also on the dip-slopes of the Yorkshire and Lincolnshire Wolds, crossing the divides between east-running dry valleys) appear to have drained from one lake to another in adjacent valleys. Sometimes the channels were deeply cut, with a typical 'railway-cutting' form, but in other instances they are hardly more than shelves on the hill-side, owing to their development right up against the ice-front in such a way that they were partly eroded in rock and partly in ice. Lateral overflows occurred characteristically, in Kendall's interpretation, in groups as follows. On some spurs a number of channels was found, giving a 'parallel sequence' (Fig. 167A). Usually such channels were developed at successively lower elevations in a down-spur direction, and seemed to mark stages in the retreat of the ice-margin and the lowering of the pro-glacial lake level as new outlets were uncovered and deepened. In other localities, channels crossing several interfluves were found to give an 'aligned sequence' (Fig. 167B). Evidently these were eroded parallel to the ice-margins, in an area where several pro-glacial lakes were dammed

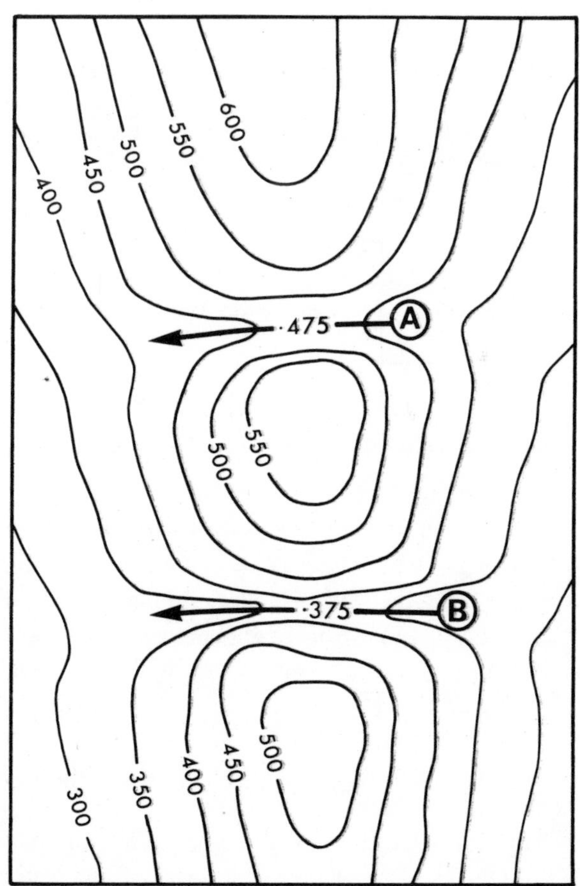

166 Glacial overflow channels

up at different levels. Ideally, in an aligned sequence the entry of each channel into a lake should be marked by a delta indicating clearly the level of that lake. This level should also be confirmed by the height of the exit channel on the far side of the lake.

From the evidence of overflow channels Kendall was able to infer not only the existence of 'Lake Pickering', as described, but also a series of pro-glacial lakes farther to the north, including the important 'Lake Eskdale' (supposedly dammed by ice entering the dale both at its western and eastern ends), which drained southwards across the main watershed to Lake Pickering by way of the great channel of Newton Dale.

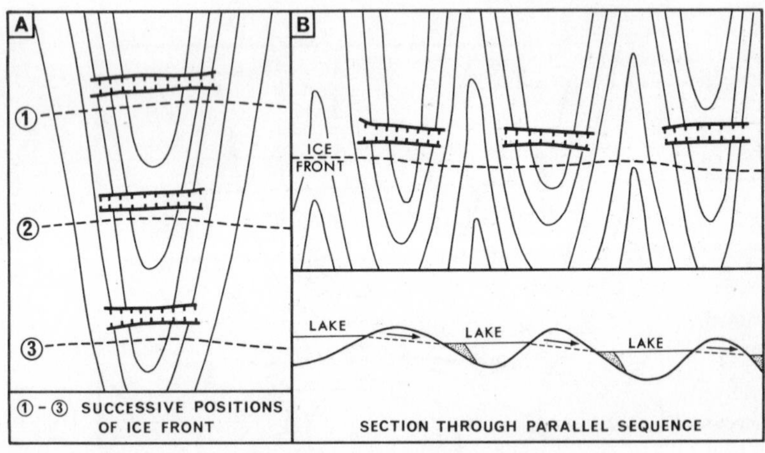

167 **Parallel and aligned sequences of overflow channels**

Sub-glacial, englacial and supra-glacial drainage

In recent years serious doubts have been cast on many of the inferences made by Kendall and others from glacial drainage channels. It is, in fact, now considered that many of these channels were not developed at or near the margins of the ice, in association with pro-glacial lakes, but by streams flowing under or within the ice or even initiated on the ice-surface. In 1949 Peel published the results of a detailed field survey of two glacial spillways (Beldon Cleugh and the East Dipton channel) in Northumberland. These had previously been studied by Dwerryhouse (1902), who had postulated the former existence in east and west Allan-dale and in the Devil's Water valley of a series of pro-glacial lakes and spillways. Peel revealed that the profiles of both channels sloped away in opposite directions from a high point (or in other words possessed an 'up-and-down' or humped form), that in Beldon Cleugh hollows containing peat up to a depth of 45 ft existed, and that the Dipton channel (draining from the Devil's Water valley) began at a level well below that of the lake assumed to have spilled through it. There seemed to be three possible explanations of the up-and-down profiles. Firstly, the 'back-slope' portions of the channels may be of secondary origin, having arisen through the incision of 'obsequent' streams since the disappearance of the lakes. However, this does not seem likely, as any existing back-draining streams are minute and do not seem capable of having cut what are often steep-walled and flat-floored trenches. Secondly, there may have been a reversal of overspill through the channels as a result of the successive creation of lakes, at different levels, on either side of a divide. However possible this may appear in theory, it is seen to be too much of a coincidence when it is known that the up-and-down form is not exceptional but rather the norm in glacial drainage channels. Thirdly, it is possible that the spillways were eroded by water able to flow up and down hill, from one pre-existing valley to another. It is conceivable that sub-glacial streams, under great hydrostatic pressure, could behave in this way. In 1949, Peel was inclined to regard this last hypothesis as 'quite untenable', but in a later paper (1956) modified his viewpoint in the light of recognition of actual sub-glacial channels of up-and-down form by the Scandinavian geomorphologist C. M. Mannerfelt.

Dwerryhouse's hypothesis of pro-glacial lakes and overflows in Northumberland was also challenged in 1958 by Sissons. Dwerryhouse had traced in some detail the supposed retreat stages of the ice damming up the lakes here. At the various stages the ice-margin was more or less straight in plan. However, in this area the valleys are up to 600 ft in depth, with the result that—as Sissons rightly emphasises—the ice would have been 600 ft thicker along the valley-lines than over the interfluves. During deglaciation, wasting back of the ice-margins would necessarily have been less effective in the valleys, so that the ice-edge would undergo a change of form (Fig. 168). In reality, downwasting of the ice by ablation would expose the interfluves at a comparatively early stage, and tongues of ice would be left in each valley. Thus Dwerryhouse's postulated parallel lines of retreat could never have existed. Indeed, even in the first instance the ice-margins were probably not straight in plan, for the simple reason that at the onset of glaciation the deep valleys would have been more easily penetrated by the ice.

Remapping of the channels in the field revealed that some tend to 'run round the hillsides'. These are presumably the true marginal channels developed when, during deglaciation, the valleys had become occupied by separate ice-tongues. The remaining channels, cutting across the interfluves, must have been developed in a totally different manner. The complete absence from this part of Northumberland of either lake shorelines or lake-bottom deposits also militates against Dwerryhouse's hypothesis, and tends to confirm Sisson's view that channels such as Beldon Cleugh were formed sub-glacially at an earlier stage of deglaciation, prior to the exposure of the interfluves by ice wasting. Finally, the remarkable lining-up of certain channels breaching divides and existing stream valleys (for example, the Rowley Burn, tributary to the Devil's Water, and the Strothers Dale channel) seems to add support to the sub-glacial hypothesis.

Evidence has accumulated from elsewhere to indicate the importance of sub-glacial streams. Sub-glacial valleys have been traced in Denmark ('tunnel valleys') and northern Germany ('Rinnentaler') which are sometimes incised to a depth of 100 metres below the level of the surrounding glacial landscape. As Schou has written:

In tunnels which they formed under the ice, the enormous masses of (melt-water) moved with such great erosive force that they were able to cut the

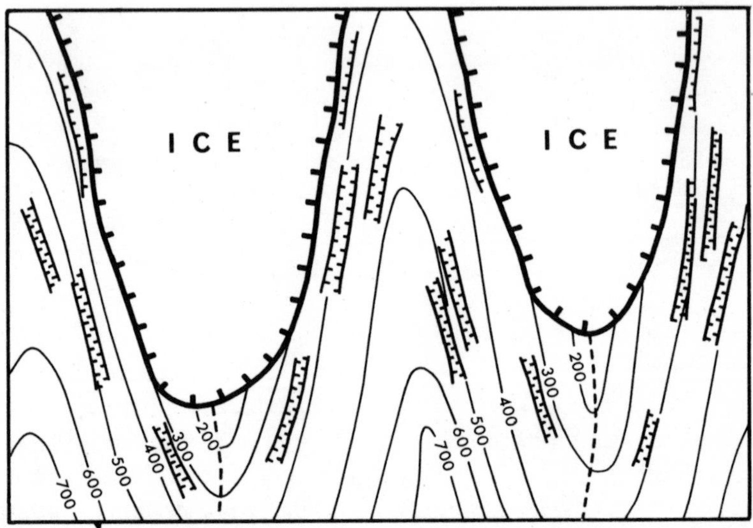

168 The form of a decaying ice-sheet and associated marginal drainage channels in an area of strong relief

tremendous sub-glacial gutters which now appear in the form of large valley systems or transgressed by the sea as in the case of the East Jutland fiords. The great thickness of the ice-cap and the resulting high water-table in the ice subjected the meltwater in the tunnels to high pressure, so that locally within these enclosed ducts it ran uphill.

Other evidence of sub-glacial streams with the ability to ignore the sub-glacial relief is afforded by eskers, the casts of such streams which for some reason have not effected erosion, which climb hill-sides and descend the far slope (p. 392).

A great deal of research during the last ten years has confirmed the dominance of sub-glacial over pro-glacial erosion in many areas. The reinterpretation of the glacial drainage system near Fishguard, Pembrokeshire, by Bowen and Gregory (1965) is an illustration of this (Fig. 169). It is clear, however, that the channels here were developed not only by meltwater initiated in tunnels at the base of the ice, but also by streams flowing in the first instance within the ice. As the ice thinned

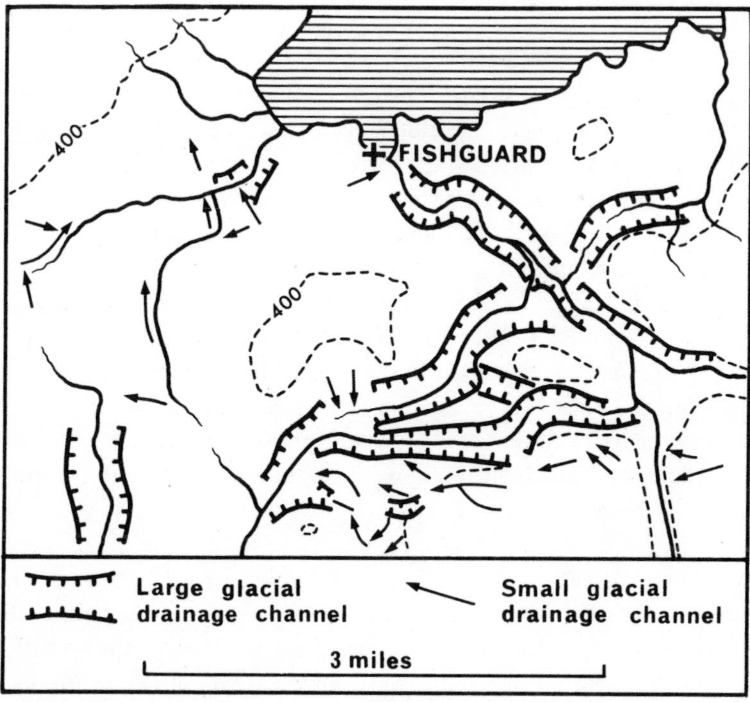

169 The pattern of glacial drainage channels near Fishguard (after K. J. Gregory and D. Q. Bowen)

during deglaciation, these streams were let down on to the underlying rock surface, where they formed discontinuous channels. The hypothesis of superimposition of englacial and even supra-glacial streams, formed by melting on the surface of the ice, has been adopted by many writers. There are, nevertheless, certain difficulties confronting the acceptance of this mechanism. One would expect superimposition of the glacial drainage system to result in many channels having a high degree of discordance to the sub-glacial relief. In reality, in most areas glacial channels cross interfluves by way of 'pre-glacial' cols (hence the entensive use of the term 'sub-glacial col gully'). Attempts have been made to explain this phenomenon in one of two ways. Those supra-glacial and englacial streams coinciding with cols are able to cut down through the ice (which is less resistant than solid rock) with greater speed than their neighbours. The water in the latter is therefore steadily abstracted by percolation through crevasses, and meltwater flow above the col is increasingly concentrated. Secondly, glacial streams which happen to be superimposed on to the slopes above a col will undergo 'uniclinal' shifting along the ice-rock contact towards the centre of the col.

Direct glacial modification of drainage

The role of diffluent and transfluent ice in the breaching of pre-glacial divides, to give cols and even deep troughs, is discussed briefly on pp. 373–4. There is no doubt that this process has operated in all the major upland areas of the British Isles affected by the Quaternary glaciations, and that in many instances the formation of the breaches was accompanied by permanent modification of the previous drainage system. Dury (1959) has identified twenty-two glacial breaches in the mountains of Donegal, north-west Ireland, and many significant displacements of the pre-glacial watershed have been effected. The high mountain ridge between Errigal Mountain and Muckish Mountain has been breached at three points by transfluent ice (Fig. 170) moving from the south-east across the Calabber valley. The Muckish Gap, immediately south-west of Muckish, has been deepened by some 400 ft, and the old watershed displaced about half a mile towards the Calabber. A valley head between Muckish and Aghla More has been converted into the remarkable high-level basin occupied by Lough Aluirg; this again lies to the south-east of the former divide. Finally, and most impressive of all, a pre-glacial col between Aghla More and Errigal has been gouged out to a depth approaching 500 ft, and is now occupied by the Altan Lough, draining northwards via the Tullaghobegly river to the coast. Into the head of the Lough now runs a system of small streams, draining the flanks of Dooish, which once formed the headwaters of the Calabber. In short, the displacement of the pre-glacial watershed here has been in the order of 1½ miles.

Examples of drainage diversion on such a local scale could be multiplied many times. However, it is relevant to comment briefly on a much more significant diversion, caused by glacial breaching and now regarded almost as a classic case, in the Scottish Highlands. The river Feshie, after running almost due east in its upper course as if to join the river Geldie, turns at an acute angle towards the west-north-west and then to the north, running through a deep glacial trough to the river Spey. Most authorities have regarded this as a simple case of piracy, in which the Feshie has taken the head off the Geldie to give a spectacular elbow of capture. However, as Linton (1949) has pointed out, two major

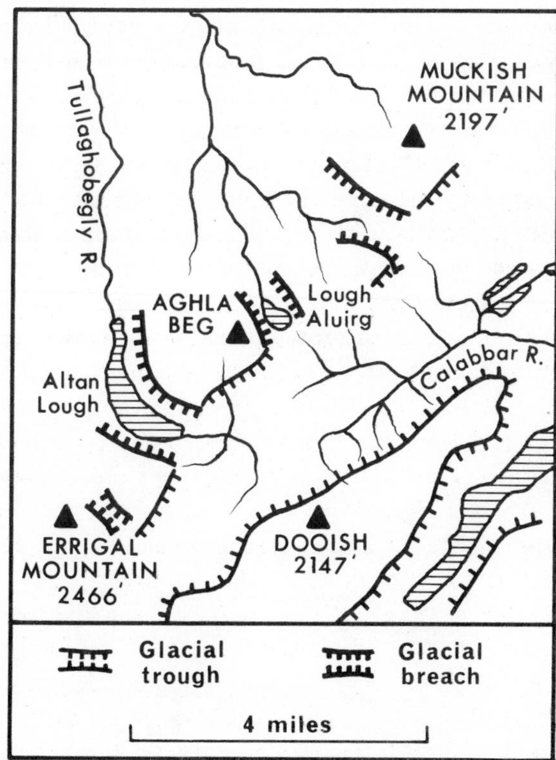

170 Glacial watershed breaching and drainage diversion, western Donegal (after G. H. Dury)

difficulties face this explanation. Firstly, there seems to be no line of geological weakness to account for the headward erosion, allied to exceptional valley deepening, of the Feshie. Secondly, the Feshie is not actually incised at the point of the 'capture', as it would have to be for the capture to occur (pp. 236–9). A special explanation of the second feature was proposed by A. Bremner. This involved assuming (i) a pre-glacial capture, (ii) the obliteration of all the normal evidence of capture during an ensuing period of glacial erosion, and (iii) a late-glacial blocking of the lower Geldie by morainic deposits in such a way that melt-water from a glacier in the present-day upper Feshie was diverted along the course of the ancient capture.

Linton, however, has convincingly reinterpreted the situation in terms of glacial diffluence. The exit of ice from the head of the Feshie-Geldie valley towards the east would have been effectively barred by glaciers moving southwards off the Cairngorms (in particular by way of the Dee valley) and northwards from An Socach, so that a basin of impeded flow would be formed. Eventually, this overspilled towards the west-north-west, probably taking advantage of a col at the head of the lower Feshie valley, and formed the Feshie trough. However, it is clear that this trough was not deepened sufficiently at its very head to cause an immediate post-glacial diversion of the drainage. At present, a fan of detritus laid down by the upper Feshie occupies a site immediately to the west of the 'elbow of capture'. The river itself has undoubtedly shifted its course over this fan, at times running eastwards down the Geldie valley, and at others into the Feshie trough. However, it now seems to have found a permanent outlet by the easiest and steepest route to the west-north-west.

12

COASTAL LANDFORMS

INTRODUCTION

Coastal landforms are of exceptional geomorphological interest for several reasons. Perhaps the most obvious is the comparative rapidity with which coastal features are changed. Few beaches remain constant, either in plan, profile or composition, for more than a short period. In response to continuously changing tidal, wave and wind conditions they are built up, combed down or even disappear temporarily. Cliffs, especially where developed in incoherent sediments, experience sudden slumps and falls, which are most noteworthy in built-up areas where roads and buildings may be severely damaged or destroyed in a few moments. As well as these day-to-day happenings, more insidious but in the long run even more important changes are wrought. The growth of large spits across estuaries and the formation of mud-banks and salt-marshes within sheltered inlets are not readily visible as processes, yet in many instances they can be shown to have occurred within little more than a century. Furthermore, long-continued but almost imperceptible cliff recession at moderate rates (less than a foot per annum) in time may threaten valuable land and property, so that the halting of such erosion becomes a pressing problem for local authorities—if rather less so than in the more catastrophic situations described above.

In fact, the coastline provides the geomorphologist with an almost unequalled opportunity for the study of *active* processes. The breaking of waves and the resultant movement of material up and down the beach; the transportation of pebbles and sand along the shore and over the sea-bed; the development of 'blow-outs' in sand-dunes; the gullying of clay cliffs by spring- and rainwater—all these, and many other processes, operate at a speed which renders observation, measurement and analysis a far more feasible task than, say, in the case of the weathering of granite or the movement of soil by creep on a slope.

In the light of this, the application of the concept of dynamic equilibrium (pp. 189–93) would seem to be more justified in the coastal field than in certain others. Features such as beaches, spits, bars and even some cliffs are without doubt the product of currently acting processes, and their forms really seem to represent a changing condition of balance between several controlling factors (aspect, weather conditions, wave action, tides, currents, supply of beach material, composition of cliffs and so on). Yet even in coastal geomorphology important examples of 'in-equilibrium' are often found. As will be seen, many features of the British coastline (particularly the cliffs and wave-cut platforms in resistant rocks) are the result of processes and/or conditions which have since been changed, mainly as a result of climatic amelioration or movement of sea-level in post-glacial times. In coastal study, therefore, the historical legacy may demand as much attention as present-day processes and their effects.

GENERAL FACTORS IN COASTAL DEVELOPMENT

The processes of wave erosion, transportation and deposition (together with allied aeolian processes, weathering, mass movements and the action of running water), which are to be discussed in detail below, play a major part in fashioning the detailed forms of any coastline. However, other more general factors need to be considered, for these determine the broader (and sometimes even the narrower) setting within which these processes function.

The trend of the coast

The trend of a coast is a fundamental factor which may influence the rate and location of erosion, the direction of longshore drifting, the formation of beaches, spits and bars, and the situation of large dune-areas. The key relationship is that between the orientation of the coastline, the direction of maximum fetch, and the direction of the prevalent and/or dominant winds. In southern England, the greatest fetch is to the south-west, and this, with the aid of the winds blowing from that direction, causes an overall drift of beach material from west to east. Individual beaches and spits clearly reflect this longshore movement. Exceptions occur (such as the spit at Sandbanks, Bournemouth), and these evidently require a special explanation (p. 459). Another striking illustration of the effect of trend is afforded by the East Anglian coast. Here the maximum fetch is towards the north and north-east, and whereas on the north Norfolk coast west of Sheringham the drift is inferred by Steers (1946) to be east–west (as indicated by Blakeney Point and similar complex beach structures), to the east of the town the movement of beach material is apparently south-eastwards. An interesting implication of this is the constant subtraction of sand and shingle from the vicinity of Sheringham itself. In fact, recent research has shown that this division of beach drifting is not actually occurring at the present time.

The influence of coastal trend on erosion can be considerable, both in a direct and indirect sense. Where the coastline turns to face at right angles the direction of dominant wave attack, erosion should in theory be more rapid simply as a result of the expenditure of maximum wave energy in a narrow zone. The ferocity of wave attack on south-west-facing coasts in Britain is an illustration of the reality of this factor. Indeed, it is fortunate that these exposed coasts are backed for the most part by extremely resistant Palaeozoic rocks, so that paradoxically the areas experiencing the most rapid erosion (p. 442) are those in comparatively sheltered but much weaker rocks farther to the east. The indirect effects are best seen where the accumulation of excessive beach material either naturally (on the windward side of a headland or estuary) or because of man's interference (through the construction of harbour-walls and groynes) leads to an accentuation of erosion to the leeward of the obstacles (Fig. 171) (p. 434).

171 The harbour mouth at West Bay, Dorset. Note the accumulation of beach shingle to the east of the harbour-walls (see p. 434) and the severe erosion of the weak clay cliffs to the west. [Eric Kay]

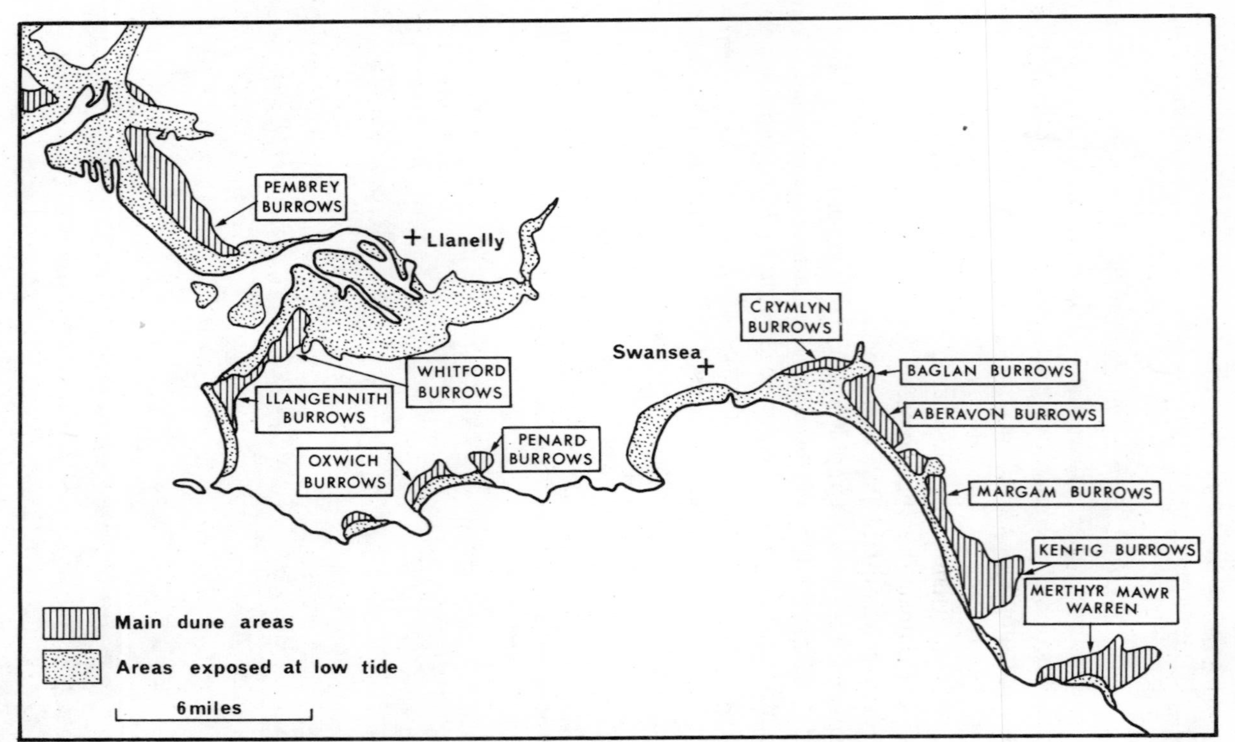

172 The sand-dune areas of South Wales

It must be emphasised, however, that exposed parts of the coast are sometimes the scene of considerable deposition. In south Wales, for example, although much of the coastline is cliffed (as in Pembrokeshire and Gower) broad expanses of sand have accumulated in estuaries and bays, even those facing the dominant winds and waves from the south-west (Fig. 172). When revealed at low tide these are affected by wind action, and in the recent past vast quantities of sand have been driven onshore to form expanses of sand-dunes, sometimes stretching up to a height of well over 100 ft O.D. Among the more important dune-areas are those at Freshwater Bay West, Towyn and Pembrey Burrows, Llangenydd Burrows in Gower, Kenfig Burrows and Merthyr Mawr Warren (Fig. 173).

Offshore gradient

This may influence coastal development both on a local and regional scale. Where the sea-floor declines gently away from the shoreline, much of the energy of the approaching waves, particularly when these are large, is dissipated by friction and wave break and resultant erosion is diminished. On the other hand, a steep offshore gradient allows far more powerful wave attack, except in the case of plunging cliffs (p. 438) where reflection rather than breaking of waves may result. In this we have the explanation of the widely observed phenomenon of shoreline straightening or 'regularisation', whereby wave erosion is concentrated on headlands and is much reduced, or even replaced by aggradation, at the heads of bays. However, such regularisation is by no means an inevitable process on all embayed shores, for many headlands are composed of very durable rocks, whereas most bays are hollowed from weaker rocks (pp. 412–18), often with the aid of stream erosion. There are indeed two conflicting tendencies: for wave action, as described, to straighten the coast, and for erosion, operating differentially on contrasting rock-types, to accentuate the existing pattern of bays and headlands. Which of these will become dominant must depend on the subtle interplay of all the factors involved in a particular area.

Even within individual bays the distribution of erosion is sometimes complex. Thus within Christchurch Bay, Hampshire, maximum erosion is concentrated at Barton-on-Sea; here the offshore gradient is appreciably steeper than a mile or two to the east, where the cliff-foot is protected by a broad shingle beach and the cliffs themselves are becoming dead. However, it is not certain that sea-floor gradient is the only factor to be reckoned with, for it so happens that the geological structure

173 Sand-dunes colonised by marram grass (*Ammophila arenaria*), Llangenydd Burrows, south Wales. [Eric Kay]

of the cliffs at Barton is itself very favourable to sub-aerial erosion and mass slumping (p. 442 and Fig. 174).

On an altogether larger scale, the offshore gradient may help to determine whether or not the coastal features as a whole are primarily erosional or depositional. For example, the abundance of constructional shore features in north Norfolk between Hunstanton and Weybourne (including the small offshore bars at Thornham, the large sand and shingle complex of Scolt Head Island, the Lodge Marsh foreland at Wells, and the Blakeney Point spit) is undoubtedly related to the shallowness of the offshore zone. This may in turn be attributed partly to extensive glacial deposition in the Quaternary and partly to the 'groyne effect' of the north Norfolk coast, which intercepts beach material being drifted southwards along the Lincolnshire coast. In post-glacial times the rising sea spread slowly across what is now the gently shelving off-shore zone. The larger waves tended to break well away from high-water mark, eroding the sea-floor and throwing up bars, which eventually became stabilised and lengthened to enclose lagoons on their landward sides. As the sea-level continued to rise, these bars were driven inwards, at the same time as the accumulation of mud in the protected lagoons led to the formation of salt-marshes. Both the features and the sequence of development of the north Norfolk coast are typically those of a so-called 'emergent' coast, despite the fact that the latest and most important event has been the post-glacial submergence.

Geological structure

Geological structure (including here rock-type) is arguably the most important single factor determining both the coastal outline in general and the detailed configuration of individual bays and headlands. The larger-scale effect is shown by the coastline of Wales. In the north, the north-east to south-west trend of the Lleyn peninsula is obviously determined by the Caledonian folding (notably by the major downfold

174 The cliff at Barton-on-Sea, Hampshire. Note the inner cliff of Barton Sand capped by plateau-gravel, and the undercliff of Barton Clay and slipped masses of sand and gravel. A major cause of erosion is the water issuing at the base of the permeable Barton Sand. Recent schemes to drain the water have led to the transformation of this particular section. [M. J. Clark]

of the Snowdon syncline); in the centre, the broad south-westerly sweep of Cardigan Bay is parallel to the axis of the Teifi anticlinorium, lying some 10 to 15 miles inland; and in the south, the west–east trend of the Pembrokeshire peninsula clearly reflects the continuation of the powerful Hercynian folds of the south Wales coalfield. Furthermore, the great inlet of the Bristol Channel, separating the uplands of south Wales and north Devon and Somerset, marks the site of a great structural and erosional depression, developed in Hercynian and early-Mesozoic times, fossilised by Permo-Triassic and younger sediments, and more recently exhumed by a combination of sub-aerial and marine processes. Again, the broadly west–east orientation of the south coast of England between Weymouth and Eastbourne appears to be a reflection of the dominant trend of the Alpine folds of the area. Such cases of large-scale structural control could be multiplied many times, but it is of more interest to examine the detailed influence of structure on coastal configuration, for no other erosional agency seems quite so sensitive as waves to slight variations of rock hardness or to the presence or absence of lines of weakness such as joints, faults and shatter-belts. This can be shown by contrasting a short stretch of cliffed coast, which will usually be characterised by an intricate series of minor headlands, irregular bays and inlets, stacks, caves and geos, with the more 'generalised' and smoothed-over relief features supported by the same geological structure a short distance inland.

The most immediate influence of rock structure may be seen in those areas where the structural grain is at right angles to the coast. A classic example is afforded by west Pembrokeshire, where the main headlands (St David's Head and St Ann's Head) are formed respectively by (i) varied igneous rocks, including ancient pre-Cambrian volcanic tuffs and younger basic lava intrusions within Ordovician sediments, and (ii) a large mass of Old Red Sandstone and the Skomer Volcanic Series of the Ordovician (Fig. 175). The intervening St Bride's Bay is eroded from a faulted synclinorium of weak Carboniferous rocks (mainly shales, with some sandstones, of the Coal Measures). However, so complex is the geological structure of Pembrokeshire that, superimposed on the overall pattern of headland and bay, is an almost infinite variety of small-scale coastal forms. On the north side of the St David's headland a series of small promontories is formed by the individual small Ordovician intrusions (for example, St David's Head proper, Pen Llechwen, Pen Clegyr, Pen Bwch Du and Strumble Head), and intervening bays (such as Aber Eiddy Bay, Traeth Llynfu and Porth Eger) are etched from

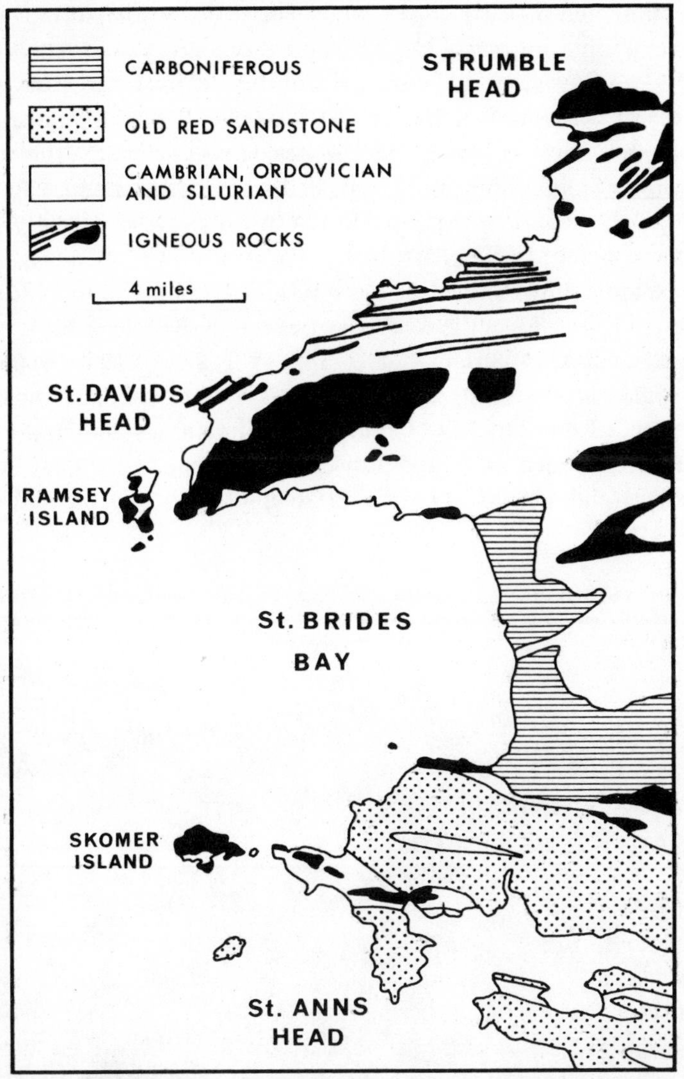

175 **Geological structure and coastal configuration, west Pembrokeshire**

unresistant shales. Even within St Bride's Bay itself the coastal features demonstrate admirably the differential erosion of the Carboniferous rocks. Most of the small headlands (for example, Settling Nose, Haroldston Chins and Rickets Head) are formed by sandstones, whereas the numerous small re-entrants mark the outcrops of easily eroded shales. A specially interesting feature is Druidston Haven, a bay developed in a faulted anticline which brings Ordovician shales to the surface over a distance of some 400 yards. On the larger promontory of St Ann's Head, the Old Red Sandstone cliffs show a considerable irregularity of outline, particularly between Mill Haven and Musselwick Bay. The many tiny inlets do not, however, mark only the weaker rocks but are often eroded along minor faults. Such fault-controlled erosion is even more apparent on the north side of the small peninsula (mainly of Ordovician volcanic rocks) leading out to Woollack Point (Fig. 176).

An example showing the features typically developed where the structural grain is parallel to the coast may be seen in the Isle of Purbeck and its western continuations, in Dorset (Fig. 177). The rocks here comprise Upper Jurassic formations (Kimmeridge Clay, Portland Sand and Stone, and Purbeck Limestone) and Cretaceous sediments (including the much attenuated Wealden Beds and Lower Greensand, Gault Clay, Upper Greensand and Chalk). The dip is steeply to the north, and slight

176 Cliffs in Silurian rocks below the Deer Park, St Ann's Head, Pembrokeshire. Note the differential erosion, producing a series of closely spaced caves, in the foreground. [Eric Kay]

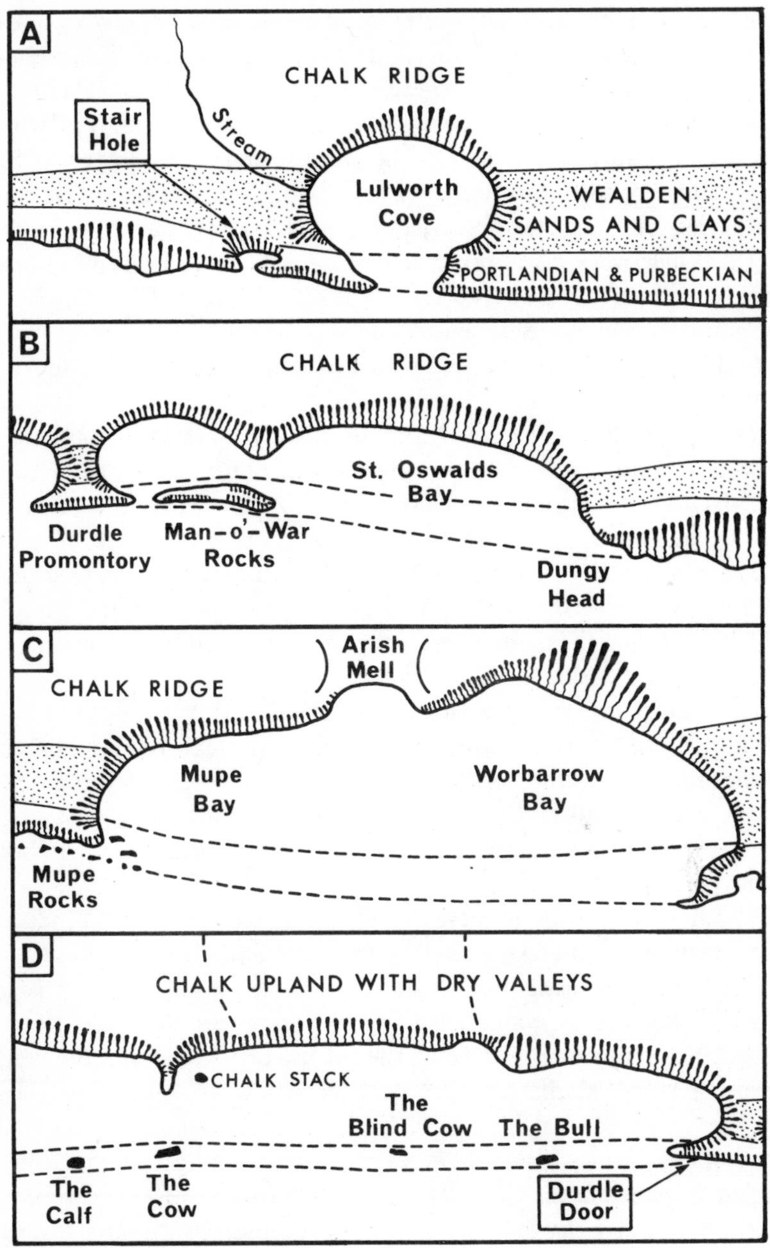

177 The development of coastal features near Lulworth, Dorset

overfolding occurs near Durdle Door. The strike is broadly west–east and the individual rock outcrops are narrow, except towards Kimmeridge Bay and farther to the east where the angle of dip is much reduced close to the crest of the Purbeck anticline (the southern limb of which has been destroyed by erosion). The Jurassic and Cretaceous rocks may be roughly grouped into four divisions, which are alternately resistant and unresistant both to sub-aerial erosion and wave attack.

The Kimmeridge Clay, preserved between Gad Cliff and Chapmans Pool, is generally unresistant, though it forms steep cliffs locally (as to the south of Swyre Head). Farther to the west the formation has been entirely removed by wave action which has exposed at the coast the next group of rocks.

The Portlandian and Purbeckian beds, mainly limestones, combine to form a discontinuous west-east ridge, which can be traced from Kimmeridge (where it is a mile inland) past Gad Cliff (where it adjoins the sea) to the Durdle promontory.

The Wealden Beds, largely weak clays and sands, give rise to a prominent strike vale in the Isle of Purbeck, around Tyneham and Steeple, which is much reduced in width westwards owing to the intensification of folding in that direction.

The Chalk forms a hog's-back in the Isle of Purbeck (for example, at Povington Hill); this is continued westwards, through Bindon Hill and Hambury Tout, to the north of which a more low-lying area of Chalk represents a structural depression, the so-called 'Purbeck foresyncline', from which a former infilling of weak Tertiary sediments has been removed.

The attack of the sea on these west–east ridges and vales has produced many features of interest, especially those resulting from the breaching of the Upper Jurassic barrier and the exposure of the unresistant Wealden Beds. It is evident that in some instances chance has determined the precise points at which the sea has penetrated, but the major breaks at Mupe-Worbarrow Bay and Lulworth Cove seem to have been initiated by past fluvial erosion. Aided by periods of rising sea-level, the waves have taken advantage of existing gaps cut (a) by the former stream responsible for the deep dry pass through the Chalk ridge at Arish Mell, and (b) by the small stream still draining southwards through West Lulworth into the Cove.

The various stages in the breaching process are represented as follows:

At Stair Hole, West Lulworth, the sea is at present beginning to break down the Portland Stone and Purbeck Limestone barrier. Small gaps and caves, probably marking joints, have been opened, and rapid attack on the soft Wealden clays, aided by sub-aerial gullying and slumping, has begun.

Lulworth Cove itself is a symmetrical, near-circular bay which has been hollowed from the Cretaceous sands and clays. At the present time, the Cove is being extended mainly westwards and eastwards, along the strike. On the inner margin the Chalk forms high cliffs, the erosion of which, although active, is less rapid.

St Oswald's Bay is an elongated double bay which has undoubtedly been formed by the amalgamation of at least two former coves comparable in origin, if not in size, with Lulworth Cove. The 'limestone barrier' has been almost destroyed between Dungy Head and the Durdle promontory, and is represented only by the elongated Man-o'-War Rocks (Fig. 178).

The much larger re-entrant of *Mupe and Worbarrow Bays* may be either a single cove which has been greatly enlarged by long-continued erosion of the Wealden clays and sands, or a series of coves which have gradually coalesced, as at St Oswald's Bay.

178 The Durdle promontory, Dorset, from St Oswald's Bay. The sea has broken through steeply dipping Purbeck Limestone to attack weak Wealden sands and clays (forming the central part of the promontory) and the Chalk (planed remnants of which are visible in the foreground). [R. J. Small]

Between the *Durdle promontory* and *Bat's Head* yet another, if in plan much more shallow, bay may again represent a number of former coves. The Jurassic limestones have been removed altogether, except for small isolated rocks exposed at low tide (the Bull and Calf Rocks) and at the fine natural arch of Durdle Door itself. Erosion of the Durdle promontory is active, both from the west and the east, and the isolation of this remaining part of the limestone barrier is a matter of time. Between Durdle Door and Bat's Head severe attack on the Chalk cliffs has begun. These are near-vertical, and are affected by frequent large falls of rock. The extent to which Bat's Head projects is a measure of the cliff retreat which has already occurred. That the erosion is differential is indicated by the fine stack at the western end of the Bay, the cave through Bat's Head (Bat's Hole), the series of caves marking a thrust fault at the base of the cliffs near Durdle Door, and the small individual bays cut into the floors of the dry valleys which hang above the beach.

Finally, at *Arish Mell* the small but pronounced bay at the dry gap through the Chalk ridge is an early stage in the breaching of the next resistant barrier. The later stages of the process cannot be seen in the Purbeck area, but are represented by the destruction of the former Chalk upland stretching from the Foreland at Swanage to the Needles of the Isle of Wight (pp. 419–21).

Inland morphology

The relief features of the area lying immediately inland can exert some measure of control over coastal forms. To take a simple example, where the sea is eroding an upland dissected by valleys running at right angles to the coast the resultant cliffs will be high and imposing at the terminations of interfluves and low or non-existent at the mouths of valleys. Once cliffs of variable height have been formed, rates of recession should —at least in theory—be directly affected. Other things being equal, a high cliff will retreat more slowly than a low cliff, simply because for a given amount of recession to be effected a greater volume of debris must be removed by the waves at the foot of the high cliff. The outcome should be an embayed coastline, with bays coincident with valley mouths and headlands with interfluves. The tendency may be accentuated by the effects of stream erosion (which can materially assist in the excavation of the bays at valley mouths) and by geological structure (for the valleys may themselves follow lines of weakness that can also be utilised by wave erosion). In reality coastal evolution does not always follow this sequence, as is shown by the Seven Sisters cliffs between Seaford and Eastbourne

179 The Seven Sisters cliffs, Sussex. Note the truncation of dip-slope dry valleys and the regularity of the coastline (see p. 409) [Eric Kay]

in Sussex (Fig. 179). The initial conditions referred to above are admirably fulfilled here, yet the cliff-line is remarkably straight in plan, with only a suspicion of bay formation opposite one or two of the hanging dry valleys; in other words the high cliffs are retreating as rapidly as the low ones. Presumably the relief factor is overridden by the absence of stream erosion and the homogeneity of the Chalk. Another relevant factor must be the tendency to regularisation of the coastline already described.

The effects of the relief and drainage pattern may be further exemplified by the coastal configuration of Hampshire and the Isle of Wight. Structurally this area is dominated by a large syncline, the Hampshire Basin, and a series of monoclines which raise Chalk and older rocks to the surface in the Isles of Purbeck and Wight. There is convincing evidence, in the form of the existing drainage pattern and remnants of gravel terraces having a gentle slope from west to east, of the former presence of a major synclinal stream, usually referred to as the Solent river (Fig. 180). This once crossed what are now Poole and Christchurch Bays and passed between the Isle of Wight and the mainland. This trunk stream was joined by tributaries both from the north (such as the Avon, Test and Itchen) and the south (the western Yar, Medina and eastern Yar, together with others of which trace has been lost).

During the Quaternary the Solent system must have been greatly affected by many changes of sea-level (pp. 421–7). During glacial periods both the main stream and the tributaries would have incised their courses deeply in response to the falling base-level, whereas in interglacials the resultant valleys were inundated by the sea to give branching estuaries. At the same time the resistant barrier to the south (particularly that of the Chalk ridge developed on the northern flanks of the Purbeck-

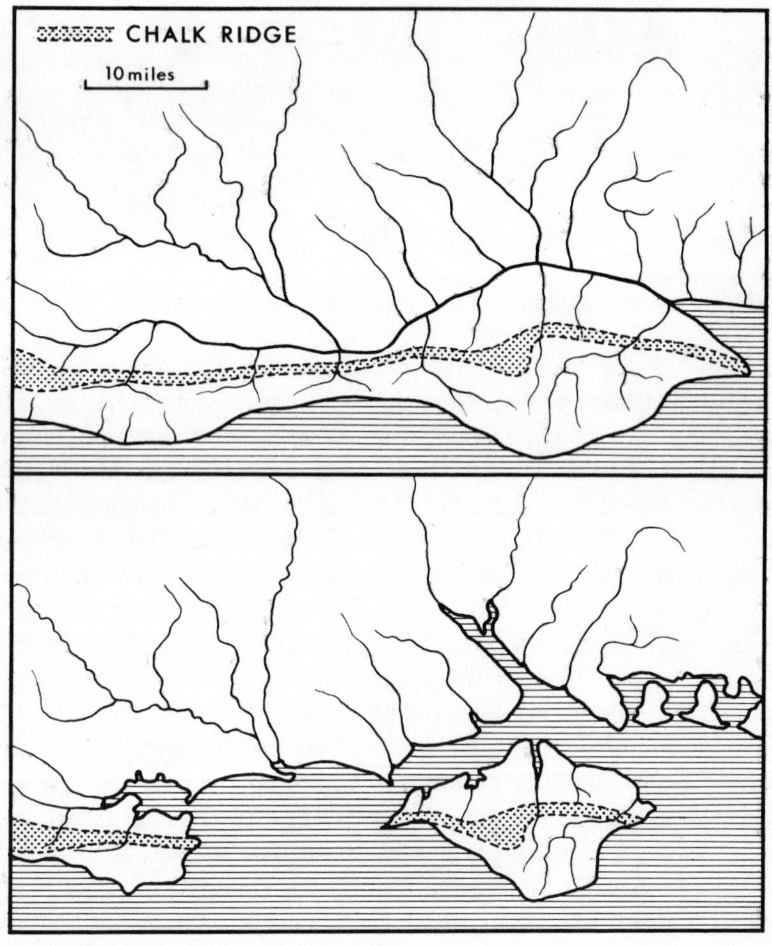

180 The Solent river

Wight anticline) seems to have been progressively reduced. Doubtless this process was aided by the presence of water-gaps, formed by tributaries of the Solent river, through which the sea could penetrate during times of rising sea-level. The major transgression of post-glacial times (pp. 423–4) saw the ultimate destruction of the Chalk ridge between the Foreland at Swanage and the Needles. The lower part of the Solent valley was drowned at this time to give the Solent and the Isle of Wight, whilst the central part of the valley was removed by a combination of inundation and marine erosion of the unprotected Tertiary sediments, forming the present Poole and Christchurch Bays.

In many other parts of the British coastline similar influences may be detected. In western Scotland the deeply penetrating and ramifying sea-lochs are obviously submerged and heavily glaciated valleys, originally the product of fluvial erosion. Furthermore, the pattern of islands here has been profoundly affected by earlier drainage-lines. Sea-bed profiles reveal numerous valleys and trenches, and the Little Minch, separating the Outer Hebrides from Skye, and the Sea of the Hebrides were once evidently occupied by an important north-north-east- to south-south-west-trending valley system. The right-bank tributaries of the trunk stream may have been responsible for initiating the narrow straits between the individual islands of the Hebrides group (for instance, North Uist, Benbecula and South Uist), though much modification has been effected by glacial erosion. An additional factor in the development of the coastal features of western Scotland has been large-scale faulting and foundering, probably in late-Tertiary times following the outpouring of vast quantities of basaltic and other lavas in Eocene and Oligocene times. A very striking feature of the map of the Outer Hebrides is their regular eastern margins (ignoring the multitude of minor indentations), and the manner in which these drop steeply into deep water in the western part of the Little Minch. As Steers has written, 'if the feature were to appear on dry land it would resemble a steep wall-like scarp'; and indeed the most likely explanation is that it has been originated by faulting and affected by later submergence.

Changes of sea-level

It will be apparent from several examples that have been discussed that changes of sea-level, whether (a) of a eustatic nature, (b) induced by isostatic depression and recovery of the earth's crust, or (c) resulting from orogenic movements, are a vital factor—perhaps on a par with geological structure—in coastal geomorphology. Relative rises often tend to exaggerate the indented character of a particular coastline, through the drowning of valleys to give ria-type inlets (Fig. 181) and of broader vales to give large bays (such as the Wash of eastern England), and through the production of islands (though it must be added that islands are not only the result of rises of sea-level, but can be the outcome of differential erosion and also coastal deposition). The formation of sea-cliffs and fringing platforms must begin again when the sea-level rises substantially, and this will be a comparatively slow process within sheltered localities (for example, in the upper reaches of drowned estuaries such as those in Cornwall and Devon and in the sea-lochs of western Scotland). Indeed, in these instances the typical features of wave erosion may be absent, and the coast will comprise sub-aerially weathered and eroded slopes dropping with little or no interruption into deep water. Falls of sea-level, on the other hand, may expose the gently sloping sea-floor, with its cover of marine sands and muds, and a more regular coastline, with few inlets and at best low cliffs cut into incoherent upraised sediments,will often be formed. Initially, the offshore gradient

181 The harbour at Solva, Pembrokeshire; a valley drowned by the post-glacial rise of sea-level. [Eric Kay]

may be so slight that shoreline sedimentation, resulting in offshore bars and banks, sand-dunes and salt-marshes, will dominate erosive processes.

However, there are serious dangers in generalising about the coastal patterns and features produced by sea-level changes. So much will depend on other factors. If the coast is an exceptionally steep one, and the sea-bed immediately offshore intricately dissected by trenches and depressions produced either by erosion or tectonic activity, a fall of sea-level will not give the 'classic' upraised coastal plain, but a coastline whose general character is but little changed from that associated with the former higher sea-level. Conversely, a transgression of the sea across a low and gently undulating land-area (perhaps one where recent glacial deposition has taken place, as in East Anglia) will produce a coastline that is comparatively simple in plan. Even where shallow valleys are submerged these may experience rapid sedimentation, as in the case of many valleys in north Norfolk, and regularisation of the coastline will quickly ensue. It is these facts, among others, which render simple classification of shorelines into 'submerged' and 'emerged', such as that proposed by Johnson (1919), of little real geomorphological value.

The study of the sea-level changes which have affected the British coastline reveals a complex and as yet imperfectly understood sequence of events. The most recent rise of sea-level has taken place in post-glacial times. Analysis of freshwater peat dredged from the bottom of the North Sea at depths approaching 200 ft reveals that in Zones IV and V (up to approximately 7000 B.C.) of the post-glacial period the sea-level had not yet fully recovered from the locking up of vast quantities of water in the Wurm ice-sheets. However, following the 'Little Ice Age' of the late-glacial period (lasting from 8800 to 8300 B.C.), the return of the bulk of this water to the sea led to the large-scale inundation of previous low-lying land-areas, the drowning of many river valleys, the final separation of England from the Continent through the formation of the Straits of Dover, and the creation of the main outline of the British coasts. Additional testimony of this 'Neolithic' or 'Flandrian' transgression lies in submerged forests (for example, the stumps of trees rooted in sub-aerial beds and covered by 30 ft of marine beds and 20 ft of estuarine beds in the Pentawan valley, Cornwall) and even by former river terraces well below present sea-level (for instance, remnants of three gravel terraces, covered by post-glacial peat and mud, found at depths of up to −98 ft O.D. in Southampton Water and the Solent). To judge from the extensive accumulation of a marine clay in the Fenlands, the sea had reached almost to its existing level by 2000–1500 B.C. (though in south-east Scotland evidence exists that the transgression culminated earlier, at about 3500 B.C.), since when minor fluctuations of the order of up to 20 ft have been superimposed on to a continued but very slow rise.

However, in Britain as a whole, the post-glacial period has encompassed not only an important marine transgression, the result of glacial eustatism, but also a noticeable isostatic recovery of the most heavily glaciated areas, in particular northern England and Scotland. Striking evidence of this is afforded by the so-called '25 ft' beach, which is generally regarded as of post-glacial age and as marking the maximum height attained by the post-glacial sea. The beach itself is represented by shingle deposits, marine clays at the sheltered heads of estuaries (as in the example of the 'carse clay' of the Forth valley), cliffs cut into unconsolidated glacial and fluvio-glacial deposits, low terraces running along the coastline for several miles, and even shingle spits and ridges (notably in the Moray Firth). Often the beach is associated with a well-marked cliff and abrasion platform cut into highly resistant ancient rocks. However, the development of these can hardly have taken place within the last few thousand years, and it is generally assumed that the 25 ft beach coincides in many places with earlier interglacial or even pre-glacial sea-levels. The term '25 ft' beach is itself something of a misnomer, for its actual height varies from sea-level (or possibly below sea-level) to above 40 ft O.D., thus indicating recent deformation. The zone of maximum isostatic recovery seems to lie in the south-west Grampians, and from this area the post-glacial beaches decline in height southwards until, south of Lancashire, they merge with the present-day beach. Thus we can infer that at some time in the post-glacial period, in those areas where the 25 ft beach is visible, the rate of isostatic recovery began to exceed the rate of eustatic sea-level rise. It is quite possible that isostatic recovery itself operated differentially, particularly on either side of ancient major faults, so that the apparent culmination of the Neolithic transgression was at different dates in different localities. Study of tidal gauges has suggested that at present Scotland is still rising isostatically at a faster rate than the eustatic rise of sea-level. Thus near Fort William sea-level is falling relative to the land at approximately 4 mm per year, whereas at Newlyn, Cornwall, it is rising at 2 mm per year.

Changes of sea-level earlier in the Quaternary are also relevant to the study of British coasts. In western Scotland the '100 ft' beach, represented by benches and shingle deposits and well developed in the western islands and on the mainland north of Skye, may have been formed during a glacial period. It is always absent from the upper parts of sea-lochs, which may have been occupied by contemporary glaciers. The '50 ft' beach (actually ranging in height from 35 to 60 ft O.D.) is far more extensive, and is associated with clays containing shells indicative of an Arctic climate. Precise dating of these two beaches has not been possible. Elsewhere in the British Isles limited planation surfaces of Quaternary age and marine origin are found at heights of up to 475 ft O.D. (for example, those on the dip-slope of the South Downs referred to on p. 256). In this context it is relevant to cite the classic work of C. Deperet on the raised shorelines of the Italian–French Riviera, in which he proposed the view that during the Quaternary the sea had occupied four main levels (the 'Sicilian' at 80–100 metres or 260–325 ft; the 'Milazzian' at 55–60 metres or 180–255 ft; the 'Tyrrhenian' at 30–35 metres or 100–120 ft; and the 'Monastirian' at 15–20 metres or 50–65 ft). More recently other workers have recognised a Low Monastirian (or 'Normannian') sea-level at 2–5 metres (7–16 ft). These five levels have been identified elsewhere, both within and outside the Mediterranean Basin, and the inference may be drawn that in the Quaternary successive eustatic falls, independent of those attributable to glacial eustatism, occurred either as the ocean basins were deepened and/or the continental land areas raised *en masse*. The pattern of change was, of course, complicated by the glaciations, and it is clear that Deperet's marine terraces represent the high sea-levels of the interglacial periods, and that the formation of each was separated by important regressions to unknown levels during the glacial advances. The two Monastirian beaches are attributed to the last interglacial, after which the Wurm, or Weichsel, glaciation resulted in a sea-level fall of some 300 ft or more; this has not been completely reversed by the post-glacial inundation.

As a result of the Quaternary fluctuations of sea-level, of which only a very simplified picture has been drawn, the products of wave action (cliffs, platforms, notches, beaches) have at times been abandoned, and in many instances fossilised by the accumulation on them of deposits of 'head', wind-blown sand, etc. (Fig. 182); subsequently they have been 'revived' by the return of the sea to approximately its former levels. An interesting illustration of this is given by the occurrence of estuarine silty clays, datable as of last interglacial age and lying a little below present sea-level, at Stone Point on the northern shore of the Solent.

182 A raised beach of late-Quaternary age, Clonque Bay, Alderney. Note the rounded beach pebbles at the base of the section. These are overlain by blown sand (formed when the sea-level fell and exposed spreads of sand on the foreshore to wind action) and a massive deposit of head (comprising angular blocks produced by frost disintegration of the former cliff). [R. J. Small]

These deposits, which were formed before the Monastirian transgression reached its maximum level, enable one to postulate an early drowning of the Solent river to give an estuary between the mainland and the Isle of Wight (pp. 419–21). More important are the numerous cliffs which are fringed by remnants of raised beaches (for example, in Devon and Cornwall and in the Gower peninsula of south Wales) (Fig. 183). It is obvious that these cliffs are not the result of wave action since the post-glacial transgression raised the sea to its existing level, but are older features in the process of being rejuvenated. Even where beach remnants are no longer preserved it is highly unlikely that post-glacial cliff recession, particularly in the very resistant Palaeozoic rocks of the west and north, has been very effective, so that in these areas too the cliffs are in essence features inherited from an earlier phase of erosion. Indeed, the conclusion may be drawn that, outside places where coastal accretion has been extensive or where cliff retreat in post-glacial times has been unduly rapid (pp. 442–3), the present-day coasts of Britain must be little different from those of the last interglacial period.

Finally, it must be stated that changes of sea-level are responsible not only for features such as drowned inlets, islands, raised platforms and raised beach deposits, but have also had some influence on constructional forms developed since the sea attained its present level. Beaches, spits and bars are usually explained in terms of currently acting processes such as waves, longshore drift, wind, etc. However, the concentration of beach materials at some points is such that other causative factors must be sought. Chesil Bank, the greatest storm beach in Britain, stretches some 16 miles from West Bay to Portland in Dorset. It has been

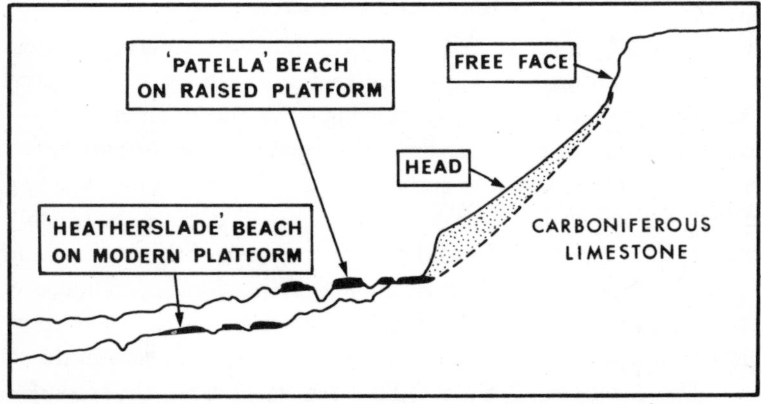

183 Raised beaches, platforms and dead cliff at Heatherslade Bay, Gower

explained by some authors as the result of west–east drift of shingle, and would thus fall into the category of a spit joining an island to the mainland ('tombolo'). However, others have argued that the great mass of pebbles in the beach has a more local provenance, either in the raised beach which must once have lain to the south (and which is still represented by a small but well-preserved deposit at Portland Bill), or in a bay mouth bar running from an extended Portland to a lost headland farther to the west in pre-Neolithic times. In the latter hypothesis the bar is regarded as being rolled inwards by the Atlantic breakers as the sea rose in post-glacial times and flooded the low-lying Jurassic clays which today form the foundation of Chesil Bank and which formerly lay behind the bar. The Fleet is considered to be the last remnant of this inundated lowland. Another example of a similar process is provided by Blakeney Point, Norfolk, which is evidently a spit which has grown from east to west. However, at present little or no shingle is arriving as a result of longshore drift from east of Weybourne. Hardy (1964) has recently suggested that the spit has been largely fashioned from a local fortuitous accumulation of shingle, derived from offshore glacial deposits and washed landwards and concentrated by breaking waves during the Neolithic transgression. That some growth westwards has since taken place is proved by the existence of laterals once marking the end of the spit, but in essence Blakeney Point has been formed by the reworking of an *in situ* mass of shingle (p. 454).

COASTAL PROCESSES

The action of sea waves

Waves breaking on the shore form by far the most important process in coastal erosion, transportation and deposition. Such waves are generated in the first instance by winds blowing over the ocean surface. The moving air exerts a frictional drag on the surface water particles, and sets up in profile a series of orbital water movements. These are largest and best developed just below the surface, and decrease noticeably in depth, until at a distance from the surface equal to half the wave length the water particle motions are barely perceptible. Again, the orbital movements are not about stationary points, but are associated with surface undulations (the waves proper) which move forward in the same direction as the generating wind. Thus sea waves are of the 'progressive oscillatory' type.

Once in existence, waves may be steepened (for instance, by air eddying against their leeward slopes), and may grow in length and height if the wind velocity is increased. In most circumstances there will be a time-lag between acceleration of wind speed and wave growth, but if the velocity remains constant for a sufficiently long period the waves will in time attain the maximum size that can be maintained without a further strengthening of the wind. It has been calculated that in a wind blowing at 8 m.p.h. a wave length of 100 ft can be formed, whereas a wind of 25 m.p.h. can produce a wave with a length of 900 ft. Needless to say a precise balance between wind strength and wave size is hardly ever achieved, because of this time-lag and the frequent changes of wind speed occurring in most meteorological situations. Normally waves travel forwards at a velocity appreciably less than that of the wind, but the reverse is often encountered, owing to the generation of the waves in another more windy area or because the local wind speed has been suddenly reduced. An interesting example of the former is the low swell sometimes observable on the Cornish coast which can be related to large storms in the Atlantic a thousand or more miles away. The decay of waves, once the wind has dropped, is very variable in rate. Short waves die out comparatively rapidly, whereas long waves are more persistent and can travel over great distances, as in the case cited. It has been suggested that, as a general rule, a wave will lose one-third of its height every time it travels a distance in miles equal to its length in feet.

The state of the sea at any place at any time thus depends to some extent on the strength of the local wind, but other factors too must be taken into consideration. These are: (a) the time during which the wind has been blowing from a particular direction, (b) the distance from the leeward shore, and (c) the presence of waves generated elsewhere. The second of these, usually termed the 'fetch', can in effect restrict the time available for wave growth. If the fetch is only 20 miles and the wave speed 5 m.p.h., there are 4 hours in which a wave has to attain maximum development. If it does not do so in this time it never will, for at the end of its journey it will be destroyed by breaking. As a general rule it may be said that wave size is limited by fetch when this is small (as in partially enclosed seas), and by wind speed and duration when the fetch is very large. Around the British Isles fetch tends to be a limiting factor in the North Sea (so far as waves with an easterly component are concerned), the English Channel (in the growth of southerly and south-easterly waves) and in St George's Channel and the Irish Sea; but it is less important on the exposed western coasts, where north-westerly, westerly and south-westerly gales are all able to generate waves of great size.

The approach of waves to the shore, and the manner of their breaking, are matters of fundamental geomorphological interest. On any beach the waves will move onshore with their crests either parallel to the shore or at an oblique angle, depending on the orientation of that beach in relation to wind direction and/or fetch. As the waves enter the shallower water they undergo a series of changes both in plan and profile. One of the most important processes is that of refraction, which affects waves approaching a straight shore obliquely or moving directly on to a deeply indented shore. When the water depth is about half the wave length, an appreciable slowing down of the wave velocity takes place, with the result that the wave tends to become more nearly parallel to the shore by the time it breaks (Figs. 184A and B). Refraction also occurs when waves pass the end of an obstacle such as a rocky headland, shingle foreland or spit. The waves tend to change direction, giving the pattern shown in Fig. 184C, and to die out gradually.

The changes that affect the profile of a wave entering shallow water involve a decrease in velocity and thus length, a marked steepening of the wave form (with the front of the wave being steeper than the rear), and an alteration in the movement of the water particles from an orbital to an elliptical or even 'to-and-fro' motion. Waves possess two types of energy: kinetic (generated by the movement of the water particles) and potential (due to the maintenance of the wave form above the average level of the sea surface). With the modifications that have been described this energy is concentrated in a narrower and narrower strip of water, until it is released by the breaking of the wave. To understand the latter process it must be realised that in a sea wave the velocity of the water particles, along their orbital paths, is invariably less than the rate of forward movement of the wave form as a whole. However, as the wave speed is retarded on approaching the shore, the orbital velocity begins to 'catch up', until a point is reached where at the crest of the wave it becomes equal to the wave velocity. Beyond this, the water particles moving in their orbits tend to leave the wave behind; or in simple terms the crest of the wave spills over and the wave form collapses. In its place, there is a swirling and highly turbulent mass of water which rushes up the beach. This 'send' or 'swash' has considerable powers of transporting sand and shingle, though its energy is quickly dissipated, especially on a steep beach, by friction, percolation and gravitational pull. The residue of water flows down the beach as the 'backwash'. Though noticeably less turbulent this also has the ability to move beach material with it, particularly since it is accelerated by gravity. As a result, the movement of shingle and sand by successive breaking waves

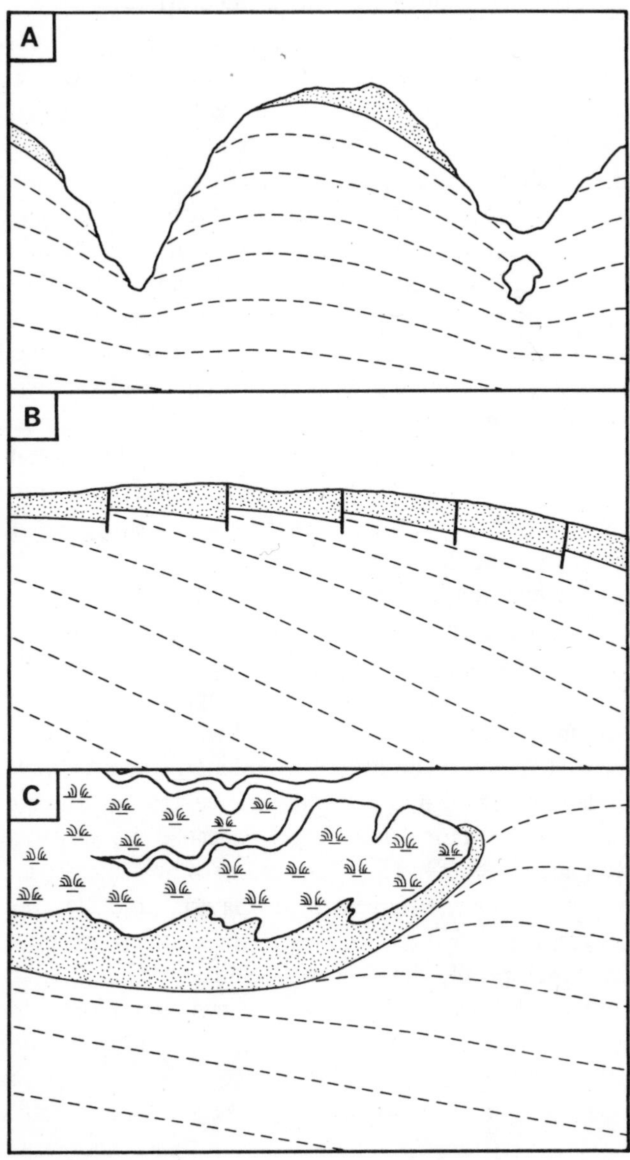

184 **Wave refraction**

up and down the beach will be approximately equal, though normally some sorting of this material is in the process effected. Thus the coarser shingle is thrown up by the vigorous swash, and the finer grit, shell and sand is carried back down by the return flow.

The movement of material up and down the beach

In the long run beach profiles are not static, but are substantially changed as either build up or combing down of the beach constituents takes place. Several attempts have been made to explain such beach 'erosion' and accretion in terms of the action of different types of wave. Two main categories of break have been recognised. 'Plunging breakers' arise when steep waves collapse suddenly with a strong overturning motion. There is comparatively little swash, but a vertical plunging of the water (producing a 'digging down' effect) and a powerful backwash with a combing down action. Such waves have been termed by Lewis (1931) 'destructive waves'. On Chesil Bank they appear to break with a high frequency (13–15 per minute) and to be associated with stormy conditions. However, Guilcher (1958) contends that plunging breakers often develop from smooth and gentle swells. 'Spilling breakers', by contrast, produce a far more powerful swash. The wave seems to break more slowly; its crest becomes turbulent and foamy, and gradually overrides the front of the wave as the beach is approached. Lewis has argued that, whereas the orbital path in plunging breakers is more nearly circular, in spilling breakers it is elliptical, with the major axis of the ellipse horizontal. Again, Lewis suggests that spilling breakers are associated with long low waves, with a comparatively low frequency (6–8 per minute), which are best developed when the wind has dropped after a storm. This too is contested by Guilcher, who attributes them to strong winds blowing onshore. In spilling breakers the powerful send is not matched by the return flow, largely because the spreading of the water over a large area of beach enhances losses by percolation. There is thus a tendency for a net movement of material upbeach to occur, so that Lewis seems justified in referring to these as 'constructive waves'. In his view beach profile changes may be directly related to alternating episodes of constructive and destructive wave action. During the former moderately long waves will build up a series of small ridges at high tide level, whereas during the latter—associated with stormy weather and high-frequency steep waves—most of the beach material will again be removed. However, the combination of unusually high water, if the strong winds are blowing onshore, and the very fierce breaking may result in a certain amount of shingle being thrown up to give a semi-permanent high beach ridge.

It is likely, however, that wave break and its nature is not the only factor determining beach change—though it may well be the most important in the movement of shingle, simply because the resultant turbulence is sufficient to move pebbles that, in normal circumstances, cannot be displaced at all by such agencies as sea currents. It is relevant to note that waves themselves are capable of causing actual water 'drift' by their passage. The orbital motion of water particles in a wave does not cause them to return exactly to their original positions. The backward movement in the troughs is in fact less than the forward movement at the wave crests, so that a net transference in the direction of wave motion is set up. In the vicinity of the beach this must be compensated by a return 'flow', which may take the form either of a seaward drift at the bed (capable of moving sand and counteracting any constructive tendency) or of closely localised 'rip' currents affecting the water at all depths. The undertow, and the resultant combing down, will be exaggerated if there are strong onshore winds, but may be nullified or reversed by offshore winds, thus causing a movement of material at the sea-bed towards the land. The observations of King and Williams (1949) indicate that as a general rule there is a net drift of sand shorewards outside the wave breakpoint, but that inside the breakpoint flat waves continue the transportation of sand up the beach while steep waves promote a downbeach movement. Under the former conditions the accumulation of sand at the margins of the send leads to the formation of a 'swash bar', whereas under the latter the concentration of material at the wave break gives rise to a 'breakpoint bar'. King and Williams base these conclusions on wave-tank experiments in the laboratory, and it is not clear whether, under natural conditions, the processes observed are responsible for, say, the 'ridge-and-runnel' forms exposed on many sandy beaches at low tide (p. 454).

The movement of material along the shore

The process of longshore drift of pebbles, grit and sand is fundamental to the formation of beaches, spits and bars, and also helps to determine the location and intensity of erosion. That such movement readily occurs is shown by a variety of evidence (including the longshore growth of spits and bars, the piling of shingle against natural and man-made obstacles (jetties and groynes), and by the tendency of many beaches to become orientated at right angles to the direction of wave approach through the transference of material to their 'far' ends), and can actually be demonstrated by experiments using coloured pebbles. However, longshore drift is a more complex process than is commonly supposed. On the south coast of Britain the predominant drift is from west to east, as a result of the prevalent and dominant waves from the south-west striking the shore at an angle. However, many apparent anomalies exist. For example, at West Bay in Dorset there is a marked deficiency of beach material to the *west* of the harbour entrance (which is bounded by two sea-walls and so forms a perfect groyne), while to the east fine shingle is abundant and the beach is set forward many yards (Fig. 171). From this may be inferred a reverse longshore movement, though possible reasons are difficult to establish. Even on spits which indicate, by their very existence, a uni-directional drift anomalous movements of shingle can be observed or are indicated by the temporary accumulation of shingle on the 'wrong side' of groynes. Usually it is possible to explain this phenomenon in terms of short-term changes in weather conditions. For example, Steers (1946) showed at Scolt Head Island that three successive days of north-east and east-north-east winds (5–10 m.p.h.) produced an average westward drift of broken bricks of 5–10 yards, whereas the next two days (of north and north-west winds blowing at 15–25 m.p.h.) resulted in an average eastward movement of 4–10 yards.

Nowadays, it is widely accepted that longshore drift is effected by swash-backwash action, but in earlier days tidal currents were invoked to explain even major shingle structures such as Dungeness or Orford Ness. In fact the precise role of currents is still not fully understood. Experiments have tended to reinforce the view that tidal currents have little transportational effect on coarser material. Thus Owens (1908, quoted in Steers, 1946) found that where sand rests in quantity on the sea-floor, all currents of up to 1·7 m.p.h. are ineffectual in moving shingle, though once this speed is exceeded the current suddenly acquires the ability to move shingle up to 3 inches in diameter over a sandy bottom. Since the sea-bed is normally quite rough, owing to the presence of ripples and sometimes large quantities of shingle, speeds considerably in excess of this are needed if effective transport of pebbles is to be achieved. In Owens's view tidal currents are normally capable of moving only fine material such as sand and mud. There may, however, be some areas in which tidal currents are notable transporters even of shingle. At the mouths of narrow estuaries (such as the Mersey and Humber) and between islands and the mainland (for example, the

Pentland Firth between the Orkneys and Caithness) the movement of sea-water is exceptionally rapid and may even cause scouring of the bed. An interesting case is the narrow western entrance to the Solent, between Hurst Castle Spit and the Isle of Wight. At spring tides the currents here approach 5 knots, and have evidently prevented further growth of the spit to the south-east, as well as causing severe erosion of the bottom, which at a distance of only a quarter of a mile from the shore lies at the astonishing depth of nearly 200 ft. The strongest flow is towards the south-west, and the great banks of shingle lying a mile or two offshore (including the 'Shingles') appear to be composed of pebbles swept from the tip of Hurst Castle Spit. The question arises as to whether any of this shingle is moved back on to the beaches of Christchurch Bay, thus giving a circulatory 'cell-like' pattern of transportation (Fig. 185). It is just conceivable that large waves approaching from the south and south-west could achieve this, by causing disturbance and displacement of pebbles on the sea-floor. In this context it is worth noting that experiments at the Hydraulics Research Station have shown (a) that 8 ft high waves are capable of moving pebbles of 1 inch diameter over other pebbles in a depth of 27 ft of water, and that (b) similar waves can move 3 inch pebbles over a hard horizontal sea-bottom in up to 37 ft of water. It is therefore possible to envisage, under

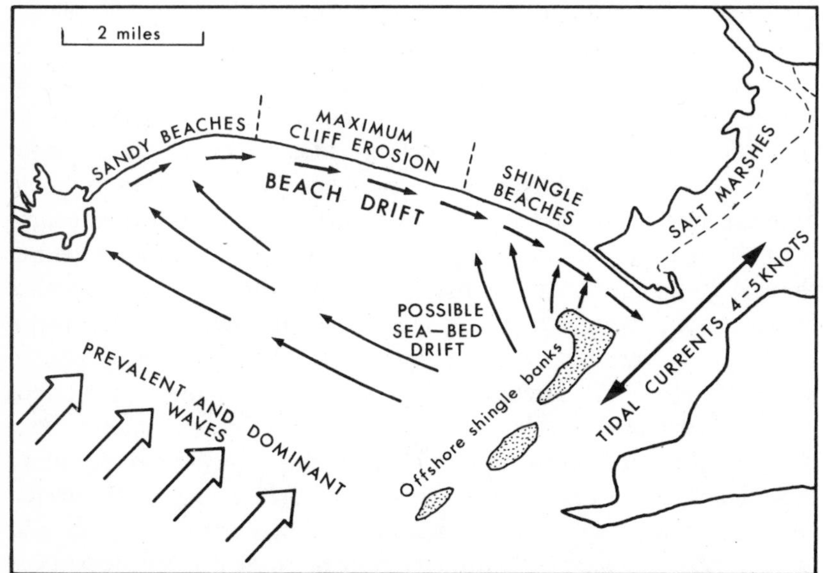

185 **The pattern of beach and sea-bed drift, Christchurch Bay, Hampshire**

favourable conditions, quite large-scale movement of shingle in moderately deep water, both along the shore and perhaps at right angles to it as well. Experiments with sea-bed drifters in Christchurch Bay have revealed well-marked onshore drifts of water during times when powerful offshore winds are blowing, and this may lead to movement of shingle on to the beaches in appreciable quantities.

The return of beach material in this way could be a very important factor in countering erosion. Where there is only a continuous drift away from the beach, the resultant deficit, unless counteracted by the effects of longshore drifting, will leave cliffed areas unprotected, and a high proportion of wave energy will be expended on the cliff-toe instead of being dissipated in the task of moving beach shingle and sand. This is shown in the many areas where groynes and jetties hold up material being moved alongshore, thus reinforcing protection at some points at the expense of others to the leeward. The latter are starved of shingle and suffer accelerated erosion. The effect may be seen on a very small scale, where individual groynes cause the piling up of shingle on one side and severe beach scour on the other, or on a much larger scale (for example, the erosion of the chalk cliffs between Brighton and Rottingdean has been attributed to the large groynes holding up shingle at Brighton itself, whilst to the east of Folkestone, at The Warren, erosion has been increased because of the large harbour-walls to the windward). Continuous longshore drift is not, however, affected by man alone. Natural accumulations of beach material at certain points are also bound to have repercussions elsewhere. A case in point is the shingle spit at Great Yarmouth, which deflects the mouth of the Yare southwards (Fig. 186). In the fourteenth century this extended some 4 miles farther to the south than at present, but breaches (which were admittedly artificial, but would probably have developed naturally in due course (p. 456)) made in the next two centuries prevented any more growth of the spit. Instead the 'dead' leeward part, deprived of its shingle supplies by these cuts, was destroyed. Much of the constituent shingle was in fact washed onshore north of Lowestoft Harbour, giving rise to Lowestoft Ness. The stretch of coast between Yarmouth and Corton, previously protected by the southern part of the Great Yarmouth Spit, now began to suffer severe erosion. A contributory factor was the failure of shingle to move past the new exit of the Yare, so that in this stretch an adequate beach could not be built up by littoral drifting. It is, in fact, frequently noticeable that to the leeward of spits the coastline is set back somewhat as a result of enhanced erosion—though it must be added

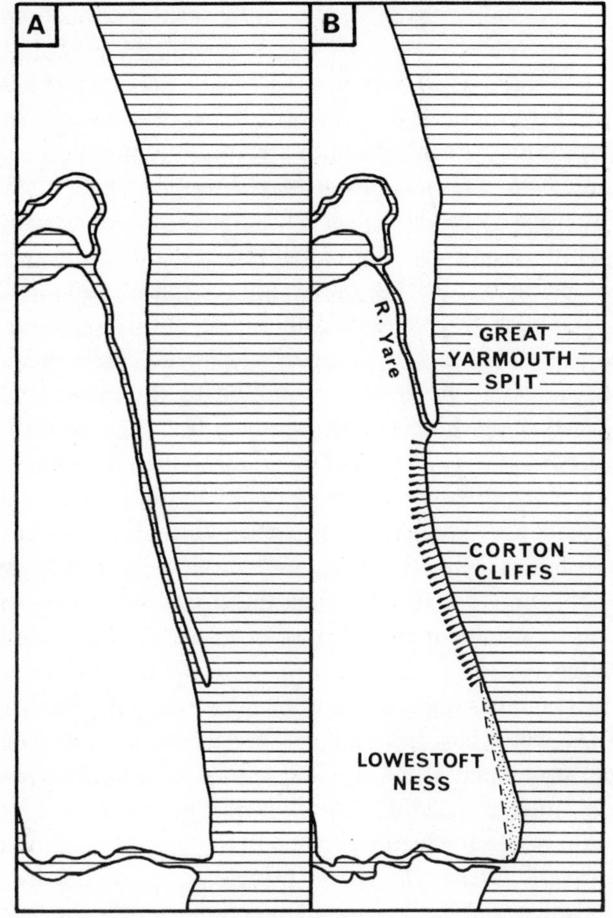

186 The development of Great Yarmouth Spit (after J. A. Steers)

that in some instances shingle does succeed in travelling from the end of the spit and causes aggradation in such a situation (p. 459).

Wave erosion

Any beach will depend in the short run for its maintenance either on the longshore drift of shingle and sand or, in certain cases, on the movement of material from the sea-bed. However, in the long run beach material must be derived from erosion of the land by waves. If such erosion did not occur the existing 'fund' would be used up as a result of attrition. It would not be adequately replaced, either in quantity or quality, by the fine mud and silt washed into the sea by the majority of present-day rivers.

Direct wave erosion is a complex process. In the first place it must be emphasised that, whilst attack is normally at its greatest at high tide level, it is by no means confined to a narrow altitudinal zone (pp. 450–1). During severe storms, wave break may throw water high up the cliff face, and this may have both corrosive effects and even achieve some corrasion—indeed on some cliffs small fissures and hollows some 20 or 30 ft above the highest level of the sea are occupied by rounded pebbles or quite large rock fragments. Furthermore, direct wave erosion can occur at any level between the highest and lowest tide-levels, and even below springs since the oscillatory motion of the incoming waves is capable of disturbing shingle on the sea-bed. In the second place it must be said that powerful waves do not invariably have a great erosive effect. Some cliffs plunge deeply into the water at their base, and lead to reflection rather than breaking of the waves. Indeed, the development of submarine platforms by corrasion may at first be a very protracted process, that is only accelerated when the erosional bevel is sufficiently wide to cause frictional retardation of waves and so render wave break more effective.

When waves break powerfully against the foot of a cliff, the sheer impact of the water may have a direct erosive effect. This is especially true of unconsolidated sands and clays (for example, the glacial drifts which form much of the cliffed coastline of East Anglia and Yorkshire). Furthermore, material which has fallen to the base of the cliff is washed away by waves; indeed this is the principal role of the sea in those areas where sub-aerial processes are largely responsible for the recession of the cliff (p. 439–42). However, it seems likely that the purely hydraulic action of waves is of little importance in hard rocks, unless these are characterised by well-developed bedding-planes, joints or small faults. It is true that the pressures exerted by wave break are considerable (up to 30 tons per square yard have been recorded), but they are also very short-lived. However, where there are cracks in the rock capable of containing air, the impact of the breaking waves causes a sudden compression of this air, which is followed by an equally rapid decompression as the wave recedes. It is believed that this process has the effect of weakening the rock structure and of opening up lines which other forms of wave attack can utilise. In this way large joint-bounded blocks can be 'quarried'. On coasts with numerous vertical faults, perhaps associated with fault-breccias giving weak zones, or major joints, narrow inlets or geos can be initiated by hydraulic action, though it is obvious that these are later broadened and extended by corrasive scour (Fig. 187). As the waves break a strong swash, carrying a large load of pebbles, rushes into these inlets and subsequently pours out again. The rocky walls are thus polished and worn away, and in many clefts a 'pot-hole' effect may be seen. Although corrasion is naturally enhanced in such situations, it can be very active on the open cliff face, and is a major factor in the shaping and extension of wave-cut platforms (pp. 450–2). Solutional effects too must not be overlooked. The many caves that characterise limestone cliffs are partly the result of carbonation-solution by underground water (the effects of which are exposed by recession of the cliff) and partly the result of corrosion and corrasion by sea-water. Wave-cut platforms too may be greatly modified by solution processes (p. 16).

Sub-aerial processes

The part played by sub-aerial processes in the development and modification of sea-cliffs is perhaps insufficiently appreciated. In the first instance it is necessary to make a distinction between cliff falls and slumps which are due to undermining of the base of the cliff (or in other words are 'marine induced') and those which are less dependent on wave action for their continued operation. Even in hard and massive rocks long-continued wave erosion must eventually create rock overhangs and promote the collapse of individual joint-bounded boulders or large masses of rock debris. The falls are naturally aided if the rock structure has previously been weakened by chemical action or the formation of ice within joints. These two processes are undoubtedly of considerable importance in the recession of chalk cliffs, whose characteristic verticality represents a neat balance between the rates of erosion of the cliff-foot by storm waves and the frequent falls of loosened chalk fragments from the upper part of the cliff face. At the other end of the scale are the cliffs in sands and clays which undergo rapid retreat because of their innate susceptibility to certain sub-aerial processes (Fig. 174). Wave action is obviously not wholly unimportant in the development of such cliffs, for it removes material which tends to collect at and protect the toe; in other words, it hinders the evolution of the cliff from a very steep unvegetated face to a stable low-angled vegetated slope.

Some of the more common processes at work on cliffs are as follows. (a) Weak sand and clay lacks the mechanical strength to sustain a high-angled face, and collapse of the cliff, frequently involving the development of rotational slip planes (p. 32), necessarily results.

187 A geo in Old Red Sandstone, Duncansby Head, Caithness, Scotland. [Eric Kay]

(b) The occurrence of sand overlying clay will lead to the formation of spring- and seepage-lines. These will sap and undermine the sands above, and lead to gullying of the clay below; furthermore, any debris lying on the cliff will be lubricated and flows of mud and sandy material down to the beach will be initiated.

(c) Heavy rainfall will lead to direct erosion of the cliff face, by sheet wash and rivulets, and can 'overload' the underlying rocks by saturation, thus causing collapse if these are mechanically weak.

(d) Wind action can have an appreciable effect on cliffs in unconsolidated sand. A honeycombing effect can sometimes be seen, as on the Sandrock and Ferruginous Sand cliffs of the southern Isle of Wight, where the sand has in the past been blown on to the top of the cliffs, giving sand-dunes at a height of nearly 200 ft above the beach.

It is a measure of the importance of these sub-aerial processes that at the base of the main cliff, which in some areas lies several hundred yards inland, there often exists an 'undercliff', comprising slipped and slumped material which is being removed comparatively slowly by wave erosion. The most impressive undercliffs (for example, those of the southern Isle of Wight to the west of Ventnor, and the Dorset coast between Seaton and Lyme Regis) are related to particular structural conditions, involving well-consolidated rocks rather than sands and clays alone. In Dorset Lower Chalk and Upper Greensand overlies Gault Clay, Lias limestones and shales and Keuper Marl, the dip being gently to the south. Water percolating through the Chalk and Greensand saturates the underlying clays, rendering them unstable. The greatest 'breakaway' occurred at Christmas 1839, when a great mass, weighing some 8 million tons, slipped seawards over the Gault Clay, leaving a deep valley (Dowland's Chasm) between itself and a new inland cliff, at its maximum over 200 ft in height (Fig. 188).

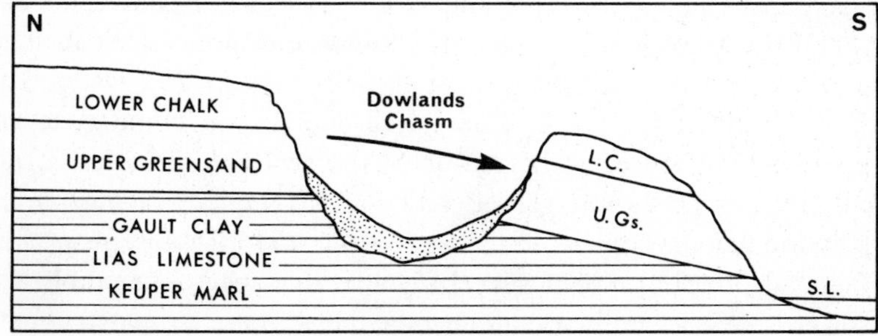

188 The landslip at Dowland's Chasm, Dorset

Cliffs which are affected by sub-aerial processes tend to recede more rapidly than those resulting mainly from marine erosion and marine-induced collapse. For one thing, they are being attacked by a number of processes instead of only one; for another, sub-aerial processes are usually at their most active in weaker rock materials. An additional, less obvious, point is that unresistant sands and clays give rise to low cliffs (often less than 100 ft and rarely more than 200 ft in height). To achieve a certain amount of retreat requires less removal of material—ultimately achieved by wave action, even where sub-aerial processes are very active—than in the case of a cliff some 500 ft high. It will be clear that in reality rates of recession will vary greatly, depending on such factors as the aspect of the cliff and its degree of exposure to large storm waves, its precise lithological composition, its structure, the presence or absence of protective shingle beaches at the foot of the cliff and so on.

Rates of cliff recession

At present many cliffs exist which appear to be undergoing no change at all. Even so there is evidence, in the existence of erosional features such as caves, geos and rock-falls, that these cliffs are 'live'. They are clearly being worn back at a rate (less than 1 inch per year) which cannot be directly observed or even inferred from successive coastal surveys spread over a number of years. In this country such cliffs are most typical of the hard Palaeozoic rocks of the west and north (such as granites, quartzites, gritstones and resistant limestones), but some younger formations (for instance, the Jurassic limestones of the Isle of Purbeck and the Weymouth area) give rise to stable cliffs. In some cases cliffs of this type have been so little affected during the past few thousand years that they preserve some record of past conditions of climate and sea-level (p. 445).

Cliffs developed in softer rocks of Mesozoic and Tertiary age, or in unconsolidated glacial deposits, usually undergo much more active retreat. Most chalk cliffs, for example, are subject to frequent if small falls, and the average annual rate of recession may approach 1 ft. This moderate figure must be set against those for the cliffs developed in Tertiary sands and clays in Hampshire and in the boulder-clay of Holderness, Yorkshire. At Barton-on-Sea, where the geological structure comprises permeable gravels and Barton Sand overlying Barton Clay and so favours seepage and gully erosion, the cliff-top retreats at an average rate of 3 ft or more a year. In Holderness, where the problem of coastal erosion is a very longstanding one, the cliffs are estimated to have receded some 2 miles since Roman times (or in other words at an average annual rate of 5–6 ft). It is often pointed out that such erosion is more than counterbalanced by deposition on spits, bars, shingle forelands, mud-flats and salt marshes, so that there is really a net gain of land from the sea at the present time. However, it needs to be emphasised that the land lost is often of high value, whereas that gained is, at least in the short run, of comparatively little value. It is for this reason that at many points local authorities are obliged to interfere with the natural processes of erosion by constructing groynes and sea-walls and undertaking more sophisticated and costly schemes of cliff drainage and stabilisation.

COASTAL FEATURES

Although much has already been written of some coastal features, in the context of discussions of general factors in coastal development and of coastal processes, it is useful to attempt a more systematic treatment. Coastal features may, for the sake of convenience, be classified as either erosional or depositional—though it must be recognised that primarily depositional forms, such as beaches and spits, are liable to erosion, and that other features, such as raised beaches consisting of shingle deposits on an abraded platform, are composite in origin.

Erosional features

The most characteristic forms due to wave erosion, sometimes allied to other processes, are cliffs and wave-cut platforms. The most simple explanation of these assumes the removal at sea-level of a wedge-shaped mass of rock, mainly by the mechanical action of breaking waves (Fig. 189). Once initiated, the cliff is driven back by a combination of basal attack, weathering, slumping, gullying and so on, and in the process may experience a steady increase of height. In the process the platform is

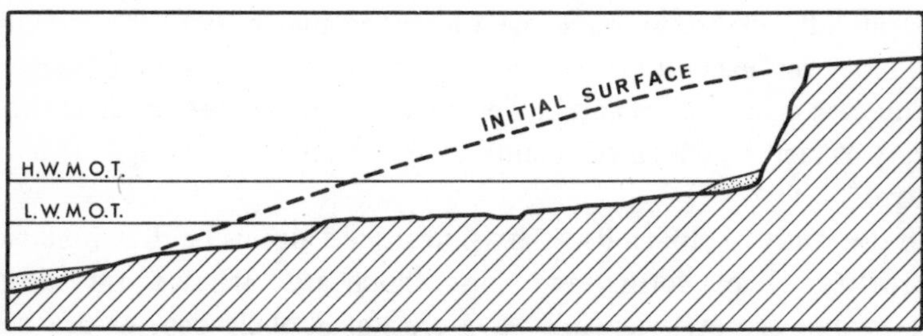

189 The development of a cliff and wave-cut platform

extended, and at the same time will be slowly lowered by the passage of waves, armed with rocks, pebbles, grit and sand derived from the erosion of the cliff itself. Initially the development of the cliff and platform will be slow, owing to the reflection of waves (p. 438) and the absence of a sufficient load for the waves to utilise, but subsequently it will be accelerated. However, later still as the wave-cut platform becomes broader and broader an increasing proportion of wave energy will be expended by passage through the shallow water overlying the platform, and attack on the cliff base will become less intense. At the same time the amount of load shed by the ever higher cliff face will also tend to minimise wave erosion. It is noticeable that even the largest wave-cut platforms around the British coasts are rarely more than a few hundred feet across; and for the reasons given on p. 451 this cannot wholly be explained as a result of the limited time for erosion since the sea attained its present level. In fact, it has been argued recently that platforms more than a half to one mile in width cannot be formed by the sea, except under conditions of rising level, owing to the great losses in efficiency of the waves passing over them.

Cliffs

Marine cliffs (the term is used rather loosely here to refer to any steep slope descending to the shore) show an almost infinite variety in profile, angle and plan, depending on the precise local conditions of lithology, jointing, structure, degree of exposure and erosional history. It is often implied in textbooks that cliff form and angle are controlled principally by the angle and direction of dip of the constituent rocks. Thus where the dip is seawards, movement of rock masses down the bedding planes is facilitated, and the resultant profile is rarely vertical. Rather the angle of the cliff face and of the dip tend to coincide, and the cliff itself is characterised by 'rock slabs' or 'boiler plates'. Where on the other hand the rocks are horizontally bedded or dip inland, very steep and often vertical cliffs are common, as even loosened masses of rock are held *in situ*. However, exceptions can always be found to these simple rules, and in any case cliff profiles and angles, as stated above, are influenced by a variety of factors, which can combine to mask the influence of structural control.

In the study of cliffs a basic division should always be drawn between 'active' (or 'live') forms, and those which are 'inactive' (or 'dead'). The former are experiencing wave erosion at their base or are being driven back by sub-aerial wash, gullying, creep and slumping, whereas the latter are (or in some cases were until very recently) isolated from the sea by shingle, sand and marsh deposits or by raised rock platforms (Fig. 190). Live cliffs are usually though not invariably characterised by free faces, whilst most dead cliffs have been converted by sub-aerial wasting into comparatively gentle rectilinear or even convex slopes. Many of the dead cliffs in Britain may be related to the changes of sea-level prior to, during and since the Wurm glacial period. In the last interglacial period the cliffs were actively eroded by the waves of the high and low Monastirian sea-levels (p. 425), but in the ensuing glacial period were abandoned. Periglacial conditions led to the wasting of the

190 Cliffs in Tertiary rocks capped by plateau-gravel, Milford-on-Sea, Hampshire. Note the broad accumulation of shingle protecting the foot of the cliff, which is becoming 'dead' and is weathering to a comparatively gentle angle. [M. J. Clark]

191 An extensive wave-cut platform and dead cliff in Carboniferous Limestone near Port Eynon, Gower, south Wales. Note the two levels of the platform: a lower level over most of the bay, and a higher level, partly covered by raised beach and head deposits, immediately at the foot of the cliff. [Eric Kay]

cliff faces by frost weathering and solifluxion, and many were converted into debris-controlled slopes at 25–35° (p. 203). Such slopes may be seen today at many points on the southern coast of Gower, in south Wales (Fig. 191). In some instances they lead upwards into unconsumed

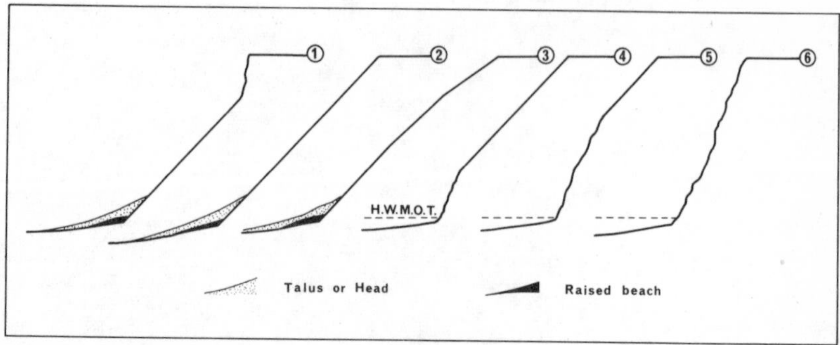

192 Stages in the rejuvenation of dead cliffs

free faces of Carboniferous Limestone; in others the slope is comparatively smooth from top to bottom. Widespread deposits of angular limestone chips, cemented into a hard carapace by percolating lime-rich water, mantle the dead cliffs, occupy former chasms eroded along faults and joints, and often spread over raised beach remnants at the foot of the cliffs.

Stages in the rejuvenation of dead cliffs may be observed in Gower and also at many places on the Devon and Cornwall coasts (Fig. 192). In the latter area three basic types of cliff profile may be identified. These are:

A *Hog's-backed cliffs.* Long steep vegetated slopes descending to sea-level and attacked by marine erosion which as yet has achieved little in the way of basal steepening. These so-called 'hog's-backed' cliffs are largely fossil features, and are best preserved on coasts sheltered from powerful wave attack.

B *Bevelled cliffs.* Vegetated slopes, often concave or convex in profile, which lead down to vertical cliffs, evidently the result of wave action at present sea-level. Such 'bevelled' cliffs are clearly the result of revival of former dead cliffs much modified by periglacial wasting (Fig. 193). In some instances the bevel comprises up to four slope facets (between 5 and 50° in angle), and these have been interpreted as the product of successive phases of marine cliffing, each followed by periods of sub-aerial slope modification. Such cliffs may therefore be regarded as 'multi-cyclic'.

C *Flat-topped cliffs.* Almost vertical cliff faces, which terminate abruptly the inland plateaus at their seaward margins. In these cases the dead cliff has been totally removed by recent retreat. They are most in evidence on very exposed sections of the coast or where particularly weak rocks outcrop (for example, the Culm Measures of parts of north Cornwall).

Cliffs such as those described are not, however, invariably the result of sub-aerial modification and subsequent revival by wave erosion. Bevelled cliffs are found on the south coast of the Isle of Purbeck between St Alban's Head and Durlston Head, but these seem to reflect lithological control, for the upper bevel is associated with easily weathered Purbeck limestones and the lower vertical cliff with the massive Portland Stone. Even in areas of lithological uniformity

193 **Bevelled cliffs in southern Alderney. [R. J. Small]**

bevelled cliffs are formed where valleys running parallel to the coast have one side destroyed by marine erosion; in such cases the bevel represents the 'inner' wall of the former valley.

A fully comprehensive classification of marine cliffs is hardly possible, but many fall into the following eight categories (modified from a scheme proposed by Guilcher).

(i) Cliffs in coherent clay, with vertical or near-vertical faces interrupted by re-entrants containing mud-flows.

(ii) Cliffs affected by large-scale landslips. These are common in structures comprising permeable rocks above, and impermeable rocks below. Usually there is an active inner cliff, produced by shearing and rock-falls, an undercliff, and an outer cliff where waves are attacking the toe of the undercliff (p. 411).

(iii) Vertical cliffs, in homogeneous rocks, with a small talus slope at the base resulting from infrequent rock-falls and a well-developed wave-cut platform. Stacks and caves are sometimes developed as a result of erosion along joints, but are far from common. Good examples include the Chalk cliffs of Dorset west of Durdle Door. The continuity of these is interrupted by hanging dry valleys which have been truncated by cliff recession.

(iv) Benched cliffs, developed in horizontal structures comprising alternate resistant and unresistant strata (for example, sandstones and clays, as in the Lower Greensand near Chale, Isle of Wight).

(v) Cliffs developed in horizontal sandstone, limestone, basalt or granite with well-marked pseudo-bedding planes (Fig. 194). These may be vertical or benched (representing the varying influences of vertical joints and bedding), and stacks and pinnacles may be common. In plan the cliffs are usually complex and irregular, with inlets and geos developed along joints and faults. Such cliffs are characteristic of the Old Red Sandstone of the Orkneys and Duncansby Head, the basalts of the Giant's Causeway, Antrim, and the granite of the 'Côte Sauvage', Quiberon, southern Brittany. Similar cliffs are developed in the Carboniferous Limestone, dipping steeply inland, near Paviland in western Gower.

(vi) Cliffs comprising vegetated slopes and a small modern sea-cliff at the base. These are the bevelled cliffs discussed above, and are to be attributed to a particular erosional history, lithological influences or the former presence of valleys.

(vii) Cliffs developed in beds dipping steeply seawards. Often the profile is controlled by an individual bedding-plane, while in other cases the cliff gradient may be less than the angle of dip, so that the profile assumes a 'saw-tooth' appearance.

(viii) 'Badland' cliffs are formed in incoherent rocks which are deeply gullied by surface rivulets and are bounded by basal detritus fans. The cliffs developed in the Lower Eocene beds at Alum Bay, Isle of Wight, fall broadly into this category.

Wave-cut platforms

Like cliffs wave-cut platforms show great variations both in detailed morphology and in the degree to which they bear the imprint of past conditions. On some coasts they are not developed at all, the cliffs plunging into deep water (p. 438); on others they are permanently or temporarily masked by beach shingle and sand. Where revealed the platforms are sometimes simple, comprising an even well-planed surface, but more usually they are morphologically complex, with numerous pinnacles, furrows, trenches, hollows and even clearly distinct levels of development. The latter inevitably raise the question of the precise height, in relation to high and low water mark, at which platforms are formed. Clearly waves are capable of eroding over a considerable vertical distance, particularly in areas such as north-east Brittany where exceptional tidal ranges are found. It is agreed by most authorities that wave attack is at its greatest at high spring tides, simply because the maximum depth of water allows the waves to reach the shore

194 Granite cliffs, Land's End, Cornwall. Note the jointing of the granite and the formation of a 'castellated' profile. [Eric Kay]

with minimum loss of energy due to friction with the bottom. In fact, coastal platforms which are being formed at the present time, in relation to existing sea-level, are generally exposed at low tide (for example, the Carstone platform off Hunstanton, Norfolk, or the chalk platform at the foot of Culver Cliff in the eastern Isle of Wight). As Steers has pointed out, it seems clear that effective cutting takes place only above mean sea-level. However, it is also evident that at low tide waves do not wholly lose their erosive ability, so that they may have the effect of 'rounding off' the edge of the platform or even of cutting a smaller lower platform seaward of the visible one. Another possibility to be considered is that of 'erosion' above high water, where spray from breaking waves wets the rock surface and accumulates in cracks and hollows, leading to chemical decay of the rock. Such 'water-level weathering' might conceivably lead to the formation of a crude bench above that produced by abrasion below high tide level. There is no doubt too that many wave-cut platforms bear the imprint of former sea-levels (Fig. 191). A typical case may be observed at Harris's Beach, Lannacombe, Cornwall. The highest platform here is at 24 ft O.D.; a middle and rather more extensive platform lies at 15 ft O.D.; and the lower main platform extends between 0 ft O.D. and low tide level. Since the mean spring tides reach only to 8–10 ft O.D. the higher levels seem explicable only in terms of wave erosion at former stands of the sea. This is confirmed by the association of the high platform in some localities with raised beach deposits (the so-called 'Patella' beach) and 'head' deposits of periglacial origin.

The detailed morphology of wave-cut platforms can be related also to factors of rock hardness, jointing, structure, chemical composition and so on. Where the rock is relatively soft and homogeneous, broad and even surfaces, with possibly some minor furrows etched out along small joint-lines, are normally formed. The platforms at the foot of the chalk cliffs between Brighton and Eastbourne illustrate this well (Fig. 3). Where, on the other hand, the rock is very hard and massively jointed (for example, the granite at Land's End) wave-cut platforms are usually poorly developed; and if they do occur are very irregular, owing to wave 'quarrying' along joints. In sedimentary rocks the angle of dip may be influential. Horizontal structures favour stepped platforms, whereas tilted and folded rocks, particularly where these have well-marked bedding-planes separating individual strata of variable resistance, will give a 'corrugated' surface, with some rock ribs standing several feet above the general level of the platform. Where igneous intrusions run at right angles to the shore (as in the case of the narrow dykes on the southern and eastern coasts of Arran), either 'rock groynes' or deep steep-walled clefts will diversify the platform. Major joints and faults, especially where the latter are allied to fault-breccias, are hollowed into narrow trenches, which may be traced across the platform into caves or geos in the cliff face. On limestones solutional effects are often striking, and the rock surface may be pitted by innumerable small corrosion hollows, separated by extremely sharp ridges and pinnacles (coastal lapiés).

Depositional features

These may be divided into three main groups: beaches, which are formed by breaking waves; sand-dunes, which are the result of wind transportation and deposition; and salt-marshes, produced by the settling out of fine mud held in suspension. Of these the first is the most important, in that beaches are ubiquitous, while dunes and marshes are more local in occurrence. In addition, beaches comprise a great variety of secondary forms (sand and shingle spits, bars, tombolos, cuspate forelands and so on).

Beaches

An idealised beach consists of two main and several minor elements (Fig. 195). The main elements are (i) the upper beach, which is often composed of coarse material such as pebbles and unbroken shells and has a steep gradient towards the sea (perhaps as high as 10–20°), and (ii) the lower beach, which is usually formed of sand, finely comminuted shells and even mud and is of very gentle gradient (as low as 2° or less). In some instances the two elements are separated by quite a sharp break of slope, but in others the less coarse material of the upper beach, with a comparatively low angle of rest, is washed downwards, with the result that the beach profile as a whole assumes a concave form. Minor elements, which are superimposed on to this generalised section, are: (i) The storm beach, which is a well-defined and semi-permanent ridge of coarse shingle standing above the level of highest spring tides. On the landward slope there may be a low and sometimes marshy depression between the edge of the shingle and the foot of the backing cliff. (ii) Beach ridges, or 'berms', which are built up by constructive waves at successive levels below that of high spring tides. On a high and steep shingle beach as many as five or six berms may be found at any one time. (iii) Beach cusps, which are small regular embayments developed

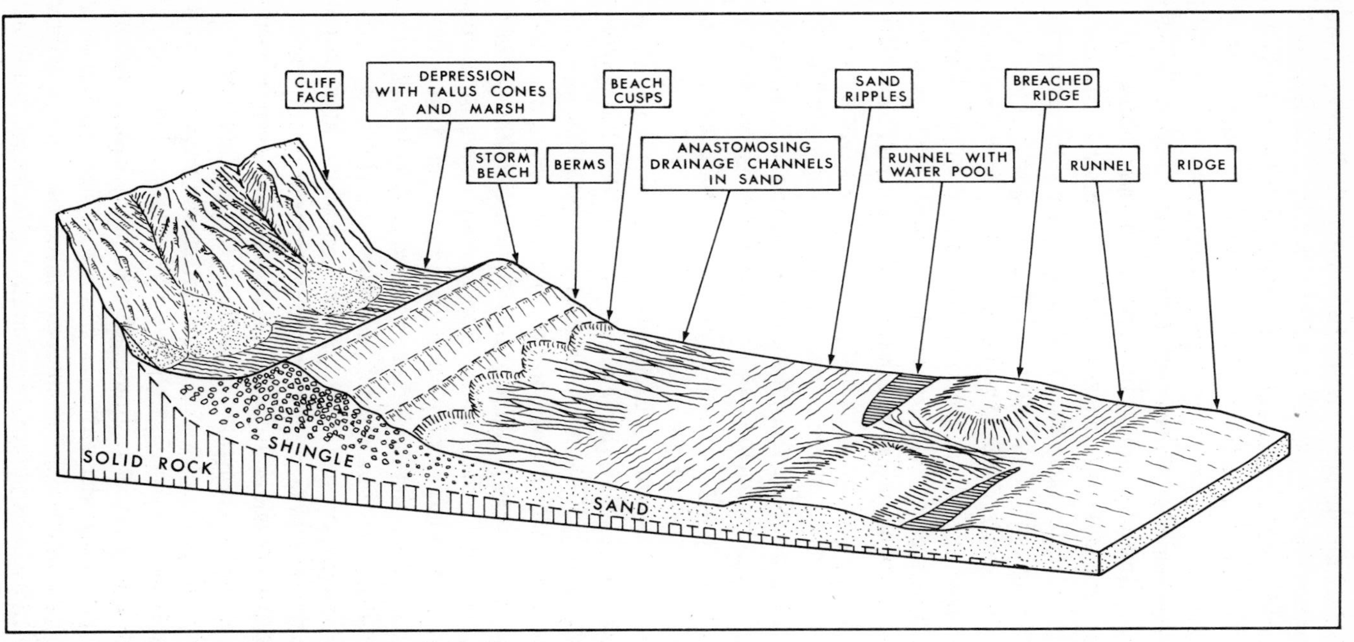

195 **The idealised features of a sand and shingle beach**

on the face of the shingle or at the junction of the steeper shingle beach and the gentler sand beach. In the latter case the re-entrants are floored by sand and small shingle, and the seaward projecting horns are of large pebbles. (iv) Patterns of ripples which are well developed on the lower beach where this is formed of sand. Symmetrical ripples, with sharp crests and rounded troughs, result from wave action and the associated forward movement of sand as the crest passes and the subsequent return with the trough. Asymmetrical ripples, with more rounded crests, are formed by tidal currents; the steeper slopes face in the direction towards which the current is running. (v) Minute and often anastomosing channels, formed in the sand by water draining from the upper beach or from the sand exposed at low tide. (vi) Ridges and runnels, which are broad and gentle rises and depressions aligned parallel to the shoreline and found towards the seaward margins of the sand beach. The runnels are usually occupied by elongated pools of sea-water, which may or may not escape via channels cut at right angles through ridges to the sea.

Whether or not all of these features will be found on any actual beach depends on several factors. One of the most important is the composition of the beach. If this is of sand only, the gradient of the upper beach is much reduced, and minor features such as berms or cusps are absent or feebly developed. Conversely, on a steep beach where shingle is especially abundant (for example, Chesil Bank), no gentle lower section including all the minor forms described above will be revealed at low water. Beach composition naturally varies from place to place. In some areas it is possible to relate this to local conditions. For example, the beaches near St Malo and Dinard in Brittany are predominantly of sand. This is to a large extent a reflection of the fact that the local rocks (granulites and schists) break down not into pebbles but directly into the constituent grains of quartz, felspar, mica, etc. However, owing to the effects of longshore drift most beaches comprise a high proportion of non-indigenous materials. An obvious example is the great shingle foreland of Dungeness, Kent. This is made up almost entirely of flint pebbles, which can only have been derived from Chalk outcrops to the west (beyond Eastbourne) or the east (beyond Folkestone); the local rocks, sandstones and clays of the Wealden Series, are capable of supplying little in the way of resistant pebbles. Many beaches in the British Isles have probably derived their constituents from the sea-floor, particularly where the latter was covered during the Quaternary by glacial deposits and outwash gravels and sands. The constructional features of north Norfolk and the southern shores of the Moray Firth can be partially explained as the outcome of reworking of such deposits (p. 428). Similarly, the magnificent sandy beaches common in south Wales, even in areas where the local rocks are limestones, may have been derived in part from outwash materials deposited below present sea-level on the northern margins of the Bristol Channel.

The many forms assumed by the beach in plan must now be discussed in some detail. Within small bays beaches follow gentle curves, of the catenary type, or occasionally approach rectilinearity. The actual orientation of the beach is normally related to the direction of approach of waves, so far as is possible within the limits set by the confining headlands. This is achieved by longshore drifting, which may have the secondary effect of sorting the beach constituents; the coarser pebbles may be carried to the leeward end of the bay and there built by the waves into a higher and steeper beach than that at the windward end. A good example of both sorting and orientation is afforded by Christchurch Bay (pp. 435–6). The beaches in the west, at Mudeford, are predominantly of sand, but there is a general increase of shingle eastwards to Milford-on-Sea and Hurst Castle Spit, which consists of a high ridge of pebbles aligned approximately parallel to the prevalent and dominant waves approaching from the south-west. Chesil Bank, situated at the eastern extremity of Lyme Bay, shows the same tendency in a more extreme form. This great bar, which becomes progressively broader and higher towards the south-east (it is 23 ft above high water mark at Abbotsbury, and 43 ft close to Portland), faces the dominant waves passing up the English Channel from the south-west. Lewis has suggested that the beach is not quite perfectly orientated, for the largest Atlantic breakers strike it at a slightly oblique angle and promote some eastward drift of pebbles. It is necessary to remember that Chesil is also affected by smaller waves generated within the Channel. At times these may approach from the south, causing a westward beach drift. Thus the present beach may represent a rough balance between two conflicting tendencies. Perhaps the most extraordinary feature of Chesil Bank is the almost perfect grading of its constituent shingle, from pea-size in the extreme west to pebbles 2–3 inches in diameter at Portland. Many explanations of this phenomenon have been given, none of them wholly convincing. The action of tidal currents has been invoked by some writers, but is no longer regarded as of any importance. An early theory, that of Sir John Coode, implied that Chesil Bank had been built south-eastwards by wind waves, and that these are capable (for reasons which are not clear) of moving large pebbles more easily than small ones, hence the grading. It seems more reasonable to suppose that the largest breakers, from the south-west, can move all grades of shingle towards Portland, but that the less powerful waves from the south can only transport the finer material back towards West Bay. However, this theory—which remains unproven—seems inadequate to account for the perfect grading. Some theories have attempted to explain a reverse grading of the shingle below low water, but since it is now known that this does not exist these have lost their point.

Discussion of the orientation and form of beaches within bays naturally leads on to a consideration of spits and bars, whose development is less confined by the pre-existing pattern of bays and headlands. The former, which are especially common on certain parts of the English coastline (for example, central southern England and East Anglia), are usually developed at points where the shoreline undergoes a sharp change of direction or, most frequently of all, at river mouths and on the windward sides of estuaries (Figs 196 and 197). In these locations shingle and sand moved along the shore by obliquely breaking waves is held up and fashioned into a projecting beach ridge. Once initiated the spit may continue to grow rapidly, particularly where there is at little depth a suitable foundation of sand- and mud-banks, salt-marshes or even drowned river terraces. It has been calculated that during the seventeenth, eighteenth and nineteenth centuries Orford Ness grew at an average rate of 15 yards per year. However, in many instances the extension of the spit is ultimately halted by such factors as a diminution in the supply of shingle, the presence of a deep-water channel, or the presence of strong river and tidal currents that sweep material from the end of the spit as quickly as it arrives. As a result some spits attain an 'equilibrium form' and thereafter experience little change (for example, Hurst Castle, Need's Ore and Calshot Spits on the northern shores of the Solent). Where conditions allow the further growth of the spit, so that a river outfall is diverted a considerable distance along the coast, the danger is that the spit will be breached, either by river flood water building up behind it, or by vigorous wave attack at a point where it is abnormally narrow. If this occurs and the breach is not healed by continued longshore drift combined with constructive waves, the far end of the spit will become dead and its constituents eventually washed onshore by wave action. The process is demonstrated by the Great Yarmouth Spit (p. 436), and in due course may affect the remarkable spit of Orford Ness, which diverts the river Alde some 11 miles to the south (Fig. 197E). The neck of the spit, near Aldeburgh, was in fact temporarily flattened during the great storm-surge of 31 January–1 February 1953. On the south coast at Mudeford, Hampshire, the sand spit across the mouth of Christchurch Harbour consists of a semi-permanent section, capped by dunes, stretching north-eastwards from Hengistbury Head, and beyond

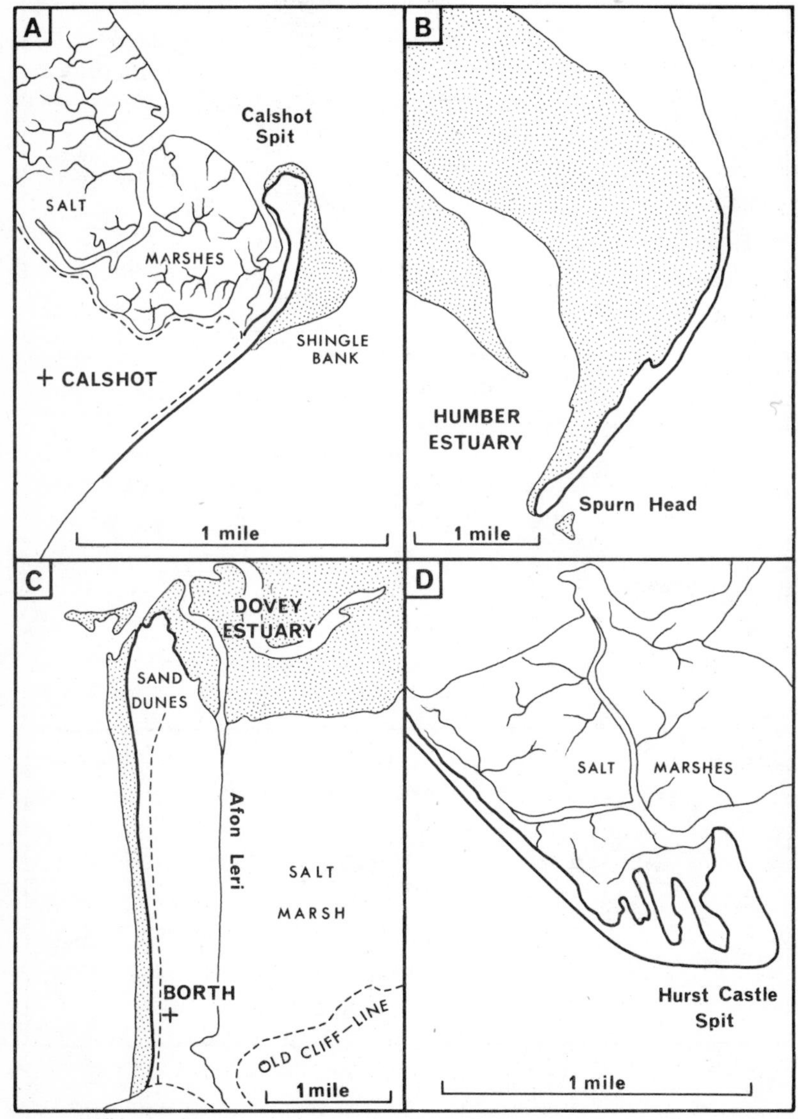

196 Selected spits: Calshot Spit (A); Spurn Head (B); Borth Spit (C); Hurst Castle Spit (D)

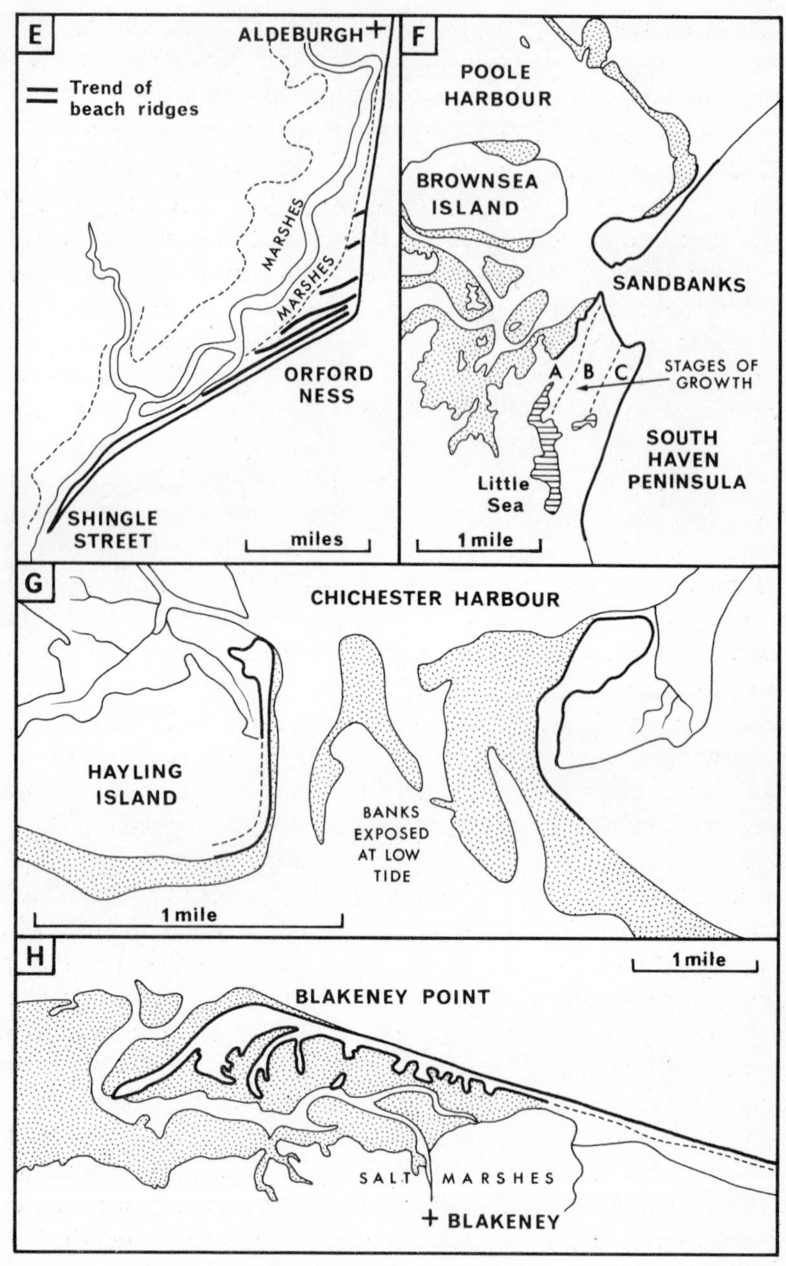

197 Selected spits: Orford Ness (E); South Haven Peninsula and Sandbanks (F); at the mouth of Chichester Harbour (G); Blakeney Point (H)

that a temporary section, which prior to the 1930s periodically grew a further distance of up to 2 miles before being broken, usually during south-east gales coinciding with high spring tides.

At the mouths of some estuaries 'double spits' are formed (Figs 197F and G). The two spits grow towards each other, but are prevented from joining by the action of strong scour by tidal currents. The spits are sometimes symmetrical, as at the mouth of Chichester Harbour, or one spit may prograde more rapidly and stand some distance to the seaward of the other (for example, at the mouth of Poole Harbour, where the spit at South Haven peninsula shows three important stages of broadening and is stepped forward by about half a mile relative to the much narrower sand spit at Sandbanks). The existence of double spits seems to imply that the longshore drift on either side of an estuary mouth can be in opposed directions. Sometimes this is feasible, as at South Haven-Sandbanks where waves from the south-east could cause beach material to move westwards from the Bournemouth area and northwards from the vicinity of Swanage. However, it is likely that in most cases there are other contributory factors. Where the estuary mouths are narrow, tidal currents will assist in the moulding of the ends of the spits (especially where these are composed of sand), producing long recurves pointing into the estuary. Again, material which crosses the outlet from the spit on the windward side may sustain the growth of the beach on the leeward side. Although river mouths and estuaries impede sand and shingle movement, they do not stop it altogether, as the ridges at Shingle Street, south of the tip of Orford Ness, clearly testify.

One of the most notable features of spits is the growth of recurves at their distal ends (Fig. 198). In the first instance, these are probably due to wave refraction (p. 430), which causes the free end of the beach to curve gently round, but once in existence are undoubtedly modified by local waves approaching from a direction counter to those responsible for the prevalent drift. This is shown by the angular breaks, in plan, of many older lateral beaches (for example, those of the central part of Scolt Head Island), and the manner in which some are orientated precisely to face the possible source of smaller waves (for instance, the main lateral at Hurst Castle Spit). In a complex shingle or sand spit the recurves, developing in each case at a former 'far point', record the main stages of growth. It is usual to find that the younger laterals curve round gently to join the main ridge of the spit, but that the older laterals meet it almost at right angles. This is due to the fact that many spits do not rely for their growth on beach material derived from some distance along

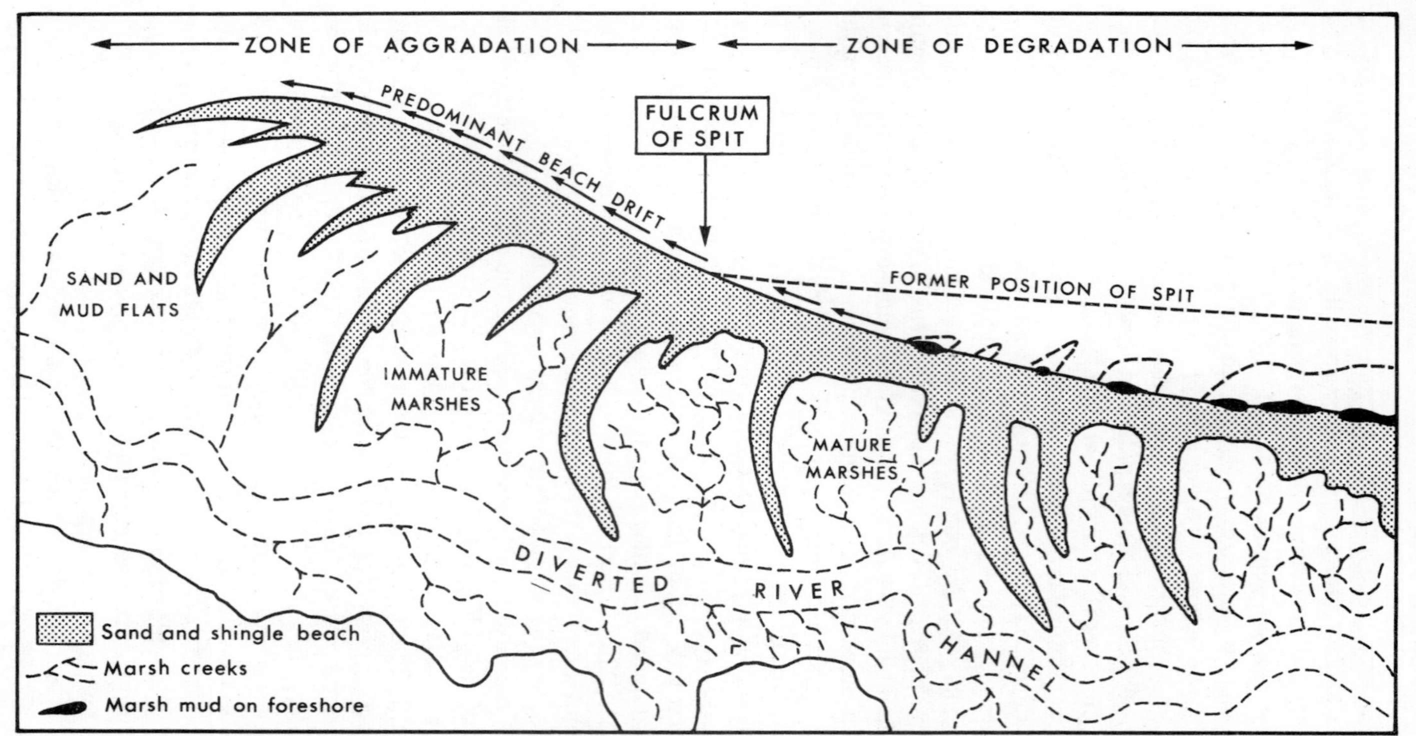

198 The development of an idealised compound spit

the coast, but are subject to erosion in their older parts and aggradation in their newer. Thus they comprise 'fixed' quantities of sand or shingle which is simply being reworked by wave action. The point at which erosion gives way to deposition has been referred to as the 'fulcrum' of the spit. In effect the spit is tending to rotate about the fulcrum, in such a way that it is (a) becoming orientated to face the oncoming waves, and (b) increasing the angle at which it meets the coast as a whole.

Offshore bars also display many of these characteristics—indeed the distinction between spits and bars is sometimes difficult to make, as a consideration of two such similar forms as Scolt Head Island (evidently a bar) and Blakeney Point (evidently a spit) will show. Bars are most common on very gently shelving coasts, particularly where the offshore zone has been the scene of extensive deposition of shingle and sand in the past (p. 428). The most difficult problem posed by bars is that of their initiation. The development of submerged bars (such as breakpoint sand bars) is well understood, but it is not clear how such features can be raised above sea-level and made permanent. The explanation of D. W. Johnson, that on a shelving coast the larger waves break well away from the shore in such a way that they erode the sea-bed and throw up what is in effect a swash bar (p. 433), is not wholly convincing. However, once formed offshore bars may be further prograded by constructive wave action or, as is more usually the case, driven landwards by large waves which overtop them and transfer beach material on to the inner side of the bar. In the latter process marsh muds that had accumulated on the land side are often exposed on the foreshore. Furthermore, offshore bars can be considerably lengthened by longshore drifting. This is admirably illustrated by The Bar at Nairn, on the southern shore of the Moray Firth. The south-western part of this consists of some eighteen gently curving shingle ridges, running approximately parallel to each other and recording the steady growth of the bar in this direction. J. A. Steers has suggested that, as little fresh material is being added to The Bar at the present day, the whole formation is undergoing a gradual shift to the south-west.

In addition to spits and bars there are numerous other beach forms produced by wave action (such as tombolos, which are sand and/or shingle accumulations linking islands to the mainland), but of these only one type merits detailed consideration. This is the large-scale prograding beach sometimes known as a 'ness' or, in its more extreme form, a 'cuspate' foreland. Sometimes such features result from the modification of spits or bars. Thus the great shingle complex of Dungeness has been explained by Lewis (1932) in terms of the re-orientation of a major spit once crossing the outer side of Romney Marsh into component sections facing south-westwards and eastwards—in each case directions of dominant wave approach. Orford Ness, essentially a spit, has also been modified in a similar fashion though to a lesser degree. Nesses are also the result of successive beaches forming on the open coast. An interesting example is Winterton Ness, in Norfolk, where sand-dunes resting on underlying shingle ridges extend for a quarter of a mile in front of the old cliff-line. Both to the north and south erosion is locally severe, and at Winterton itself the cliffs were washed by the sea up to the beginning of the eighteenth century. The change from erosion to deposition seems to have been sudden, and possibly to have been related to the movement onshore of an offshore bank. Farther to the south Lowestoft Ness is derived from sand and shingle washed onshore after the breaching of the Great Yarmouth Spit (p. 436).

Sand-dunes

Dunes are characteristic of all coastal areas where large expanses of sand are revealed at low tide, dried by the sun and wind, and blown onshore by the prevailing or even occasional winds. Among the most important dune areas in Britain are the Lancashire coast north of Liverpool (notably at Southport, much of which is built on land washed by the sea in the eighteenth century), parts of the coast of West Wales (for example, the large forelands of Morfa Harlech and Morfa Dyffryn), much of the south Wales coast (p. 409), and parts of the Devon and Cornwall coasts (for example, Braunton Burrows and Penhale Sands, near Newquay). These are all affected by prevailing westerly and south-westerly winds, and most are associated with river estuaries in which large banks and spreads of sand have been deposited. It is noticeable that on the south coast of England large dune-areas are rare, and that on the east coasts of England and Scotland the largest dunes occur on stretches that face generally northwards (such as the dune-belt of north Norfolk between Hunstanton and Weybourne, and the great dune-mass of the Culbin Sands). A partial explanation may be that in the latter areas there is some opportunity for winds with a westerly component to affect the beach.

Sand-dunes require a foundation on which to grow, and this may be provided in a number of ways, for instance by (i) prograding beaches at the head of a sandy bay, (ii) shingle and sand spits and bars, (iii) low-lying marshy areas near the coast, and (iv) by the cliff face, if this is suitably degraded, the slopes of estuaries and even the plateau-top a short distance inland. In the growth and stabilisation of dunes, whatever their situation, certain types of vegetation play a leading role—indeed in their absence drifting and totally formless spreads of sand alone would exist. Dunes are initiated when the beach or small mounds of sand lying on it are colonised by two species of grass, namely *Agropyrum junceum*, or sea-couch grass, and *Ammophila arenaria*, or marram (Fig. 173). These grasses, whose ramifying root systems hold the sand *in situ*, trap further sand, with the result that individual small mounds grow and coalesce. This is especially true of marram, which grows in tufts and indeed depends for its continued healthy growth on fresh supplies of sand brought by the wind. Under favourable circumstances the dunes will grow steadily in height, to 20–50 ft or even more, and individual dunes will amalgamate to give a continuous dune-line backing the beach.

The actual forms assumed by dunes will vary immensely. Often the pattern seems quite chaotic, but sometimes developing dunes are of a parabolic shape, with the concave (in plan) side facing the wind; in other words, a parabolic dune is the reverse of the desert barchan (p. 302). This has been explained in terms of the more effective stabilisation of the lower ends of the dune by marram, at a time when the central high and less vegetated part is migrating inland. Many dunes are characterised by 'blow-outs'. Once the vegetation cover is breached (for example, by wave attack on the front of the dune), the unconsolidated sand is easily blown away, and a large hollow or 'corridor' through the dune-line can be quickly formed by deflation. As a result, new dunes can be formed inside the old dune-line, and in this way the dune-area may be extended inland. Alternatively, where the sand supply is exceptionally favourable, new dunes can be added on the seaward side, and in this way a number of parallel and individual dune-ridges can result. However, when this happens the inner dunes, cut off from the sand source, will eventually suffer serious decay (into remanié dunes) or complete destruction.

Few features of the coast give such an impression of impermanence as dunes. Those of the east coast of England suffered badly from direct wave erosion during the storm-surge of 1953, and many breaches would not have been easily repaired but for the planting of marram and the construction of brushwood barriers to accelerate the replenishment of the sand. In the past many British dune-areas have experienced even greater vicissitudes. The large-scale movement of sand inland, to destroy settlements and churches, has been noted in many localities.

The best-known and most striking case is that of the Culbin Sands, which overwhelmed the Culbin Estate after 1695 and completely covered an area of rich cultivated land. In the words of Steers, this was a 'fertile estate before 1700, a desert in the nineteenth century'.

Salt-marshes

In most sheltered coastal localities (within estuaries, behind spits and bars, in the compartments of recurved spits, and in certain deeply penetrating bays) sedimentation to give mud-flats and vegetated marshes is an important process. The sediments themselves are brought in suspension and gradually settle out, particularly at times of slack water. Often the amount of mud in estuaries (for example, that of the Severn) is so great as to pose a problem of origin, for it is far more concentrated than in the river upstream or in the sea itself. The first stage in the development of a marsh involves the uneven deposition of mud, usually on a basement of sand. Accumulation is naturally greatest at the inner edges of the area flooded at high tide, but away from the shore a thinner layer of mud, with an irregular surface that guides the drainage of water as the tide ebbs, will also slowly be formed. An important event is the colonisation of these muddy areas by halophytes (notably *Salicornia*, or marsh samphire, and *Suaeda maritima*). Although only annuals, these have an effect on the rate of sedimentation, especially where they succeed in forming a close vegetation covering over the young marsh. As the latter is built up, other plants such as *Aster tripolium* (sea-aster), *Limonium vulgare* (sea-lavender) and *Puccinellia maritima* (common marsh grass) give rise to a dense vegetation growth rising some 6 inches above the surface of the mud. Trapping of sediment, by the often fleshy foliage, is now greatly accelerated, the marsh grows upwards at a rate approaching 1 cm a year, and the system of drainage creeks becomes firmly established. Along the banks of the latter another species, the bushy *Halimione portulacoides*, may thrive, and by trapping mud in large quantities give rise to a slight levée effect. In time, the rate of marsh growth of necessity slows down, if only because the number of inundations is reduced, until finally it is covered only at the highest spring tides. At this stage the fully mature marsh (now colonised by plants such as *Juncus maritimus* and *Artimesia maritima*) may be reclaimed by the building of a protective sea-wall.

This whole sequence of events, and in particular the order of plant colonisation, is a generalised one, and will vary according to local conditions. The major exception to be noted, in this country, is the

199 Salt-marsh development, Newtown Marshes, Isle of Wight. Note the clumps of *Spartina townsendii* colonising bare mud. [Eric Kay]

remarkable dominance of many of the south coast marshes by the species *Spartina townsendii* (Fig. 199). Since it was first found at Hythe, in Southampton Water, in 1870 this has spread throughout the estuaries of the region, becoming dominant in Poole Harbour by the 1920s. It has also been introduced elsewhere, on account of its unrivalled powers of trapping mud. In the words of Tansley 'no other species of salt-marsh plant, in north-western Europe at least, has anything like so rapid and so great an influence in gaining land from the sea'.

To the geomorphologist one of the most interesting features of marshes is their intricately meandering creek-systems (Fig. 200). These develop in the first place purely as drainage channels, and are the result not so much of erosion as of non-deposition (for mud accumulation is only effective on the flat vegetated interfluves). However, they can be greatly modified by erosion during the ebbing of the tide. Small 'headwater' creeks are extended by a process similar to knickpoint recession, the banks of meanders are undercut and collapse, and even

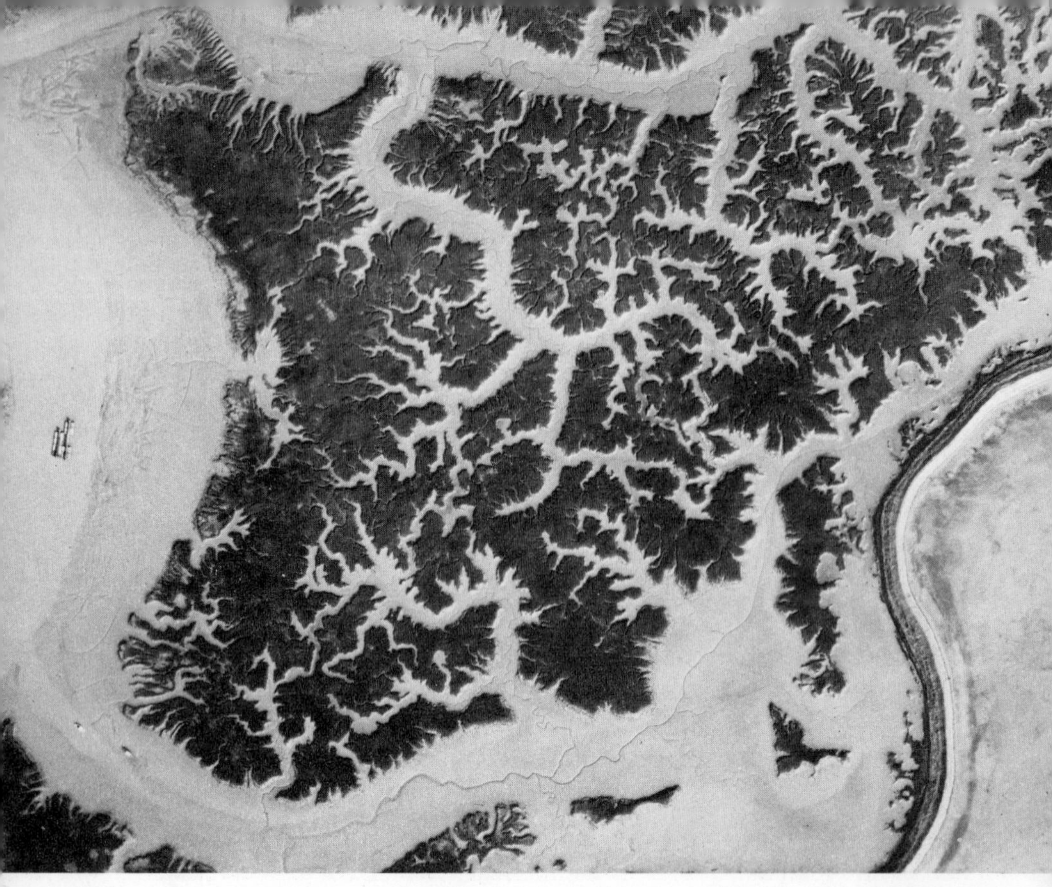

200 Aerial view of salt-marshes, Southampton Water. Note the ramifying creek patterns. [Aerofilms Ltd]

small-scale captures and abstractions can occur. The smaller creeks, drained by the smaller and least effective ebb streams, can become increasingly constricted, mainly as a result of the activity of *Halimione portulacoides*, and even blocked at some points. Elongated 'channel pans', contrasting with the rounded and shallow depressions, unoccupied by vegetation and representing points at which deposition has been ineffective, are the end-product of this process of creek dismemberment.

Marshes depend for their growth upon protection, and if this is removed erosion of the marsh mud is usually rapid. A common cause is the inward migration of a bar or spit, over the marsh which formed in its lee. Once revealed on the seaward side the mud is easily removed by wave attack. It is noticeable that even in protected estuaries marshes are often in a state of decay. Instead of the continuous and flat layer of vegetated mud, there are small mesa-like remnants and fallen masses. Probably the outer marsh is attacked and reduced through lateral erosion by ebbing tidewater near the mouths of the larger creeks, as well as by normal wave action. Occasionally the failure of the marsh vegetation itself can be the cause. An outstanding instance of this is the 'die-back', in the last thirty years, of *Spartina townsendii* in the Lymington and Beaulieu estuaries of Hampshire, and the resultant decay of the marshes here.

SELECT BIBLIOGRAPHY

CHAPTER 1

Davis, W. M. (1909). *Geographical Essays*. Boston.

Lawson, A. C. (1915). Epigene profiles of the desert. *Univ. Calif. Dept Geol.* Pub. No. 9, 23–48.

Lewis, W. V. (1944). Stream trough experiments and terrace formation. *Geol. Mag.* **81**, 241–53.

Lewis, W. V. and Miller, M. M. (1955). Kaolin model glaciers. *J. Glaciol.* **2**, 533–8.

Peltier, L. C. (1950). The geographic cycle in periglacial regions as it is related to climatic geomorphology. *Ann. Ass. Am. Geogr.* **40**, 214–36.

Penck, W. (1953). *The morphological analysis of landforms*. (Trans. from *Die Morphologie Analyse*. Stuttgart 1924, by H. Czech and K. C. Boswell.) London.

Russell, R. J. (1949). Geographical geomorphology. *Ann. Ass. Am. Geogr.* **39**, 1–11.

Strahler, A. N. (1950). Equilibrium theory of erosional slopes, approached by frequency distribution analysis. *Amer. J. Sci.* **248**, 673–96 and 800–14.

Waters, R. S. (1958). Morphological mapping. *Geography*, **43**, 10–17.

Wood, A. (1942). The development of hillside slopes. *Proc. Geol. Ass. Lond.* **53**, 128–40.

Wooldridge, S. W. (1958). The trend of geomorphology. *Trans. Inst. Brit. Geogr.* **25**, 29–35.

Wooldridge, S. W. and Linton, D. L. (1955). *Structure, surface and drainage in south-east England*. London.

CHAPTER 2

Bates, R. E. (1939). Geomorphic history of the Kickapoo region, Wisconsin. *Bull. geol. Soc. Amer.* **50**, 819–80.

Baulig, H. (1948). Le problème des méandres. *Bull. Soc. Belge d'Etudes Geog.* **17**, 103–43.

Brown, E. H. (1952). The river Ystwyth, Cardiganshire: a geomorphological analysis. *Proc. Geol. Ass. Lond.* **63**, 244–69.

Corbel, J. (1959). Erosion en terrain calcaire. *Ann. de Geogr.* **68**, 97–120.

Crickmay, C. H. (1933). The later stages of the cycle of erosion. *Geol. Mag.* **70**, 337–47.

Davis, W. M. (1909). *Geographical essays*. Boston.

Dury, G. H. (1958). Tests of a general theory of misfit streams. *Trans. Inst. Brit. Geogr.* **25**, 105–18.
Dury, G. H. (1966). *Essays in geomorphology* (ed. G. H. Dury). London.
Gilbert, G. K. (1909). The convexity of hilltops. *J. Geol.* **17**, 344–51.
Green, J. F. N. (1934). The river Mole: its physiography and superficial deposits. *Proc. Geol. Ass. Lond.* **45**, 35–69.
Horton, R. E. (1945). Erosional development of streams and their drainage basins; hydrophysical approach to quantitative morphology. *Bull. geol. Soc. Amer.* **56**, 275–370.
Kesseli, J. (1941). The concept of the graded river. *J. Geol.* **49**, 561–88.
Kidson, C. (1953). The Exmoor storm and the Lynmouth floods. *Geography.* **38**, 1–9.
Leighly, J. B. (1934). Turbulence and the transportation of rock debris by streams. *Geogr. Rev.* **24**, 453–64.
Leopold, L. B. (1953). Downstream change of velocity in rivers. *Amer. J. Sci.* **251**, 606–24.
Leopold, L. B. and Wolman, M. G. (1960). River meanders. *Bull. geol. Soc. Amer.* **71**, 769–94.
Leopold, L. B., Wolman, M. G. and Miller, J. P. (1964). *Fluvial processes in geomorphology*. San Francisco and London.
Mackin, J. H. (1948). Concept of the graded river. *Bull. geol. Soc. Amer.* **59**, 463–512.
Rastall, R. H. (1944). Rainfall, rivers and erosion. *Geol. Mag.* **81**, 39–44.
Reiche, P. (1950). A survey of weathering processes and products. *Univ. of New Mex. Pub. in Geol.* No. 3.
Schumm, S. A. (1956). Evolution of drainage systems and slopes in badlands at Perth Amboy, New Jersey. *Bull. geol. Soc. Amer.* **67**, 597–646.

CHAPTER 3

Clayton, K. M. (1966). The origin of the landforms of the Malham area. *Field Studies.* **2**, 359–84.
Cotton, C. A. (1942). *Geomorphology*. London.
Hack, J. T. (1960). Interpretation of erosional topography in humid temperate regions. *Amer. J. Sci.* **258**, 80–97.
Pitty, A. F. (1965). A study of some escarpment gaps in the southern Pennines. *Trans. Inst. Brit. Geogr.* **37**, 127–45.
Scarth, A. (1966). The physiography of the fault-scarp between the Grand Limagne and the Plateaux des Dômes, Massif Central. *Trans. Inst. Brit. Geogr.* **38**, 25–40.
Small, R. J. (1961). The morphology of chalk escarpments. *Trans. Inst. Brit. Geogr.* **29**, 71–90.
Small, R. J. (1964). In *A survey of Southampton and its region* (ed. F. J. Monkhouse). Southampton.
Small, R. J. (1964). The escarpment dry valleys of the Wiltshire Chalk. *Trans. Inst. Brit. Geogr.* **34**, 33–52.
Trueman, A. E. (1949). *Geology and scenery in England and Wales.* Harmondsworth.

CHAPTER 4

Baulig, H. (1928). *Le plateau central de la France*. Paris.

King, L. C. (1948). A theory of bornhardts. *Geogr. J.* **112**, 83–7.

King, L. C. (1958). Correspondence on the problem of tors. *Geogr. J.* **124**, 289–91.

Linton, D. L. (1955). The problem of tors. *Geogr. J.* **121**, 470–87.

Marres, P. (1935). *Les Grands Causses*. Tours.

Palmer, J. and Neilson, R. A. (1962). The origin of granite tors, Dartmoor, Devonshire. *Proc. Yorks. Geol. Soc.* **33**, 315–40.

Sweeting, M. M. (1950). Erosion cycles and limestone caves in the Ingleborough district. *Geogr. J.* **116**, 63–78.

Trueman, A. E. (1949). *Geology and scenery in England and Wales*. Harmondsworth.

Waters, R. S. (1957). Differential weathering in oldlands. *Geogr. J.* **123**, 503–13.

Waters, R. S. (1964). In D. Brunsden *et al.* The denudation chronology of parts of south-western England. *Field Studies*. **2**, 115–32.

Waters, R. S. (1964). The Pleistocene legacy to the geomorphology of Dartmoore, in *Dartmoor Essays*.

CHAPTER 5

Chorley, R. J. (1962). Geomorphology and General Systems Theory. *U.S. Geol. Surv. Prof. Paper*, 500–B.

Cotton, C. A. (1942). *Climatic accidents in landscape making*. Christchurch.

Cotton, C. A. (1961). The theory of savanna planation. *Geography*. **46**, 89–101.

Crickmay, C. H. (1933). The later stages of the cycle of erosion. *Geol. Mag.* **70**, 337–47.

Davis, W. M. (1909). *Geographical essays*. Boston.

Hack, J. T. (1960). Interpretation of erosional topography in humid temperate regions. *Amer. J. Sci.* **258**, 80–97.

Johnson, D. W. (1919). *Shore processes and shoreline development*. New York.

King, L. C. (1948). A theory of bornhardts. *Geogr. J.* **112**, 83–7.

King, L. C. (1950). The study of the world's plainlands. *Quart. J. geol. Soc. Lond.* **106**, 101–31.

King, L. C. (1962). *The morphology of the Earth*. Edinburgh.

Peltier, L. C. (1950). The geographic cycle in periglacial regions as it is related to climatic geomorphology. *Ann. Ass. Am. Geogr.* **40**, 214–36.

Penck, W. (1953). *The morphological analysis of landforms*. (Trans. from *Die Morphologie Analyse*, Stuttgart 1924, by H. Czech and K. C. Boswell.) London.

Pugh, J. C. (1966) in *Essays in geomorphology* (ed. G. H. Dury). London.

Ruxton, B. P. and Berry, L. (1957). Weathering of granite and associated erosional features in Hong Kong. *Bull. geol. Soc. Amer.* **68**, 1263–90.

Saunders, E. M. (1921). The cycle of erosion in a karst region (after Cvijić). *Geogr. Rev.* **11**, 593–604.

Strahler, A. N. (1952). Dynamic basis of geomorphology. *Bull geol. Soc. Amer.* **63**, 923–38.

Thomas, M. F. (1966). Some geomorphological implications of deep weathering patterns in crystalline rocks in Nigeria. *Trans. Inst. Brit. Geogr.* **40**, 173–98.

CHAPTER 6

Baulig, H. (1940). Le profil d'équilibre des versants. *Ann. de Geog.* **49**, 81–97.

Clark, M. J. (1965). The form of chalk slopes. *Southampton Research Series in Geography.* **2**, 3–34.

Cotton, C. A. (1952). The erosional grading of convex and concave slopes. *Geogr. J.* **118**, 197–204.

Fenneman, N. M. (1908). Some features of erosion by unconcentrated wash. *J. Geol.* **16**, 746–54.

Gilbert, G. K. (1909). The convexity of hilltops. *J. Geol.* **17**, 344–51.

Horton, R. E. (1945). Erosional development of streams and their drainage basins; hydrophysical approach to quantitative morphology. *Bull. geol. Soc. Amer.* **56**, 275–370.

King, L. C. (1957). The uniformitarian nature of hillslopes. *Trans. Roy. Geol. Soc. Edinburgh.* **17**, 81–102.

Lawson, A. C. (1932). Rainwash erosion in humid regions. *Bull. geol. Soc. Amer.* **43**, 703–24.

Penck, W. (1953). *The morphological analysis of landforms.* (Trans. from *Die Morphologie Analyse*, Stuttgart 1924, by H. Czech and K. C. Boswell.) London.

Savigear, R. A. (1952). Some observations on slope development in south Wales. *Trans. Inst. Brit. Geogr.* **18**, 31–51.

Schumm, S. A. (1956). Evolution of drainage systems and slopes in badlands at Perth Amboy, New Jersey. *Bull. geol. Soc. Amer.* **67**, 597–646.

Strahler, A. N. (1950). Equilibrium theory of erosional slopes, approached by frequency distribution analysis. *Amer. J. Sci.* **248**, 673–96 and 800–14.

Wood, A. (1942). The development of hillside slopes. *Proc. Geol. Ass. Lond.* **53**, 128–40.

Young, A. (1960). Soil movement by denudational processes on slopes. *Nature.* **188**, 120–22.

Young, A. (1963). Some field observations of slope form and regolith, and their relation to slope development. *Trans. Inst. Brit. Geogr.* **32**, 1–29.

CHAPTER 7

Brown, E. H. (1957). The physique of Wales. *Geogr. J.* **122**, 208–30.

Coleman, A. (1958). The terraces and antecedence of a part of the river Salzach. *Trans. Inst. Brit. Geogr.* **25**, 119–34.

Davis, W. M. (1895). On the development of certain English rivers. *Geogr. J.* **5**, 128–46.

Horton, R. E. (1945). Erosional development of streams and their drainage basins; hydrophysical approach to quantitative morphology. *Bull. geol. Soc. Amer.* **51**, 275–375.

Jones, O. T. (1951). The drainage systems of Wales and the adjacent regions. *Quart. J. geol. Soc. Lond.* **107**, 201–25.

Jones, R. O. (1939). The evolution of the Neath-Tawe drainage. *Proc. Geol. Ass. Lond.* **50**, 530–66.

Linton, D. L. (1930). Notes on the development of the western part of the Wey drainage system. *Proc. Geol. Ass. Lond.* **41**, 160–74.

Linton, D. L. (1951). Midland drainage. *The Advancement of Science.* **7**, 449–56.

Pinchemel, P. (1954). *Les plaines de craie.* Paris.

Schumm, S. A. (1956). Evolution of drainage systems and slopes in badlands at Perth Amboy, New Jersey. *Bull. geol. Soc. Amer.* **67**, 597–646.

Sissons, J. B. (1954). The erosion surfaces and drainage system of south-west Yorkshire. *Proc. Yorks. Geol. Soc.* **29**, 305–42.

Small, R. J. (1962). A short note on the origin of the Devil's Dyke, near Brighton. *Proc. Geol. Ass. Lond.* **73**, 187-92.

Sparks, B. W. (1949). The denudation chronology of the dipslope of the South Downs. *Proc. Geol. Ass. Lond.* **60**, 165–215.

Strahler, A. N. (1957). Quantitative analysis of watershed geometry. *Trans. Amer. Geophys. Union.* **38**, 913–20.

Wager, L. R. (1937). The Arun river drainage pattern and the rise of the Himalaya. *Geogr. J.* **89**, 239–49.

Wooldridge, S. W. and Linton, D. L. (1955). *Structure, surface and drainage in south-east England.* London.

Yates, E. M. (1963). The development of the Rhine. *Trans. Inst. Brit. Geogr.* **32**, 65–87.

Zernitz, E. R. (1932). Drainage systems and their significance. *J. Geol.* **40**, 498–521.

CHAPTER 8

Baulig, H. (1936). The changing sea-level. *Trans. Inst. Brit. Geogr.* **3**.

Brown, E. H. (1960). *The relief and drainage of Wales.* Cardiff.

Bury, H. (1910). On the denudation of the western end of the Weald. *Quart. J. geol. Soc. Lond.* **66**, 640–92.

Clark, M. J., Lewin, J. and Small, R. J. (1967). The sarsen stones of the Marlborough Downs and their geomorphological implications. *Southampton Research Series in Geography.* **4**, 3–40.

Clayton, K. M. (1953). The denudation of part of the middle Trent basin. *Trans. Inst. Brit. Geogr.* **19**, 25–36.

Everard, C. (1956). Erosion platforms on the borders of the Hampshire Basin. *Trans. Inst. Brit. Geogr.* **22**, 33–46.

Jones, O. T. (1931). Some episodes in the geological history of the Bristol Channel region. *Report of Brit. Assn. Adv. Science.* 57–82.

King, L. C. (1950). The study of the World's plainlands. *Quart. J. geol. Soc. Lond.* **106**, 101–31.

Linton, D. L. (1951). Problems of Scottish scenery. *Scott. Geogr. Mag.* **69**, 65–85.

Wooldridge, S. W. (1950). The upland plains of Britain; their origin and geographical significance. *The Advancement of Science*. **7**, 162–75.

Wooldridge, S. W. (1952). The changing physical landscape of Britain. *Geogr. J.* **118**, 297–308.

Wooldridge, S. W. and Linton, D. L. (1955). *Structure, surface and drainage in south-east England*. London.

CHAPTER 9

Bagnold, R. A. (1941). *The physics of blown sand and desert dunes*. London.

Balchin, W. G. V. and Pye, N. (1955). Piedmont profiles in the arid cycle. *Proc. Geol. Ass. Lond.* **66**, 167–81.

Blackwelder, E. (1933). The insolation hypothesis of rock weathering. *Amer. J. Sci.* **26**, 97–113.

Bryan, K. and McCann, F. (1936). Successive pediments and terraces of the upper Rio Puerco in New Mexico. *J. Geol.* **44**, 145–72.

Davis, W. M. (1938). Sheetfloods and streamfloods. *Bull. geol. Soc. Amer.* **49**, 1337–1416.

Griggs, D. T. (1936). The factor of fatigue in rock weathering. *J. Geol.* **44**, 781–96.

Holmes, D. Chauncey (1955). Geomorphic development in humid and arid regions; a synthesis. *Amer. J. Sci.* **253**, 357–90.

Howard, A. D. (1942). Pediments and the pediment pass problem. *Journ. Geomorph.* **5**, 3–31 and 95–136.

Johnson, D. W. (1932). Rock planes of arid regions. *Geogr. Rev.* **22**, 656–65.

King, L. C. (1948). A theory of bornhardts. *Geogr. J.* **112**, 83–7.

Lawson, A. C. (1915). Epigene profiles of the desert. *Univ. Calif. Dept. Geol.* Pub. No. 9, 23–48.

Madigan, C. T. (1936). The Australian sand-ridge deserts. *Geogr. Rev.* **26**, 205–27.

McGee, W. J. (1897). Sheetflood erosion. *Bull. geol. Soc. Amer.* **8**, 87–112.

Peel, R. F. (1941). Denudational landforms of the central Libyan Desert. *Journ. Geomorph.* **5**, 3–23.

Peel, R. F. (1960). Some aspects of desert geomorphology. *Geography*. **45**, 241–62.

Peel, R. F. (1966). The landscape in aridity. *Trans. Inst. Brit. Geogr.* **38**, 1–23.

CHAPTER 10

Bryan, K. (1946). Cryopedology—the study of frozen ground and intensive frost action with suggestions on nomenclature. *Amer. J. Sci.* **244**, 622–42.

Bull, A. J. (1940). Cold conditions and landforms in the South Downs. *Proc. Geol. Ass. Lond.* **51**, 63–70.

Chambers, M. J. G. (1966) (1967). Investigations of patterned ground at Signy Island, South Orkney Islands. *Brit. Antarct. Surv. Bull.* **9**, 21–40: **10**, 71–83: **12**, 1–22.

Clark, M. J., Lewin, J. and Small, R. J. (1967). The sarsen stones of the Marlborough Downs and their geomorphological implications. *Southampton Research Series in Geography*. **4**, 3–40.

Corbel, J. (1961). Morphologie périglaciaire dans l'Arctique. *Ann. de Geogr.* **70**, 1–24.

Embleton, C. and King, C. A. M. (1968). *Glacial and periglacial geomorphology*. London.

Jenness, J. L. (1952). Erosive forces in the physiography of western Arctic Canada. *Geogr. Rev.* **42**, 238–52.

Kerney, M. P., Brown, E. H. and Chandler, T. J. (1964). The Late-glacial and Post-glacial history of the Chalk escarpment near Brook, Kent. *Phil. Trans. Roy. Soc. Series B.* **248**, 135–204.

Lewis, W. V. (1939). Snowpatch erosion in Iceland. *Geogr. J.* **94**, 153–61.

Ollier, C. D. and Thomasson, A. J. (1957). Asymmetrical valleys of the Chiltern Hills. *Geogr. J.* **123**, 71–80.

Peltier, L. C. (1950). The geographic cycle in periglacial regions as it is related to climatic geomorphology. *Ann. Ass. Am. Geogr.* **40**, 214–36.

Pissart, A. (1963). Les traces de 'Pingos' du Pays de Galles (Grande-Bretagne) et du Plateau des Hautes Fagnes (Belgique). *Zeit. Geomorph.* **7**, 147–65.

Rapp, A. (1960). Recent development of mountain slopes in Karkevagge and surroundings, northern Scandinavia. *Geogr. Annaler.* **42**, 65–200.

Sharp, R. P. (1942). Soil structures in the St. Elias Range, Yukon Territory. *Journ. Geomorph.* **5**, 274–301.

Taber, S. (1943). Perennially frozen ground in Alaska; its origin and history. *Bull. geol. Soc. Amer.* **54**, 1433–548.

Washburn, A. L. (1956). Classification of patterned ground and review of suggested origins. *Bull. geol. Soc. Amer.* **67**, 823–65.

Williams, J. E. (1949). Chemical weathering at low temperatures. *Geogr. Rev.* **39**, 129–35.

Williams P. J. (1959). An investigation into processes occurring in solifluction. *Amer. J. Sci.* **257**, 481–90.

CHAPTER 11

Battey, M. H. (1960). Geological factors in the development of Veslgjuvbotn and Vesl-Skautbotn. *Investigations on Norwegian cirque glaciers*, R.G.S. Research Series No. 4, 11–24.

Battle, W. B. R. and Lewis, W. V. (1951). Temperature observations in bergschrunds and their relationship to cirque erosion. *J. Geol.* **59**, 537–45.

Bowen, D. Q. and Gregory, K. J. (1965). A glacial drainage system near Fishguard, Pembrokeshire. *Proc. Geol. Ass. Lond.* **76**, 275–81.

Bremner, A. (1915). The capture of the Geldie by the Feshie. *Scott. Geogr. Mag.* **31**, 589–96.

Brown, J. C. (1959). The sub-glacial surface in east Hertfordshire and its relation to the valley pattern. *Trans. Inst. Brit. Geogr.* **26**, 37–50.

Carol, H. (1947). Formation of roches moutonnées. *J. Glaciol.* **1**, 57–9.

Dury, G. H. (1959). A contribution to the geomorphology of central Donègal. *Proc. Geol. Ass. Lond.* **70**, 1–27.

Dwerryhouse, A. R. (1902). The glaciation of Teesdale, Weardale and the Tyne valley and their tributary valleys. *Quart. J. geol. Soc. Lond.* **58**, 572–608.

Embleton, C. and King, C. A. M. (1968). *Glacial and periglacial geomorphology*. London.

Flint, R. F. (1957). *Glacial and Pleistocene geology*. New York.

Garwood, E. J. (1910). Features of Alpine scenery due to glacial protection. *Geogr. J.* **36**, 310–39.

Johnson, W. D. (1899). An unrecognised process in glacial erosion. *Science.* **9**, 106.

Kendall, P. F. (1902). A system of glacier lakes in the Cleveland Hills. *Quart. J. geol. Soc. Lond.* **58**, 471–571.

Lewis, W. V. (1940). The function of meltwater in cirque formation. *Geogr. Rev.* **30**, 64–83.

Lewis, W. V. (1947). Valley steps and glacial valley erosion. *Trans. Inst. Brit. Geogr.* **13**, 19–44.

Lewis, W. V. (1954). Pressure release and glacial erosion. *J. Glaciol.* **2**, 417–22.

Linton, D. L. (1949). Some Scottish river captures re-examined. *Scott. Geogr. Mag.* **65**, 123–31.

Linton, D. L. (1951). Watershed breaching by ice in Scotland. *Trans. Inst. Brit. Geogr.* **17**, 1–16.

Linton, D. L. (1963). The forms of glacial erosion. *Trans. Inst. Brit. Geogr.* **33**, 1–28.

Matthes, F. E. (1930). Geologic history of the Yosemite valley. *U.S. Geol. Surv. Prof. Paper*, No. 160.

McCall, J. G. (1952). The internal structure of a cirque glacier. *J. Glaciol.* **2**, 122–30.

Peel, R. F. (1949). A study of two Northumbrian spillways. *Trans. Inst. Brit. Geogr.* **15**, 75–89.

Peel, R. F. (1956). The profiles of glacial drainage channels. *Geogr. J.* **122**, 483–7.

Sissons, J. B. (1958). Sub-glacial stream erosion in southern Northumberland. *Scott. Geogr. Mag.* **74**, 163–74.

Sissons, J. B. (1958). The deglaciation of part of east Lothian. *Trans. Inst. Brit. Geogr.* **25**, 59–77.

Sparks, B. W. (1957). The evolution of the relief of the Cam valley. *Geogr. J.* **123**, 188–207.

Sparks, B. W. and West, R. G. (1964). The drift landforms around Holt, Norfolk. *Trans. Inst. Brit. Geogr.* **35**, 27–35.

West, R. G. and Donner, J. J. (1956). The glaciations of East Anglia and the East Midlands: a differentiation based on stone-orientation measurements of the tills. *Quart. J. geol. Soc. Lond.* **112**, 69–91.

CHAPTER 12

Guilcher, A. (1958). *Coastal and submarine morphology*. London.

Hardy, J. R. (1964). The movement of beach material and wave action near Blakeney Point, Norfolk. *Trans. Inst. Brit. Geogr.* **34**, 53–69.

Johnson, D. W. (1919). *Shore processes and shoreline development*. New York.

King, C. A. M. (1959). *Beaches and coasts*. London.

King, C. A. M. and Williams, W. W. (1949). The formation and movement of sand bars by wave action. *Geogr. J.* **113**, 70–85.

Lewis, W. V. (1932). The formation of Dungeness foreland. *Geogr. J.* **80**, 309–24.

Steers, J. A. (1946). *The coastline of England and Wales*. Cambridge.

Steers, J. A. (1953). *The sea coast*. London.

Steers, J. A. (1960). *The coast of England and Wales in pictures*. Cambridge.

Zenkovich, V. P. (1967). *Processes of coastal development*. (Trans. from Russian and ed. J. A. Steers.) Edinburgh.

Special number on the vertical displacement of shorelines in Highland Britain. *Trans. Inst. Brit. Geogr.* **39** (1966).

ABBREVIATIONS

(The abbreviations of journals in the bibliography follow the World List)

Amer. J. Sci.	American Journal of Science.
Ann. Ass. Am. Geogr.	Annals of the Association of American Geographers.
Ann. de. Geogr.	Annales de Geographie.
Brit. Antarct. Surv. Bull.	British Antarctic Survey Bulletin.
Bull. geol. Soc. Amer.	Bulletin of the Geological Society of America.
Bull. Soc. Belge d'Etudes Geog.	Bulletin Société Belge d'Etudes Geographie.
Geogr. J.	Geographical Journal.
Geogr. Annaler	Geographiske Annaler.
Geol. Mag.	Geological Magazine.
Geogr. Rev.	Geographical Review.
J. Geol.	Journal of Geology.
Journ. Geomorph.	Journal of Geomorphology.
J. Glaciol.	Journal of Glaciology.
Phil. Trans. Roy. Soc. Series B	Philosophical Transactions of the Royal Society Series B.
Proc. Geol. Ass. Lond.	Proceedings of the Geologists Association of London.
Proc. Yorks. Geol. Soc.	Proceedings of the Yorkshire Geological Society.
Quart. J. geol. Soc. Lond.	Quarterly Journal of the Geological Society of London.
R.G.S. Research Series	Royal Geographical Society Research Series.
Report of Brit. Ass. Adv. Science	Report of the British Association for the Advancement of Science.
Scott. Geogr. Mag.	Scottish Geographical Magazine.
Trans. Amer. Geophys. Union	American Geophysical Union Transactions.
Trans. Inst. Brit. Geogr.	Transactions of the Institute of British Geographers.
Trans. Roy. Geol. Soc. Edinburgh	Transactions of the Royal Geological Society of Edinburgh.
Univ. Calif. Dept. Geol.	University of California Department of Geology.
Univ. of New Mex. Pub. Geol.	University of New Mexico Publication in Geology.
U.S. Geol. Surv. Prof. Paper	United States Geological Survey Professional Paper.
Zeit. Geomorph.	Zeitschrift für Geomorphologie.

INDEX